ANALYTICAL METHODS FOR MILK AND MILK PRODUCTS

Volume 1

Sampling Methods and Chemical and Compositional Analysis

Analytical Methods for Milk and Milk Products, 3-volume set
ISBN: 978-1-77491-305-5 (hbk)
ISBN: 978-1-77491-306-2 (pbk)

Volume 1: Sampling Methods and Chemical and Compositional Analysis
ISBN: 978-1-77491-307-9 (hbk)
ISBN: 978-1-77491-308-6 (pbk)
ISBN: 978-1-03267-573-2 (ebk)

Volume 2: Physicochemical Analysis of Concentrated, Coagulated and Fermented Products
ISBN: 978-1-77491-309-3 (hbk)
ISBN: 978-1-77491-310-9 (pbk)
ISBN: 978-1-03267-581-7 (ebk)

Volume 3: Microbiological Analysis, Isolation, and Characterization
ISBN: 978-1-77491-741-1 (hbk)
ISBN: 978-1-77491-740-4 (pbk)
ISBN: 978-1-03267-607-4 (ebk)

Innovations in Agricultural and Biological Engineering

ANALYTICAL METHODS FOR MILK AND MILK PRODUCTS

Volume 1

Sampling Methods and Chemical and Compositional Analysis

Edited by

Megh R. Goyal, PhD, PE

N. Veena, PhD

Santosh Kumar Mishra, PhD

AAP | APPLE ACADEMIC PRESS

First edition published 2024

Apple Academic Press Inc.
1265 Goldenrod Circle, NE,
Palm Bay, FL 32905 USA

760 Laurentian Drive, Unit 19,
Burlington, ON L7N 0A4, CANADA

CRC Press
2385 NW Executive Center Drive,
Suite 320, Boca Raton FL 33431

4 Park Square, Milton Park,
Abingdon, Oxon, OX14 4RN UK

Library and Archives Canada Cataloguing in Publication

CIP data on file with Canada Library and Archives

Library of Congress Cataloging-in-Publication Data

CIP data on file with US Library of Congress

ISBN: 978-1-77491-307-9 (hbk)
ISBN: 978-1-77491-308-6 (pbk)
ISBN: 978-1-03267-573-2 (ebk)

ABOUT THE BOOK SERIES: INNOVATIONS IN AGRICULTURAL AND BIOLOGICAL ENGINEERING

Under this book series, Apple Academic Press Inc. is publishing book volumes over a span of 8–10 years in the specialty areas defined by the American Society of Agricultural and Biological Engineers (<asabe.org>). Apple Academic Press Inc. aims to be a principal source of books in agricultural and biological engineering. We welcome book proposals from readers in areas of their expertise.

The mission of this series is to provide knowledge and techniques for agricultural and biological engineers (ABEs). The book series offers high-quality reference and academic content on agricultural and biological engineering (ABE) that is accessible to academicians, researchers, scientists, university faculty and university-level students, and professionals around the world.

Agricultural and biological engineers ensure that the world has the necessities of life, including safe and plentiful food, clean air and water, renewable fuel and energy, safe working conditions, and a healthy environment by employing knowledge and expertise of the sciences, both pure and applied, and engineering principles. Biological engineering applies engineering practices to problems and opportunities presented by living things and the natural environment in agriculture.

ABE embraces a variety of the following specialty areas (<asabe.org>): aquaculture engineering, biological engineering, energy, farm machinery and power engineering, food, and process engineering, forest engineering, information, and electrical technologies, soil, and water conservation engineering, natural resources engineering, nursery, and greenhouse engineering, safety, and health, and structures and environment.

For this book series, we welcome chapters on the following specialty areas (but not limited to):

1. Academia to industry to end-user loop in agricultural engineering;
2. Agricultural mechanization;
3. Aquaculture engineering;
4. Biological engineering in agriculture;

5. Biotechnology applications in agricultural engineering;
6. Energy source engineering;
7. Farm-to-fork technologies in agriculture;
8. Food and bioprocess engineering;
9. Forest engineering;
10. GPS and remote sensing potential in agricultural engineering;
11. Hill land agriculture;
12. Human factors in engineering;
13. Impact of global warming and climatic change on agriculture economy;
14. Information and electrical technologies;
15. Irrigation and drainage engineering;
16. Micro-irrigation engineering;
17. Nanotechnology applications in agricultural engineering;
18. Natural resources engineering;
19. Nursery and greenhouse engineering;
20. Potential of phytochemicals from agricultural and wild plants for human health;
21. Power systems and machinery design;
22. Robot engineering and drones in agriculture;
23. Rural electrification;
24. Sanitary engineering;
25. Simulation and computer modeling;
26. Smart engineering applications in agriculture;
27. Soil and water engineering;
28. Structures and environment engineering;
29. Waste management and recycling;
30. Any other focus areas.

For more information on this series, readers may contact:

Megh R. Goyal, PhD, PE
Book Series Senior Editor-in-Chief: Innovations in Agricultural and Biological Engineering

OTHER BOOKS ON AGRICULTURAL AND BIOLOGICAL ENGINEERING BY APPLE ACADEMIC PRESS, INC.

Management of Drip/Trickle or Micro Irrigation
Megh R. Goyal, PhD, PE, Senior Editor-in-Chief
Evapotranspiration: Principles and Applications for Water Management
Megh R. Goyal, PhD, PE and Eric W. Harmsen, PhD Editors

Book Series: Research Advances in Sustainable Micro Irrigation
Senior Editor-in-Chief: Megh R. Goyal, PhD, PE

- **Volume 1:** Sustainable Micro Irrigation: Principles and Practices
- **Volume 2:** Sustainable Practices in Surface and Subsurface Micro Irrigation
- **Volume 3:** Sustainable Micro Irrigation Management for Trees and Vines
- **Volume 4:** Management, Performance, and Applications of Micro Irrigation Systems
- **Volume 5:** Applications of Furrow and Micro Irrigation in Arid and Semi-Arid Regions
- **Volume 6:** Best Management Practices for Drip Irrigated Crops
- **Volume 7:** Closed Circuit Micro Irrigation Design: Theory and Applications
- **Volume 8:** Wastewater Management for Irrigation: Principles and Practices
- **Volume 9:** Water and Fertigation Management in Micro Irrigation
- **Volume 10:** Innovation in Micro Irrigation Technology

Book Series: Innovations and Challenges in Micro Irrigation
Senior Editor-in-Chief: Megh R. Goyal, PhD, PE

- Engineering Interventions in Sustainable Trickle Irrigation: Water Requirements, Uniformity, Fertigation, and Crop Performance
- Management Strategies for Water Use Efficiency and Micro Irrigated Crops: Principles, Practices, and Performance

- Micro-Irrigation Engineering for Horticultural Crops: Policy Options, Scheduling, and Design
- Micro-Irrigation Management: Technological Advances and Their Applications
- Micro-Irrigation Scheduling and Practices
- Performance Evaluation of Micro-Irrigation Management: Principles and Practices
- Potential of Solar Energy and Emerging Technologies in Sustainable Micro-Irrigation
- Principles and Management of Clogging in Micro-Irrigation
- Sustainable Micro-Irrigation Design Systems for Agricultural Crops: Methods and Practices

Book Series: Innovations in Agricultural and Biological Engineering
Senior Editor-in-Chief: Megh R. Goyal, PhD, PE

- Advanced Research Methods in Food Processing Technologies
- Advances in Food Process Engineering: Novel Processing, Preservation and Decontamination of Foods
- Advances in Green and Sustainable Nanomaterials: Applications in Energy, Biomedicine, Agriculture, and Environmental Science
- Advances in Sustainable Food Packaging Technology
- Analytical Methods for Milk and Milk Products, 3-volume set:
    - o Volume 1: Sampling Methods and Chemical and Compositional Analysis
    - o Volume 2: Physicochemical Analysis of Concentrated, Coagulated and Fermented Products
    - o Volume 3: Microbiological Analysis, Isolation, and Characterization
- Biological and Chemical Hazards in Food and Food Products: Prevention, Practices, and Management
- Bioremediation and Phytoremediation Technologies in Sustainable Soil Management, 4-volume set:
    - o Volume 1: Fundamental Aspects and Contaminated Sites
    - o Volume 2: Microbial Approaches and Recent Trends
    - o Volume 3: Inventive Techniques, Research Methods, and Case Studies
    - o Volume 4: Degradation of Pesticides and Polychlorinated Biphenyls
- Dairy Engineering: Advanced Technologies and Their Applications

- Developing Technologies in Food Science: Status, Applications, and Challenges
- Emerging Technologies in Agricultural Engineering
- Engineering Interventions in Agricultural Processing
- Engineering Interventions in Foods and Plants
- Engineering Practices for Agricultural Production and Water Conservation: An Interdisciplinary Approach
- Engineering Practices for Management of Soil Salinity: Agricultural, Physiological, and Adaptive Approaches
- Engineering Practices for Milk Products: Dairyceuticals, Novel Technologies, and Quality
- Enzyme Inactivation in Food Processing: Technologies, Materials, and Applications
- Field Practices for Wastewater Use in Agriculture: Future Trends and Use of Biological Systems
- Flood Assessment: Modeling and Parameterization
- Food Engineering: Emerging Issues, Modeling, and Applications
- Food Process Engineering: Emerging Trends in Research and Their Applications
- Food Processing and Preservation Technology: Advances, Methods, and Applications
- Food Technology: Applied Research and Production Techniques
- Functional Dairy Ingredients and Nutraceuticals: Physicochemical, Technological, and Therapeutic Aspects
- Handbook of Research on Food Processing and Preservation Technologies, 5-volume set:
    - Volume 1: Nonthermal and Innovative Food Processing Methods
    - Volume 2: Nonthermal Food Preservation and Novel Processing Strategies
    - Volume 3: Computer-Aided Food Processing and Quality Evaluation Techniques
    - Volume 4: Design and Development of Specific Foods, Packaging Systems, and Food Safety
    - Volume 5: Emerging Techniques for Food Processing, Quality, and Safety Assurance
- Modeling Methods and Practices in Soil and Water Engineering
- Nanotechnology and Nanomaterial Applications in Food, Health, and Biomedical Sciences

- Nanotechnology Applications in Agricultural and Bioprocess Engineering: Farm to Table
- Nanotechnology Applications in Dairy Science: Packaging, Processing, and Preservation
- Nanotechnology Horizons in Food Process Engineering, 3-volume set:
  - Volume 1: Food Preservation, Food Packaging and Sustainable Agriculture
  - Volume 2: Scope, Biomaterials, and Human Health
  - Volume 3: Trends, Nanomaterials, and Food Delivery
- Novel and Alternative Methods in Food Processing: Biotechnological, Physicochemical, and Mathematical Approaches
- Novel Dairy Processing Technologies: Techniques, Management, and Energy Conservation
- Novel Processing Methods for Plant-Based Health Foods: Extraction, Encapsulation and Health Benefits of Bioactive Compounds
- Novel Strategies to Improve Shelf-Life and Quality of Foods: Quality, Safety, and Health Aspects
- Phytochemicals and Medicinal Plants in Food Design: Strategies and Technologies for Improved Healthcare
- Processing of Fruits and Vegetables: From Farm to Fork
- Processing Technologies for Milk and Milk Products: Methods, Applications, and Energy Usage
- Quality Control in Fruit and Vegetable Processing: Methods and Strategies
- Scientific and Technical Terms in Bioengineering and Biological Engineering
- Soil and Water Engineering: Principles and Applications of Modeling
- Soil Salinity Management in Agriculture: Technological Advances and Applications
- State-of-the-Art Technologies in Food Science: Human Health, Emerging Issues and Specialty Topics
- Sustainable and Functional Foods from Plants
- Sustainable Biological Systems for Agriculture: Emerging Issues in Nanotechnology, Biofertilizers, Wastewater, and Farm Machines
- Sustainable Green Nanomaterials: Synthesis, Characterization, and Engineering Applications
- Sustainable Green Nanomaterials: Synthesis, Characterization, and Engineering Applications

- Sustainable Nanomaterials for Biomedical Engineering: Impacts, Challenges, and Future Prospects
- Sustainable Nanomaterials for Biosystems Engineering: Trends in Renewable Energy, Environment, and Agriculture
- Technological Interventions in Dairy Science: Innovative Approaches in Processing, Preservation, and Analysis of Milk Products
- Technological Interventions in Management of Irrigated Agriculture
- Technological Interventions in the Processing of Fruits and Vegetables
- Technological Processes for Marine Foods, From Water to Fork: Bioactive Compounds, Industrial Applications, and Genomics
- The Chemistry of Milk and Milk Products: Physicochemical Properties, Therapeutic Characteristics, and Processing Methods

ABOUT SENIOR EDITOR-IN-CHIEF

Megh R. Goyal, PhD, PE

Retired Professor in Agricultural and Biomedical Engineering, University of Puerto Rico, Mayaguez Campus; Senior Technical Editor-in-Chief, Biomedical Engineering and Agricultural Science, Apple Academic Press, Inc.

Megh R. Goyal, PhD, PE, is, currently a retired professor of agricultural and biomedical engineering from the General Engineering Department at the College of Engineering at the University of Puerto Rico–Mayaguez Campus (UPRM); and Senior Acquisitions Editor and Senior Technical Editor-in-Chief for Agricultural and Biomedical Engineering for Apple Academic Press Inc. During his professional career, he has worked as a Soil Conservation Inspector; Research Assistant at Haryana Agricultural University and Ohio State University; Research Agricultural Engineer/Professor at the Department of Agricultural Engineering of UPRM; and Professor of Agricultural and Biomedical Engineering in the General Engineering Department of UPRM. He spent a one-year sabbatical leave in 2002–2003 at the Biomedical Engineering Department of Florida International University, Miami, USA.

Dr. Goyal was the first agricultural engineer to receive the professional license in agricultural engineering from the College of Engineers and Surveyors of Puerto Rico. In 2005, he was proclaimed the "Father of Irrigation Engineering in Puerto Rico for the Twentieth Century" by the American Society of Agricultural and Biological Engineers, Puerto Rico Section, for his pioneering work on micro irrigation, evapotranspiration, agroclimatology, and soil and water engineering.

During his professional career of 52 years, he has received many awards, including Scientist of the Year, Membership Grand Prize from the American Society of Agricultural Engineers Campaign, Felix Castro Rodriguez Academic Excellence Award, Man of Drip Irrigation by the Mayor of Municipalities of Mayaguez/Caguas/Ponce and Senate/Secretary of Agriculture of ELA, Puerto Rico, and many others. He has been recognized as one of the experts "who rendered meritorious service for the development of [the] irrigation sector in India" by the Water Technology Centre of Tamil Nadu

Agricultural University in Coimbatore, India, and ASABE who bestowed on him the 2018 Netafim Microirrigation Award.

Dr. Goyal has authored more than 200 journal articles and edited more than 120 books. AAP has published many of his books, including *Management of Drip/Trickle or Micro Irrigation; Evapotranspiration: Principles and Applications for Water Management;* ten-volume set on *Research Advances in Sustainable Micro Irrigation.* He has also authored the textbooks *Elements of Agroclimatology* (Spanish) by UNISARC, Colombia, and two *Bibliographies on Drip Irrigation.* Dr. Goyal has also developed several book series with AAP, including Innovations in Agricultural & Biological Engineering; Innovations and Challenges in Micro Irrigation; and Innovations in Plant Science for Better Health: From Soil to Fork.

Dr. Goyal received his BSc degree in Engineering from Punjab Agricultural University, Ludhiana, India, and his MSc and PhD degrees from the Ohio State University, Columbus, Ohio, USA. He also earned a Master of Divinity degree from the Puerto Rico Evangelical Seminary, Hato Rey, Puerto Rico, USA.

ABOUT THE EDITORS

N. Veena, PhD
Associate Professor (Food Safety and Quality), DSLD-College of Horticultural Engineering and Food Technology, University of Horticultural Sciences (UHS), Devihosur, Haveri, Karnataka, India.

N. Veena, PhD, received her BTech degree in Dairy Technology from Karnataka Veterinary, Animal and Fisheries Sciences University, Bidar, Karnataka, India and her MTech and PhD degrees in Dairy Chemistry from ICAR-National Dairy Research Institute, Karnal, Haryana, India. She has worked as an Assistant Professor in the Department of Dairy Chemistry at the College of Dairy Science and Technology, Guru Angad Dev Veterinary and Animal Sciences University (GADVASU), Ludhiana, Punjab, India, from March 2014 to March 2023.

Dr. Veena's area of expertise includes fortification of milk, development and characterization of functional dairy foods, and nanoencapsulation of nutraceuticals and their delivery. She is the recipient of junior research and institutional fellowships during her master's and doctoral programs from the ICAR-National Dairy Research Institute, Karnal, Haryana, India. She has been involved in different institutional and externally funded research projects such as Milkfed Punjab, the Indian Council of Agricultural Research (ICAR), Indian Department of Science and Technology, and RKVY as principal and co-principal investigator. She has published research papers in peer-reviewed journals, edited books, book chapters, popular articles and practical manuals. She has developed the e-learning content for postgraduate dairy chemistry courses under National Higher Agricultural Education Project (NAHEP). Dr. Veena has presented her research work at various national and international seminars and conferences and received several awards for best oral and poster presentations. Venus International Foundation awarded her a Young Women Researcher in Dairy Chemistry for her contribution and achievement in the discipline of Veterinary and Animal Sciences at the 8th Annual Women's Meet held on 4th March 2023, Chennai, India. She has membership in various professional bodies, such as life membership in the Indian Dairy Association, Association of Food Scientists and Technologists (India), Dairy Technology Society of India, Swedish South Asian Network on Functional Foods, etc.

Santosh Kumar Mishra, PhD
Assistant Professor, Department of Dairy Microbiology, College of Dairy Science and Technology, Guru Angad Dev Veterinary and Animal Sciences University, Ludhiana, Punjab, India

Santosh Kumar Mishra, PhD, is working as an Assistant Professor in the Department of Dairy Microbiology, College of Dairy Science and Technology, Guru Angad Dev Veterinary and Animal Sciences University, Ludhiana, Punjab.

He received his BTech degree in Dairy Technology from Maharashtra Animal and Fisheries Sciences University, Nagpur, India; his MSc and PhD degrees from National Dairy Research Institute, Karnal, Haryana, India.

He is currently working in the area of functional foods and dairy products incorporating live probiotics, and technology of functional lactic cultures for fermented and non-fermented dairy products. He served the dairy industry as a Quality Assurance Executive at Mother Dairy, New Delhi, India. In India, he has handled externally funded projects granted by the Department of Science and Technology (DST), Ministry of Food Processing Industries (MoFPI), and University Grants Commission (UGC) as principal or co-principal investigator. He received several awards for best papers and posters/presentations. He is the recipient of junior and senior research fellowships during his master's and doctoral programs at the National Dairy Research Institute, Karnal, Haryana. He was a university gold medalist during his graduate program and also received the best thesis award from his university for his PhD research. Recently he got an Award of Honor at the international conference sponsored by the Partap College of Education, Ludhiana, in association with the International Professionals Development Association, UK. He is a member of various scientific societies: life member of SASNET-Fermented Foods, Anand; and member of the Indian Dairy Association, New Delhi. He has published several research, review, and popular articles in national and international journals. He has also published five international edited books, several book chapters, and teaching reviews in various training programs. He recently completed a young scientist project by DST, SEED Department, Government of India, New Delhi, on isolation and characterization of novel oxalate degrading lactic acid bacteria for potential probiotic management of kidney stones.

CONTENTS

CONTRIBUTORS

S. Athira
Assistant Professor, Department of Food Chemistry, College of Food Technology, Kerala Veterinary, and Animal Sciences University, Chalakkudy, Thrissur, Kerala, India

Mitul Bumbadiya
Assistant Professor, Dairy Chemistry Department, College of Dairy Science and Post-Graduate Institute of Dairy Education and Research, Kamdhenu University, Amreli, Gujarat, India

Sudheendra V. K. Chaganti
Assistant Professor, Dairy Microbiology Department, College of Dairy Science and Post-Graduate Institute of Dairy Education and Research, Kamdhenu University, Amreli, Gujarat, India

Partha P. Debnath
Assistant Professor, Department of Dairy Technology, West Bengal University of Animal and Fishery Sciences (WBUAFS), Mohanpur, West Bengal, India

Hemant Gawande
Assistant Professor, Dairy Chemistry Department, College of Dairy Technology, Maharashtra Animal and Fishery Sciences University (MAFSU), Warud (Pusad), Yavatmal, Maharashtra, India

Saurabh Gosewade
Research Scientist (NPD), Schreiber Dynamix Dairies, Baramati, Maharashtra, India

Megh R. Goyal
Retired Faculty in Agricultural and Biomedical Engineering from College of Engineering at University of Puerto Rico – Mayaguez Campus; and Senior Technical Editor-in-Chief in Agricultural and Biomedical Engineering for Apple Academic Press Inc., Rincon, USA

Chitra Gupta
Assistant Professor, Dairy Chemistry Department, College of Dairy Technology,Maharashtra Animal and Fishery Sciences University (MAFSU), Warud (Pusad),Yavatmal, Maharashtra, India

Tanmay Hazra
Assistant Professor, Dairy Chemistry Department, College of Dairy Science and Post-Graduate Institute of Dairy Education and Research, Kamdhenu University, Amreli, Gujarat, India

Amit K. Jain
Assistant Professor, Dairy Chemistry Department, SMC College of Dairy Science, Kamdhenu University, Anand, Gujarat, India

Payal Karmakar
PhD Research Scholar, Division of Dairy Chemistry, ICAR–National Dairy Research Institute (NDRI), Karnal, Haryana, India

Anuradha Kumari
Assistant Professor, Dairy Chemistry Department, Sanjay Gandhi Institute of Dairy Technology, Bihar Animal Sciences University, Patna, Bihar, India

Ronit Mandal
PhD Research Scholar, Faculty of Land and Dairy Systems, University of British Columbia (UBC), Vancouver, British Columbia, Canada

Santosh K. Mishra
Assistant Professor, Department of Dairy Microbiology, College of Dairy Science and Technology, Guru Angad Dev Veterinary and Animal Sciences University (GADVASU), Ludhiana, Punjab, India

Satishkumar C. Parmar
Assistant Professor, Dairy Chemistry Department, SMC College of Dairy Science, Kamdhenu University, Anand, Gujarat, India

Shriyesh Patel
Assistant Professor, Dairy Chemistry Department, SMC College of Dairy Science, Kamdhenu University, Anand, Gujarat, India,

Prasad Patil
Assistant Professor, Dairy Microbiology Department, College of Dairy Technology, Maharashtra Animal and Fishery Sciences University (MAFSU), Warud (Pusad), Yavatmal, Maharashtra, India

M. P. Rahila
Assistant Professor, Department of Dairy Chemistry, College of Dairy Science and Technology, Kerala Veterinary and Animal Sciences University, Pookode, Wayanad, Kerala, India

Suvartan Ranveer
Assistant Professor, Dairy Chemistry Department, Warner College of Dairy Technology, Allahabad, Uttar Pradesh, India

Kuntal Roy
Assistant Professor, Faculty of Dairy Technology, West Bengal University of Animal and Fishery Sciences (WBUAFS), Mohanpur, Nadia, West Bengal, India

Pravin Sawale
Assistant Professor, Dairy Technology Department, College of Dairy Technology, Maharashtra Animal and Fishery Sciences University (MAFSU), Warud (Pusad), Yavatmal, Maharashtra, India

Ahesanvarish Shaikh
Associate Professor, Dairy Chemistry Department, SMC College of Dairy Science, Kamdhenu University, Anand, Gujarat, India

Sneha Shende
Assistant Professor, Dairy Chemistry Department, College of Dairy Technology, Maharashtra Animal and Fishery Sciences University (MAFSU), Warud (Pusad), Yavatmal, Maharashtra, India

Rohit G. Sindhav
Assistant Professor, Dairy Technology Department, College of Dairy Science and Post-Graduate Institute of Dairy Education and Research, Kamdhenu University, Amreli, Gujarat, India

Rinku Sundar
Guest Lecturer, College of Dairy Science and Technology, Kerala Veterinary and Animal Sciences University, Pookode, Wayanad, Kerala, India

N. Veena
Associate Professor, Department of Food Safety and Quality, DSLD - College of Horticultural Engineering and Food Technology, University of Horticultural Sciences (UHS), Devihosur, Haveri, Karnataka

ABBREVIATIONS

$(NH_4)_2C_2O_4$	ammonium oxalate
AA	ascorbic acid
AAS	atomic absorption spectroscopy
AGMARK	agriculture marketing
$AgNO_3$	silver nitrate
Al_2O_3	aluminum oxide
AMF	anhydrous milk fat
ANSI	American National Standards Institute
AOAC	Association of Official Analytical Chemists
AR	analytical reagents
ARRF	average of all RRF
ASTM	American Society of Testing and Material
AV	absorbance value
BCG	bromocresol green
BDI	Bureau of Dairy Industries
BHT	Butylated Hydroxytoluene
BIS	Bureau of Indian Standards
BR	butyro-refractometer
BTB	bromothymol blue
C_2H_5OH	ethanol
Ca	calcium
CA	citric acid
CF	conversion factor
CH_2Cl_2	methylene chloride
CH_3CN	acetonitrile
CH_3COOH	acetic acid
CH_3OH	methanol
CHR	corrected hydrometer reading
CO_2	carbon dioxide
COB	clot-on-boiling
Conc.	concentration
cP	centipoise
Cu	copper
$CuSO_4$	copper sulfate

dBA	decibel
DF	dilution factor
DMAB	dimethyl amino benzaldehyde
DMF	dimethylformamide
EM	evaporated milk
ESI	electrospray ionization
eV	electron volt
FAMEs	fatty acids methyl esters
FAO	Food and Agricultural Organization
FAs	fatty acids
FAS	ferrous ammonium sulfate
FFAs	free fatty acids
FID	flame ionization detector
FP	freezing point
FR	flow rate
f_r	response factor
FSSAI	Food Safety Standard Authority of India
FSSRI	Food Safety and Standards Regulations of India
GLC	gas-liquid chromatogram
H_2O	water
H_2O_2	hydrogen peroxide
H_2SO_4	sulfuric acid
H_3BO_3	boric acid
HCl	hydrochloric acid
HDPE	high density polyethylene
HNO_3	nitric acid
HPLC	high performance liquid chromatography
IDF	International Dairy Federation
IS	internal standard
ISI	Indian Standardization Institute
ISO	International Organization for Standardization
ISS	Indian Standard Specification
IU	International Units
K_2CrO_4	potassium chromate
KBr	potassium bromide
kg	kilogram
KH_2PO_4	potassium dihydrogen orthophosphate
$KMnO_4$	potassium permanganate
KOH	potassium hydroxide

kPa	kilopascal
KSCN	potassium thiocyanate
kV	kilovolt
KV	Kirschner value
$LaCl_3$	lanthanum trichloride
LB	Liebermann-Burchard
LC	liquid chromatography
LMSS	labeled melamine stock solution
LMWS	labeled melamine working solution
M	molar or molarity
Max	maximum
MB	methylene blue
Mg	magnesium
mg	milligram
Min.	minimum
ml	milliliter
mmol	millimole
mN/m	millinewtons per meter
MPG	mobile phase gradient
MRL	maximum residue limit
MS	mass spectrometry
MSNF	milk solid-not-fat
MSS	melamine stock solution
MTBE	methyl-tert-butyl ether
MWA	microwave-assisted
MWS	melamine working solution
N	normality or normal
N/m	Newton per meter
Na_2SO_4	sodium sulfate
NABL	National Accreditation Board of Testing and Calibration Laboratories
NaCl	sodium chloride
NAD	nicotinamide adenine dinucleotide
NaOH	sodium hydroxide
NCN	non-casein nitrogen
ng	nanogram
NH_4OH	ammonia solution/ammonium hydroxide
NIH	National Institute of Health
NPN	non-protein nitrogen

OC	organochlorine
OPA	orthophosphoric acid
PA	peak area
Pa.s	pascal second
PAR	peak area ratio
psi	pounds per square inch
PTA	phosphor-tungstic acid
PTFE	polytetrafluorethylene
PV	peak volume
PV	Polenske value
RA	retinyl acetate
RI	refractive index
RM	Reichert-Meissel
RP	retinyl palmitate
RP	reverse phase
RP-LC	reverse phase-liquid chromatography
rpm	revolutions per minute
RRF	relative response factor
RRT	relative retention time
RT	retention time
SACN	Scientific Advisory Committee on Nutrition
SCM	sweetened condensed milk
SDS	solvent delivery system
SG	specific gravity
SNF	solids-not-fat
SRM	selected reaction monitoring
SSS	standard stock solution
ST	surface tension
SV	saponification value
TCA	trichloroacetic acid
TLC	thin layer chromatography
TN	total nitrogen
TR	transition reaction
TS	total solids
USDA	U.S. Department of Agriculture
USM	unsaponifiable matter
UV	ultraviolet
v/v	volume/volume
W	watt

w/w	weight/weight
WHO	World Health Organization
WSS	working standard solution
Zn	zinc
μg	microgram
μl	microliter
μm	micrometer

PREFACE

Milk is an essential commodity entering trade and required in everyday life as an article of food. Dairy food analysis is very important to meet the permissible limits provided by national and international standards, to assess physicochemical and sensory attributes, to provide information on nutritive value, to assess the keeping quality, to prevent fraud, and to ensure freedom from contaminants, toxins, anti-nutrients, pathogens, spoilage-causing bacteria, etc.

A wide range of analytical methods has been used in dairy products analysis while considering the complexity and for assessing various physical, chemical, and biological properties. Among all the foods, milk products are probably the most diverse group. Each of these milk products has its own unique properties and therefore demands the application of precise analytical methods. Proximate analysis of products is of fundamental importance to consumers, manufacturers, and regulatory bodies. However, in certain cases, more detailed information on composition is required, such as for product safety or authentication, nutritional or toxicological purpose, technological reasons, etc., depending on the aim of the dairy food analysis.

This book volume aims to be a reference book for the analysis of milk and milk products with a description of the main analytical techniques and methodologies and their application to the compounds involved in nutritional and technological quality. The book consists of two main sections. The Part 1: Sampling Methods and Chemical Analysis of Milk – of this book highlights the standard methods used for calibration of different glasswares, sampling procedure of milk and milk products, physicochemical and compositional aspects, and assessment of quality of raw milk intend for processing and manufacturing. Part 2: Compositional Analysis of Frozen and Fat-Rich Products – of this book deals with physicochemical and compositional analysis of different dairy products, including cream, butter, butter oil, clarified fat (ghee), ice cream, and frozen desserts. Each of the laboratory exercises includes the following: introduction, objective, principle of method, chemicals, and apparatus required, sample preparation, experimentation, data collection sheet and calculations, and resource materials.

This book volume on "Analytical Methods for Milk and Milk Products" is a 3-volume set announced in the book series on 'Innovations in

Agricultural and Biological Engineering' by www.appleacademicpress.com. It attempts to illustrate the wide range of analytical techniques and methodologies for assessing the physical, chemical, and microbiological properties and to ensure nutritional or technological quality and safety of milk and milk products. The first book volume on "Analytical Methods for Milk and Milk Products: Sampling Methods and Chemical and Compositional Analysis" emphasizes the calibration of glasswares, sampling techniques, and physicochemical, compositional, and overall quality aspects of raw and processed milk has been described. This volume also covers the physicochemical and compositional analysis of fat-rich dairy products (such as cream, butter, butter oil, and clarified fat) and ice cream and frozen desserts. The contents have been planned to meet the requirements for students, researchers, scientists, food analysts, food analysis or research laboratory, and personnel involved in the area of milk and milk products analysis.

This book has exceeded our revelation and anticipations due to the support of all contributing authors to this book, who have been most valuable in this compilation. The names of contributing authors are mentioned in their respective chapters. The editorial team appreciates authors for their proficiency, pledge, and perseverance, without which the work would not have seen the light of day. The effort of contributors and the editorial team will make this volume more informative for readers from different dairy processing industries and academic institutions.

Sincere thanks to the editorial staff at Apple Academic Press, Inc. for their cooperation, valuable assistance, great support, and guidance in the right direction all the way through this project. We also extend our deepest appreciation to the Production Staff of Apple Academic Press, Inc. for the electronic publishing of the entire document and for the exceptional production of this book volume. We appeal to the reader to suggest your feedback that may benefit in improving the subsequent edition of this book.

We offer the deepest gratitude and indebtedness to our friends and families for their constant support and encouragement during the course of editing this book volume. Finally, and most importantly, we would like to commend our spouses, Subhadra, Mohan Kumar, and Anamika, for their understanding and patience throughout this project.

—Editors

PART I

SAMPLING METHODS AND CHEMICAL ANALYSIS OF MILK

CHAPTER 1

CALIBRATION OF DIFFERENT GLASSWARES FOR ANALYTICAL METHODS

CHITRA GUPTA, SNEHA SHENDE, M. P. RAHILA, and SAURABH GOSEWADE

ABSTRACT

The calibration of glassware is important in the various chemical, biological, microbiological, and pharmaceutical industries. The accuracy, reliability, and reproducibility of the result of any analysis are also dependent upon the accuracy and calibration of the glassware used in the analysis. Glassware is divided into two categories, i.e., type A and type B, depending upon the accuracy of calibration and tolerance values. This chapter describes the methods used to calibrate various volumetric glasswares (pipette, burette, volumetric flask, measuring cylinder), butyrometer, thermometer, and hydrometer.

1.1 CALIBRATION OF VOLUMETRIC GLASSWARE

1.1.1 INTRODUCTION

Calibration is the process of standardization of various glasswares and instruments to determine the deviation from the standards and finding out the appropriate correction factors. It is also known as setting and correcting a measuring device using dependably known and unvarying measures. It is a two-step process, and in the first step, the relationship between measurement uncertainties (quantitative values) and measurement standards is established. In the second step, information is utilized for obtaining the measurement results; however, in general, the first step is only considered as calibration.

Analytical Methods for Milk and Milk Products, Volume 1: Sampling Methods and Chemical and Compositional Analysis. Megh R. Goyal, N. Veena, and Santosh Kumar Mishra (Eds.)

Various volumetric glasswares, lactometer, butyrometer, and instruments need to be authentically calibrated before their use in dairy and other industries. Two methods, i.e., (a) reference method and (b) comparison method are generally used to calibrate the glasswares and instruments [2].

Different IS and ISO methods [2] were used by the National Accreditation Board of Testing and Calibration Laboratories (NABL) for the calibration of different volumetric apparatus, i.e., single mark pipette, graduated pipette, burette, volumetric flask, and measuring cylinders. Volumetric glasswares are precisely calibrated using the gravimetric method. The temperature of distilled water and barometric pressure are two important parameters for converting the weight into volume. However, barometric fluctuations are negligible as this parameter can be fixed based upon the location. Weighing balance should be pre-calibrated and temperature fluctuations must be properly controlled, monitored, and recorded. Calibration of volumetric apparatus using gravimetric method involves various uncertainty factors, i.e., calibration and performance of weighing balance, weighing error, distilled water evaporation during weighing, and operator proficiency [14].

1.1.2 PRINCIPLE

The gravimetric method is more precise for the calibration of volumetric apparatus (glasswares), because the equipment is not available for quantitative measurement of very less/minute volume. In this method, the mass volume of distilled water (physical properties are known) dispensed through the volumetric apparatus is weighed with highly precise or pre-calibrated weighing balance and based upon the density converted into the volume value. The weight of the water dispensed or contained by volumetric glassware is dependent upon the distill water temperature, density of water, density of air, density of weight by which weighing balance is calibrated and coefficient of cubical thermal expansion of glasswares material [14]. The coefficient for cubical thermal expansion for borosilicate glass 3.3 is 9.9×10^{-6}°C^{-1}, for borosilicate glass 5.0 is 15×10^{-6}°C^{-1} and for soda lime glass is 27×10^{-6}°C^{-1}. Other factors are affected by the temperature of analysis and atmospheric pressure [8].

1.1.3 REQUIREMENTS OF MATERIALS

1. **Laboratory Requirements:** Noise level in the laboratory should be less than 60 dBA. Illumination shall be maintained from 250 to 500 lux, on a working bench. It is provided with regulated and

uninterrupted power supply with regulation level ±2% or better, and frequency variation ±2.5 Hz or better. Effective mains earthing should be provided to avoid electric shock. Temperature of the laboratory should be maintained at 18 to 23°C and stable for 1 h at ±1°C and relative humidity of the laboratory should be maintained at 40–60% and stable at ±10°C for 4 h [14]. Under Indian conditions, temperature in the laboratory is 27±2°C and relative humidity 65±5%. Excessive air in-flow in the laboratory should also be avoided [8].

2. **Pre-Calibrated Weighing Balance:** This of different resolution is used based on the volume of different glasswares [14], which will be calibrated; and requirements are given in Table 1.1.

TABLE 1.1 Recommended Weighing Balances Used for the Calibration of Various Volumetric Apparatus

Volume (V) Under Test	Resolution of Weighing Balance (mg)	Repeatability (mg)	Linearity (mg)
100 μl < V ≤ 10 ml	0.1	0.2	0.2
10 ml < V ≤ 1,000 ml	1	1	2
1,000 ml < V ≤ 2,000 ml	10	10	20
V > 2,000 ml	100	100	200

Note: Nominal volume may be used to choose the balance, for practical purposes.

3. **Distilled Water:** Water purity is very important during volumetric calibration of glasswares. Water is used for two purposes, i.e., as: (i) cleaning fluid; and (ii) standard reference material. Water used for these purposes should be clean and pure. Dirty water could form the deposits, soil in the interior surfaces and result in volumetric error. Physical contaminants should be removed through the filtration. Despite the impurities, ratio of density value of clean water versus pure water is essentially same, however, gravimetric calibration of volumetric glasswares does not use the ratios, hence, impurity of water resulted in significant error in the volumetric calibration results. Pure water cannot be produced only by the filtration, however there is a need to remove the chemical contaminants also. These can be removed using distillation, ion exchange/water softener, electrodialysis, continuous electro deionization, reverse osmosis or combination thereof. Distilled water with electrical conductivity of ≤ 5 μS/cm is most suitable for gravimetric calibration of volumetric glasswares [3].

4. **Receiving Vessel:** Beaker or conical flask.
5. **Thermometer.**

1.1.4 CALIBRATION OF PIPETTES

A pipette is volumetric glassware consisting of a narrow tube with graduation marks in which solution is sucked by suction and solution is retained by closing the upper end (Figure 1.1). It is used in all chemicals, biological, and medical laboratories to deliver the desired quantity of liquid. It should be well annealed and made up of stout-walled transparent glass tubing and free from any visible defects and internal strains [5]. Glass pipettes are of two types, i.e., (i) one mark or fixed volume pipette and (ii) graduated or variable volume pipette. One mark pipette or fixed volume has a single graduation mark and able to deliver one single quantity, however, graduated, or variable volume pipette has various graduation mark and deliver the variable volumes (Figure 1.1).

Graduated pipettes can be further divided into two categories, i.e., (i) serological or blow-out pipette and (ii) Mohr or drain-out pipette. Serological pipettes are the graduated pipette in which graduation marks are up to the tips, hence total deliverable volume is up to tip, however, in Mohr pipettes, graduation marks are slightly above the tip of the pipette, hence tip should not be empty during delivery of volume (Figures 1.1 and 1.2).

American society of testing and material (ASTM) has divided the glassware into two categories based on relative tolerance [1], i.e., class A and class B glasswares. Class A designate that it is complying with the necessary construction and accuracy requirements. Class B glasswares are the general-purpose instruments, which has the similar design as class A apparatus with twice volumetric tolerance [1]. Table 1.2 indicates maximum permissible error on capacities of pipette [5, 13]. Delivery time is the time in seconds (s) required for the movement of the meniscus from graduation mark to the mark at which it comes to rest in the tip of pipette [5, 13] as shown in Table 1.3.

1.1.4.1 PROCEDURE FOR PIPETTE CALIBRATION

Cleaning of the pipette and other glasswares used in the analysis is very important. Wash the glassware to remove the adhering dirt particles and grease. The presence of grease and dirt affects the delivery volume of the

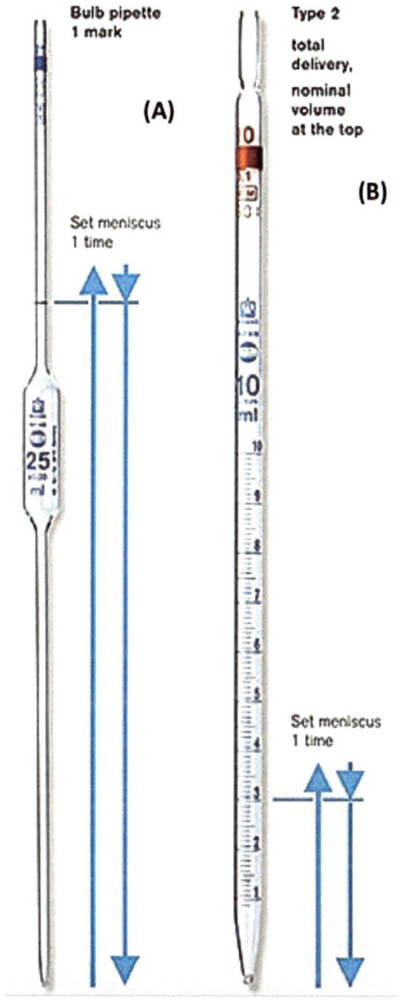

FIGURE 1.1 (A) One mark pipette; and (B) graduated pipette.

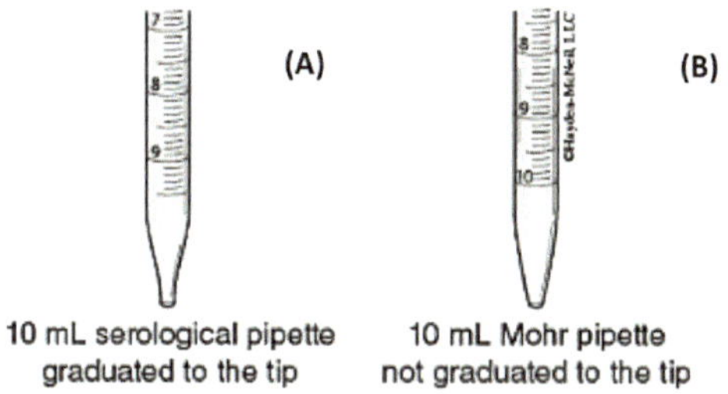

FIGURE 1.2 (A) Serological pipette; and (B) Mohr pipette.

TABLE 1.2 Maximum Permissible Error for Different Capacities of Pipette [5, 13]

Nominal Capacity, cm^3 (ml)	Maximum Permitted Error, cm^3 (ml)	
	Class A	Class B
0.5	±0.005	±0.01
1	±0.007	±0.015
2	±0.010	±0.020
5	±0.015	±0.030
10	±0.020	±0.040
20	±0.030	±0.060
25	±0.030	±0.060
50	±0.050	±0.100
100	±0.080	±0.160
200	±0.100	±0.200

TABLE 1.3 Delivery Time (in Seconds) for Pipettes [5, 13]

Nominal Capacity, cm^3 (ml)	Delivery Time (s)				Permissible Difference between Observed and Inscribed Delivery Times (s), Max.
	Class A		Class B		
	Min.	Max.	Min.	Max.	
0.5	10	20	7	20	2
1	10	20	7	20	2
2	10	25	7	25	2
5	15	30	10	30	3
10	15	40	10	40	3
20	25	50	20	50	4
25	25	50	20	50	4
50	30	60	20	60	5
100	40	60	30	60	5
200	50	70	40	70	5

pipette. Fill the distilled water into pipette to a short distance above the graduation marks, retaining the water into the pipette by pressing a finger on the top of the pipette. Hold the pipette in the vertical position and adjust the meniscus to the graduation line. Set the meniscus in a way that the upper edge of the graduation line is horizontally tangential to the lowest level of the meniscus. Bring the glass vessel in contact with the tip of the pipette to remove any droplet of water adhered on the tip or wipe it with the filter paper.

Note down the delivery time with the help of the stopwatch. Slightly inclined the tip of the pipette touching the inside surface of another cleaned and tarred (pre-weighed) glass vessel. Allow the water to run out slowly by reducing the pressure of the finger so as to bring the water surface to rest with the lowest point of the meniscus in horizontal plane with the top edge of the graduation mark. Remove the finger from the top of the pipette and allow the pipette to drain within 15 seconds. Tip of the pipettes should remain in contact with the internal surface of receiving vessel for 3 more seconds after visible outflow has ceased. Determine the weight of the water, thus delivered. Carry out all the operations at room temperature. Observe the temperature of water and calculate the volume delivered by the pipettes using the formulae given in the calculation applying the necessary correction factors [4, 5, 8, 13].

1.1.5 CALIBRATION OF BURETTE

Burette is a uniform bore glass tube with fine graduation and stopcock at the bottom, used especially in laboratory procedure for accurate fluid dispensing and measurement (Figure 1.3). It should be well annealed and made up of stout-walled transparent glass tubing and free from any visible defects and internal strains [6, 11].

On the basis of the accuracy of burettes, these are divided into two classes, i.e., Class A and Class B. Class A is further divided into two classes, i.e., Class A and Class AS. Class A and class AS burette are of high accuracy and class B burette is of low accuracy. Waiting time is the time lapse between the apparent completion of liquid delivery from burette and noting the final volume of liquid delivered. Waiting time is not specified for class A and class B burette, however, 30 sec waiting time has been specified for class AS burettes. Maximum permitted error in burette is the error at any point on the scale (Table 1.4), and the maximum permissible difference between the errors at any two points [6, 11]. Delivery time is the time in second required for the movement of the meniscus from graduation mark to the mark at which it comes to rest in the tip of the pipette (Table 1.5).

1.1.5.1 PROCEDURE FOR CALIBRATION OF BURETTE

Burette and other glasswares used in the analysis should be properly cleaned to remove the adhering grease and dirt particles. Stopcock of burette should

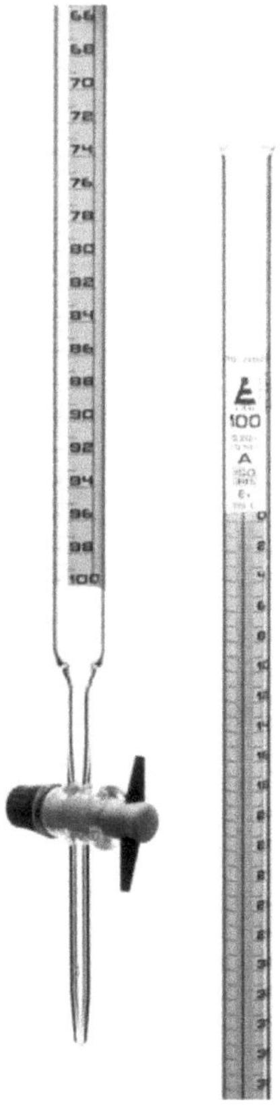

FIGURE 1.3 Burette.

TABLE 1.4 Maximum Permissible Error in the Burette [6, 11]

Nominal Capacity (ml)	Subdivision (ml)	Maximum Permissible Error (ml)	
		Class A and Class AS	Class B
1	0.01	±0.006	±0.01
2	0.01	±0.01	±0.02
5	0.01	±0.01	±0.02
5	0.02	±0.01	±0.02
10	0.02	±0.02	±0.05
10	0.05	±0.03	±0.05
25	0.05	±0.03	±0.05
25	0.10	±0.05	±0.10
50	0.10	±0.05	±0.10
100	0.20	±0.10	±0.20

TABLE 1.5 Delivery Time (in Seconds) for Various Classes of Burettes [6, 11]

Nominal Capacity	Subdivision	Delivery Time (Seconds)					
		Class A		Class B		Class AS	
ml	ml	Min.	Max.	Min.	Max.	Min.	Max.
1	0.01	20	50	20	50	–	–
2	0.01	15	45	10	45	8	20
5	0.01	20	75	20	65	15	25
5	0.02	20	75	20	65	15	25
10	0.02	75	95	40	95	35	45
10	0.05	75	95	45	75	35	45
25	0.05	70	100	30	70	35	45
25	0.10	35	75	30	70	35	45
50	0.10	50	100	40	100	35	45
100	0.20	60	100	30	100	–	–

be checked for lubrication and free movement. Clamp the burette in a vertical position on the burette stand and one beaker or test tube which can hold the thermometer. Fill the same water in the beaker or test tube, which is filled in burette for recording the temperature of water. If the top of the burette is not provided with the filling device, place the funnel for filling the burette with distilled water. Note down the temperature of distilled water, preferably it should be at room temperature. Fill the burette with distilled water in a way that there should not be any presence of air bubbles in the stopcock or jet. Check the burette for any leakage through the tip or stopcock and remove any liquid droplet on the tip. Burette is filled and drained various times to condition the burette.

Adjust the meniscus to the zero-graduation mark and record the delivery time, i.e., time required for running of liquid from zero graduation mark to lowest graduation mark when stopcock is fully open and tip is not in contact with the receiving vessel. Again fill the burette to the zero graduation mark or any other graduation mark. Note down the weight of the empty receiving vessel (I_E). Fully open the stopcock and deliver the quantity under test in the receiving vessel. Tip of the burette should be in contact with the internal surface of the receiving flask. Slow or restrict the flow through the stopcock near to the graduation line being calibrated and precisely adjust the meniscus to the graduation line. Remove the receiving flask from the tip of the burette and again read the meniscus. Weigh the receiving flask with water and note down the weight (I_L). Record the temperature of water, air pressure and air temperature and calculate the volume dispensed by the burette using the formulae given in the calculation. Repeat the same procedure for calibration of another interval. Record all the readings in triplicate for more accurate results [4, 6, 8, 11].

1.1.6 CALIBRATION OF VOLUMETRIC FLASK

A volumetric flask is a type of volumetric glassware used for the preparation of chemical solutions of different concentrations (Figure 1.4). It is a flat bottom bulb with elongated neck consisting of one graduation mark on the neck, which is calibrated to hold a specific quantity of liquid. Volumetric flasks are designed to contain a specific quantity of liquid at a specific temperature, not the volume to be dispensed. Quantity contained and temperature at which it is calibrated is written on the glasswares. Volumetric flasks are of two types, i.e., narrow-necked and wide-necked flasks [9, 10].

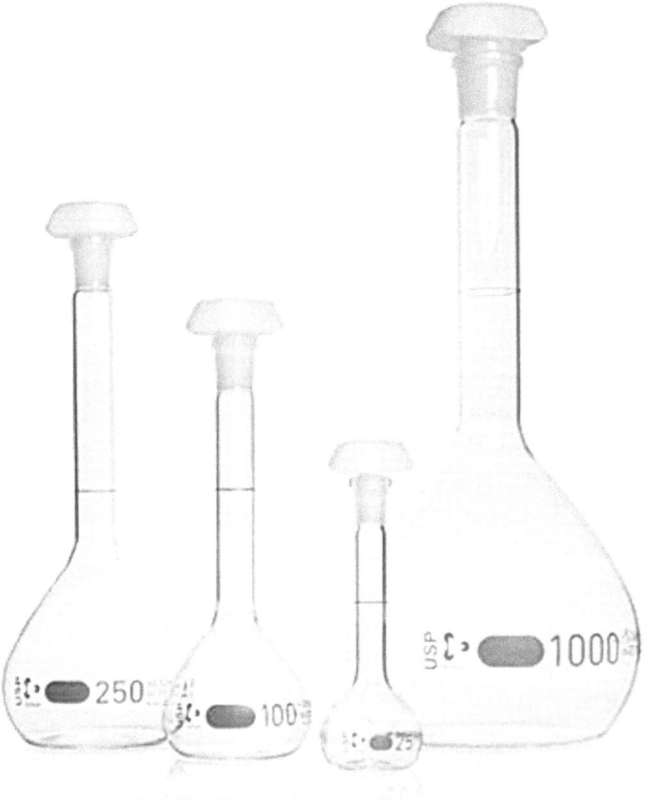

FIGURE 1.4 Volumetric flask of different capacities.

Like other glasswares, high accuracy volumetric flask is class A and less accuracy volumetric flask is class B. The capacity of the volumetric flask (in ml) is defined as the volume of water at 20°C, contained by the flask at 20°C, when filled to the graduation mark. When reference temperature is 27°C, then this value shall be substituted for 20°C [9, 10]. Table 1.6 indicates maximum permissible error for narrow-necked and wide-necked flasks [9, 10].

TABLE 1.6 Maximum Permissible Error for Narrow-Necked and Wide-Necked Flasks [9, 10]

Nominal Capacity (ml)	Maximum Permissible Error (ml)			
	Narrow-Necked Flask		Wide-Necked Flask	
	Class A	Class B	Class A	Class B
1	±0.025	±0.050	–	–
2	±0.025	±0.050	–	–
5	±0.025	±0.050	±0.040	±0.080
10	±0.025	±0.050	±0.040	±0.080
20	±0.040	±0.080	±0.060	±0.120
25	±0.040	±0.080	±0.060	±0.120
50	±0.060	±0.120	±0.100	±0.200
100	±0.100	±0.200	–	–
200	±0.150	±0.300	–	–
250	±0.150	±0.300	–	–
500	±0.250	±0.500	–	–
1,000	±0.400	±0.800	±0.600	±1.200
2,000	±0.600	±1.200	–	–
5,000	±1.200	±2.400	–	–

1.1.6.1 PROCEDURE FOR CALIBRATION OF VOLUMETRIC FLASK

Volumetric flasks are calibrated to contain a specific quantity of liquid. Prior to calibration, the volumetric flask is properly cleaned in a way that there should be the uniform drainage of water without any adherence of water droplets on the interior surface. Reproducible weight should be observed after washing and drainage of the water. Volumetric flask should be properly cleaned and dry prior to calibration. Use specific resolution pre-calibrated weighing balance based upon the capacity of the volumetric flask as specified in Table 1.1. Weigh and note down the weight of volumetric flask with stopper as I_E. Place a funnel in the neck and transfer the distilled water a few mm below the graduation line, in a way that it should not wet the neck above the graduation mark. Wait for two minutes and carefully adjust the meniscus of the distilled water up to the graduation mark and check for the presence of air bubbles. Wait for few minutes until all air bubbles subsides. Carefully check the internal surface above the graduation mark and outside of volumetric flask for the absence of water droplets. Stopper the flask and weigh it. Note down the weight of filled volumetric flask with stopper as I_L. Record the temperature of water, air pressure and air temperature and calculate the volume dispensed by the volumetric flask using the formulae given in the calculation. Repeat the same procedure and record the readings in triplicate for more accurate results [4, 8–10].

1.1.7 CALIBRATION OF MEASURING CYLINDER

Measuring cylinder, graduated cylinder or mixing cylinder is graduated glassware to accurately measure the volume of liquid. These are narrow cylinder with minimum circumference. Each marked line on the measuring cylinder indicates the volume of liquid being measured. Measuring cylinders are less accurate than pipette and volumetric flask, however, more accurate than beaker and conical flask. Measuring cylinder are of three types: (i) tall form with spouted neck (Type 1a); (ii) tall form with stoppered flask (Type 1b); and (iii) squat form with spouted neck (Type 2). Types 1a and 1b measuring cylinder are of both class A and class B grades and type 2 measuring cylinder is of class B grade only [7, 12]. Subdivision and permissible error of all the three types of measuring cylinder are shown in Table 1.7.

TABLE 1.7 Subdivision and Maximum Permissible Error in Measuring Cylinders [7, 12]

Nominal Capacity (ml)	Type 1a and Type 1b			Type 2	
	Subdivision (ml)	Maximum Permissible Error (ml)		Subdivision (ml)	Maximum Permissible Error (ml)
		Class A	Class B		Class B
5	0.1	±0.05	±0.1	0.5	±0.2
10	0.2	±0.1	±0.2	1.0	±0.3
25	0.5	±0.25	±0.5	1.0	±0.5
50	1.0	±0.5	±1.0	1.0 or 2.0	±1.0
100	1.0	±0.5	±1.0	2.0	±1.0
250	2.0	±1.0	±2.0	5.0	±2.0
500	5.0	±2.5	±5.0	10.0	±5.0
1,000	10.0	±5.0	±10.0	20.0	±10.0
2,000	20.0	±10.0	±20.0	50.0	±20.0

1.1.7.1 PROCEDURE FOR CALIBRATION OF MEASURING CYLINDER

Measuring cylinders are calibrated for delivering the measured volume of liquid. Prior to calibration, the measuring cylinder is properly cleaned in a way that there should be the uniform drainage of water without any adherence of water droplets on the interior surface. It should be conditioned by filling and holding the distilled water for 30 s and then emptying the measuring cylinder without splashing. Use specific resolution pre-calibrated weighing balance based upon the capacity of the measuring cylinder. Fill the measuring cylinder with distilled water up to the desired graduation mark and note down the weight of the filled measuring cylinder as I_L. Transfer the water in the other container without splashing. Completely empty the measuring cylinder and hold it in a vertical position for 10 s. Note down the weight of the empty measuring cylinder as I_E. Record the temperature of water, air pressure, and air temperature and calculate the volume dispensed by the measuring using the formulae given in the calculation. Repeat the same procedure and record the readings in triplicate for more accurate results [4, 7, 8, 12].

1.1.8 CALCULATION

Volume delivered by volumetric glassware at the reference temperature (Vt_R) [8] is as follows:

$$Vt_R = (I_L - I_E) \times (p_W - p_A)^{-1} \times (1 - p_A/p_B) \times [1 - \gamma\,(t_1 - t_R)]$$

$$V_{20} = (I_L - I_E) \times Z$$

$$Z = (p_W - p_A)\text{–}1 \times (1 - p_A/p_B) \times [1 - \gamma\,(t_1 - 20)] \quad (1)$$

where; t_R is the reference temperature (20°C); t_1 is the temperature of water during testing; Vt_R is the volume of water delivered by volumetric glassware at reference temperature (volume at 20°C); I_L is the weight of vessel with water (in g); I_E is the weight of empty vessel (in g); p_A is the density of air (g/ml) at temperature and atmospheric pressure of laboratory; p_B is the density of balance weight (g/ml) with which weighing balance is calibrated, generally p_B is the 8.0 g/ml; p_W is the density of water at t_1°C (g/ml); γ is the coefficient of cubical thermal expansion of the material with which volumetric apparatus is manufactured; and Z is the conversion value dependent upon the type of glass used for manufacturing glassware, temperature of analysis and density of air.

➢ ***Note:*** Z-value can be obtained from appendices A1, A2, or A3.

If volumetric glassware is used at different temperature other than reference temperature (i.e., 20°C), then the volume of volumetric instrument at t_2 is calculated as follows:

$$Vt_2 = V_{20}\,[1 + \gamma\,(t_2 - 20)] \quad (2)$$

where; Vt_2 is the volume of water delivered at t_2°C.

1.1.9 DATA COLLECTION SHEET

Title: Calibration of Volumetric glasswares

Location and Organization: ____________________________

Participants: ____________________________

Date (Month/Day/Year): ____________________________

Observations:

Parameter	Repeat 1	Repeat 2	Repeat 3	Average
Weight of vessel with water (I_L)				
Weight of empty vessel (I_E)				
Thermal expansion of glassware material (γ)				
Temperature of water (t_1)				
Density of water at t_1°C (p_W)				
Density of air (p_A)				
Density of balance weight (p_B)				
Conversion value (Z)				
Delivery time				

Comments: ____________________________

1.1.10 APPENDICES FOR Z-FACTORS

➢ **Note:** The standard atmosphere (symbol: atm) is a unit of pressure defined as 101,325 Pa (1,013.25 hPa; 1,013.25 mbar), which is equivalent to 760 mm Hg, or 29.9212 inches Hg, or 14.696 psi.

APPENDIX-1 Z-Factor for Borosilicate Glass 3.3

Temperature (°C)	Atmospheric Pressure or Barometric Pressure (hPa or mbar)							
	850	880	910	940	970	1,000	1,030	1,060
15.0	1.00185	1.00188	1.00191	1.00195	1.00198	1.00201	1.00204	1.00207
15.2	1.00188	1.00191	1.00194	1.00197	1.00201	1.00204	1.00207	1.00210
15.4	1.00191	1.00194	1.00197	1.00200	1.00203	1.00207	1.00210	1.00213
15.6	1.00194	1.00197	1.00200	1.00203	1.00206	1.00209	1.00213	1.00216
15.8	1.00196	1.00200	1.00203	1.00206	1.00209	1.00212	1.00216	1.00219
16.0	1.00199	1.00203	1.00206	1.00209	1.00212	1.00215	1.00218	1.00222
16.2	1.00202	1.00206	1.00209	1.00212	1.00215	1.00218	1.00221	1.00225
16.4	1.00205	1.00209	1.00212	1.00215	1.00218	1.00221	1.00225	1.00228
16.6	1.00209	1.00212	1.00215	1.00218	1.00221	1.00224	1.00228	1.00231
16.8	1.00212	1.00215	1.00218	1.00221	1.00224	1.00228	1.00231	1.00234
17.0	1.00215	1.00218	1.00221	1.00224	1.00228	1.00231	1.00234	1.00237
17.2	1.00218	1.00221	1.00225	1.00228	1.00231	1.00234	1.00237	1.00240
17.4	1.00222	1.00225	1.00228	1.00231	1.00234	1.00237	1.00241	1.00244
17.6	1.00225	1.00228	1.00231	1.00234	1.00238	1.00241	1.00244	1.00247
17.8	1.00228	1.00231	1.00235	1.00238	1.00241	1.00244	1.00247	1.00250
18.0	1.00232	1.00235	1.00238	1.00241	1.00244	1.00247	1.00251	1.00254
18.2	1.00235	1.00238	1.00241	1.00245	1.00248	1.00251	1.00254	1.00257
18.4	1.00239	1.00242	1.00245	1.00248	1.00251	1.00254	1.00258	1.00261
18.6	1.00242	1.00245	1.00249	1.00252	1.00255	1.00258	1.00261	1.00264
18.8	1.00246	1.00249	1.00252	1.00255	1.00258	1.00262	1.00265	1.00268
19.0	1.00249	1.00253	1.00256	1.00259	1.00262	1.00265	1.00268	1.00272
19.2	1.00253	1.00256	1.00259	1.00263	1.00266	1.00269	1.00272	1.00275
19.4	1.00257	1.00260	1.00263	1.00266	1.00270	1.00273	1.00276	1.00279
19.6	1.00261	1.00264	1.00267	1.00270	1.00273	1.00276	1.00280	1.00283
19.8	1.00265	1.00268	1.00271	1.00274	1.00277	1.00280	1.00283	1.00287
20.0	1.00268	1.00272	1.00275	1.00278	1.00281	1.00284	1.00287	1.00290
20.2	1.00272	1.00275	1.00279	1.00282	1.00285	1.00288	1.00291	1.00294
20.4	1.00276	1.00279	1.00283	1.00286	1.00289	1.00292	1.00295	1.00298
20.6	1.00280	1.00283	1.00287	1.00290	1.00293	1.00296	1.00299	1.00302
20.8	1.00284	1.00287	1.00291	1.00294	1.00297	1.00300	1.00303	1.00306
21.0	1.00288	1.00292	1.00295	1.00298	1.00301	1.00304	1.00307	1.00310
21.2	1.00288	1.00292	1.00295	1.00298	1.00301	1.00304	1.00307	1.00310
21.4	1.00297	1.00300	1.00303	1.00306	1.00309	1.00312	1.00315	1.00319
21.6	1.00301	1.00304	1.00307	1.00310	1.00313	1.00316	1.00320	1.00323
21.8	1.00305	1.00308	1.00311	1.00314	1.00318	1.00321	1.00324	1.00327
22.0	1.00309	1.00313	1.00316	1.00319	1.00322	1.00325	1.00328	1.00331
22.2	1.00314	1.00317	1.00320	1.00323	1.00326	1.00329	1.00332	1.00336
22.4	1.00318	1.00321	1.00324	1.00327	1.00331	1.00334	1.00337	1.00340

APPENDIX-1 (Continued)

Temperature	Atmospheric Pressure or Barometric Pressure (hPa or mbar)							
(°C)	850	880	910	940	970	1,000	1,030	1,060
22.6	1.00322	1.00326	1.00329	1.00332	1.00335	1.00338	1.00341	1.00344
22.8	1.00327	1.00330	1.00333	1.00336	1.00339	1.00343	1.00346	1.00349
23.0	1.00331	1.00335	1.00338	1.00341	1.00344	1.00347	1.00350	1.00353
23.2	1.00336	1.00339	1.00342	1.00345	1.00348	1.00352	1.00355	1.00358
23.4	1.00341	1.00344	1.00347	1.00350	1.00353	1.00356	1.00359	1.00362
23.6	1.00345	1.00348	1.00351	1.00354	1.00358	1.00361	1.00364	1.00367
23.8	1.00350	1.00353	1.00356	1.00359	1.00362	1.00365	1.00368	1.00372
24.0	1.00354	1.00358	1.00361	1.00364	1.00367	1.00370	1.00373	1.00376
24.2	1.00359	1.00362	1.00365	1.00369	1.00372	1.00375	1.00378	1.00381
24.4	1.00364	1.00367	1.00370	1.00373	1.00376	1.00379	1.00383	1.00386
24.6	1.00369	1.00372	1.00375	1.00378	1.00381	1.00384	1.00387	1.00390
24.8	1.00374	1.00377	1.00380	1.00383	1.00386	1.00389	1.00392	1.00395
25.0	1.00379	1.00382	1.00385	1.00388	1.00391	1.00394	1.00397	1.00400
25.2	1.00383	1.00387	1.00390	1.00393	1.00396	1.00399	1.00402	1.00405
25.4	1.00388	1.00392	1.00395	1.00398	1.00401	1.00404	1.00407	1.00410
25.6	1.00393	1.00397	1.00400	1.00403	1.00406	1.00409	1.00412	1.00415
25.8	1.00398	1.00402	1.00405	1.00408	1.00411	1.0044	1.00417	1.00420
26.0	1.00404	1.00407	1.00410	1.00413	1.00416	1.00419	1.00422	1.00425
26.2	1.00409	1.00412	1.00415	1.00418	1.00421	1.00424	1.00427	1.00430
26.4	1.00414	1.00417	1.00420	1.00423	1.00426	1.00429	1.00432	1.00435
26.6	1.00419	1.00422	1.00425	1.00428	1.00431	1.00434	1.00438	1.00441
26.8	1.00424	1.00427	1.00430	1.00433	1.00437	1.00440	1.00443	1.00446
27.0	1.00430	1.00433	1.00436	1.00439	1.00442	1.00445	1.00448	1.00451
27.2	1.00435	1.00438	1.00441	1.00444	1.00447	1.00450	1.00453	1.00456
27.4	1.00440	1.00443	1.00446	1.00449	1.00452	1.00456	1.00459	1.00462
27.6	1.00446	1.00449	1.00452	1.00455	1.00458	1.00461	1.00464	1.00467
27.8	1.00451	1.00454	1.00457	1.00460	1.00463	1.00466	1.00469	1.00472
28.0	1.00456	1.00460	1.00463	1.00466	1.00469	1.00472	1.00475	1.00478
28.2	1.00462	1.00465	1.00468	1.00471	1.00474	1.00477	1.00480	1.00483
28.4	1.00467	1.00471	1.00474	1.00477	1.00480	1.00483	1.00486	1.00489
28.6	1.00473	1.00476	1.00479	1.00482	1.00485	1.00488	1.00491	1.00494
28.8	1.00479	1.00482	1.00485	1.00488	1.00491	1.00494	1.00497	1.00500
29.0	1.00484	1.00487	1.00490	1.00493	1.00497	1.00500	1.00503	1.00506
29.2	1.00490	1.00493	1.00496	1.00499	1.00502	1.00505	1.00508	1.00511
29.4	1.00496	1.00499	1.00502	1.00505	1.00508	1.00511	1.00514	1.00517
29.6	1.00501	1.00504	1.00508	1.00511	1.00514	1.00517	1.00520	1.00523
29.8	1.00507	1.00510	1.00513	1.00516	1.00519	1.00522	1.00526	1.00529
30.0	1.00513	1.00516	1.00519	1.00522	1.00525	1.00528	1.00531	1.00534

APPENDIX-2 Factor *Z* for Borosilicate Glass 5.0

Temperature (°C)	Atmospheric Pressure or Barometric Pressure (hPa or mbar)							
	850	880	910	940	970	1,000	1,030	1,060
15.0	1.00188	1.00191	1.00194	1.00197	1.00200	1.00203	1.00207	1.00210
15.2	1.00190	1.00193	1.00197	1.00200	1.00203	1.00206	1.00209	1.00213
15.4	1.00193	1.00196	1.00199	1.00203	1.00206	1.00209	1.00212	1.00215
15.6	1.00196	1.00199	1.00202	1.00205	1.00208	1.00212	1.00215	1.00218
15.8	1.00199	1.00202	1.00205	1.00208	1.00211	1.00214	1.00218	1.00221
16.0	1.00201	1.00205	1.00208	1.00211	1.00214	1.00217	1.00221	1.00224
16.2	1.00204	1.00208	1.00211	1.00214	1.00217	1.00220	1.00223	1.00227
16.4	1.00207	1.00211	1.00214	1.00217	1.00220	1.00223	1.00226	1.00230
16.6	1.00210	1.00214	1.00217	1.00220	1.00223	1.00226	1.00229	1.00233
16.8	1.00213	1.00217	1.00220	1.00223	1.00226	1.00229	1.00232	1.00236
17.0	1.00217	1.00220	1.00223	1.00226	1.00229	1.00232	1.00236	1.00239
17.2	1.00220	1.00223	1.00226	1.00229	1.00232	1.00235	1.00239	1.00242
17.4	1.00223	1.00226	1.00229	1.00232	1.00236	1.00239	1.00242	1.00245
17.6	1.00226	1.00229	1.00232	1.00236	1.00239	1.00242	1.00245	1.00248
17.8	1.00229	1.00233	1.00236	1.00239	1.00242	1.00245	1.00248	1.00251
18.0	1.00233	1.00236	1.00239	1.00242	1.00245	1.00248	1.00252	1.00255
18.2	1.00236	1.00239	1.00242	1.00246	1.00249	1.00252	1.00255	1.00258
18.4	1.00239	1.00243	1.00246	1.00249	1.00252	1.00255	1.00258	1.00262
18.6	1.00243	1.00246	1.00249	1.00252	1.00256	1.00259	1.00262	1.00265
18.8	1.00246	1.00250	1.00253	1.00256	1.00259	1.00262	1.00265	1.00269
19.0	1.00250	1.00253	1.00256	1.00259	1.00263	1.00266	1.00269	1.00272
19.2	1.00254	1.00257	1.00260	1.00263	1.00266	1.00269	1.00272	1.00276
19.4	1.00257	1.00260	1.00264	1.00267	1.00270	1.00273	1.00276	1.00279
19.6	1.00261	1.00264	1.00267	1.00270	1.00273	1.00277	1.00280	1.00283
19.8	1.00265	1.00268	1.00271	1.00274	1.00277	1.00280	1.00283	1.00287
20.0	1.00268	1.00272	1.00275	1.00278	1.00281	1.00284	1.00287	1.00290
20.2	1.00272	1.00275	1.00278	1.00282	1.00285	1.00288	1.00291	1.00294
20.4	1.00276	1.00279	1.00282	1.00285	1.00289	1.00292	1.00295	1.00298
20.6	1.00280	1.00283	1.00286	1.00289	1.00292	1.00296	1.00299	1.00302
20.8	1.00284	1.00287	1.00290	1.00293	1.00296	1.00300	1.00303	1.00306
21.0	1.00288	1.00291	1.00294	1.00297	1.00300	1.00304	1.00307	1.00310
21.2	1.00292	1.00295	1.00298	1.00301	1.00304	1.00308	1.00311	1.00314
21.4	1.00296	1.00299	1.00302	1.00305	1.00308	1.00312	1.00315	1.00318
21.6	1.00300	1.00303	1.00306	1.00309	1.00313	1.00316	1.00319	1.00322
21.8	1.00304	1.00307	1.00310	1.00314	1.00317	1.00320	1.00323	1.00326
22.0	1.00308	1.00311	1.00315	1.00318	1.00321	1.00324	1.00327	1.00330
22.2	1.00313	1.00316	1.00319	1.00322	1.00325	1.00328	1.00331	1.00334
22.4	1.00317	1.00320	1.00323	1.00326	1.00329	1.00332	1.00336	1.00339

APPENDIX-2 (Continued)

Temperature	Atmospheric Pressure or Barometric Pressure (hPa or mbar)							
(°C)	850	880	910	940	970	1,000	1,030	1,060
22.6	1.00321	1.00324	1.00327	1.00331	1.00334	1.00337	1.00340	1.00343
22.8	1.00326	1.00329	1.00332	1.00335	1.00338	1.00341	1.00344	1.00347
23.0	1.00330	1.00333	1.00336	1.00339	1.00342	1.00345	1.00349	1.00352
23.2	1.00334	1.00337	1.00341	1.00344	1.00347	1.00350	1.00353	1.00356
23.4	1.00339	1.00342	1.00345	1.00348	1.00351	1.00354	1.00357	1.00361
23.6	1.00343	1.00346	1.00350	1.00353	1.00356	1.00359	1.00362	1.00365
23.8	1.00348	1.00351	1.00354	1.00357	1.00360	1.00363	1.00366	1.00370
24.0	1.00352	1.00356	1.00359	1.00362	1.00365	1.00368	1.00371	1.00374
24.2	1.00357	1.00360	1.00363	1.00366	1.00369	1.00373	1.00376	1.00379
24.4	1.00362	1.00365	1.00368	1.00371	1.00374	1.00377	1.00380	1.00383
24.6	1.00366	1.00370	1.00373	1.00376	1.00379	1.00382	1.00385	1.00388
24.8	1.00371	1.00374	1.00377	1.00380	1.00384	1.00387	1.00390	1.00393
25.0	1.00376	1.00379	1.00382	1.00385	1.00388	1.00391	1.00395	1.00398
25.2	1.00381	1.00384	1.00387	1.00390	1.00393	1.00396	1.00399	1.00402
25.4	1.00386	1.00389	1.00392	1.00395	1.00398	1.00401	1.00404	1.00407
25.6	1.00391	1.00394	1.00397	1.00400	1.00403	1.00406	1.00409	1.00412
25.8	1.00396	1.00399	1.00402	1.00405	1.00408	1.00411	1.00414	1.00417
26.0	1.00400	1.00404	1.00407	1.00410	1.00413	1.00416	1.00419	1.00422
26.2	1.00405	1.00409	1.00412	1.00415	1.00418	1.00421	1.00424	1.00427
26.4	1.00411	1.00414	1.00417	1.00420	1.00423	1.00426	1.00429	1.00432
26.6	1.00416	1.00419	1.00422	1.00425	1.00428	1.00431	1.00434	1.00437
26.8	1.00421	1.00424	1.00427	1.00430	1.00433	1.00436	1.00439	1.00442
27.0	1.00426	1.00429	1.00432	1.00435	1.00438	1.00441	1.00444	1.00447
27.2	1.00431	1.00434	1.00437	1.00440	1.00443	1.00447	1.00450	1.00453
27.4	1.00436	1.00439	1.00443	1.00446	1.00449	1.00452	1.00455	1.00458
27.6	1.00442	1.00445	1.00448	1.00451	1.00454	1.00457	1.00460	1.00463
27.8	1.00447	1.00450	1.00453	1.00456	1.00459	1.00462	1.00465	1.00468
28.0	1.00452	1.00455	1.00458	1.00462	1.00465	1.00468	1.00471	1.00474
28.2	1.00458	1.00461	1.00464	1.00467	1.00470	1.00473	1.00473	1.00479
28.4	1.00463	1.00466	1.00469	1.00472	1.00475	1.00478	1.00482	1.00485
28.6	1.00469	1.00472	1.00475	1.00478	1.00481	1.00484	1.00487	1.00490
28.8	1.00474	1.00477	1.00480	1.00483	1.00486	1.00489	1.00492	1.00496
29.0	1.00480	1.00483	1.00486	1.00489	1.00492	1.00495	1.00498	1.00501
29.2	1.00485	1.00488	1.00491	1.00494	1.00497	1.00501	1.00504	1.00507
29.4	1.00491	1.00494	1.00497	1.00500	1.00503	1.00506	1.00509	1.00512
29.6	1.00496	1.00500	1.00503	1.00506	1.00509	1.00512	1.00515	1.00518
29.8	1.00502	1.00505	1.00508	1.00511	1.00514	1.00517	1.00520	1.00524
30.0	1.00508	1.00511	1.00514	1.00517	1.00520	1.00523	1.00526	1.00529

APPENDICE-3 Factor *Z* for Soda-Lime Glass

Temperature (°C)	Atmospheric Pressure or Barometric Pressure (hPa or mbar)							
	850	880	910	940	970	1,000	1,030	1,060
15.0	1.00194	1.00197	1.00200	1.00203	1.00206	1.00210	1.00213	1.00216
15.2	1.00196	1.00199	1.00202	1.00206	1.00209	1.00212	1.00215	1.00218
15.4	1.00199	1.00202	1.00205	1.00208	1.00211	1.00214	1.00218	1.00221
15.6	1.00201	1.00204	1.00207	1.00211	1.00214	1.00217	1.00220	1.00223
15.8	1.00204	1.00207	1.00210	1.00213	1.00216	1.00220	1.00223	1.00226
16.0	1.00206	1.00209	1.00213	1.00216	1.00219	1.00222	1.00225	1.00229
16.2	1.00209	1.00212	1.00215	1.00218	1.00222	1.00225	1.00228	1.00231
16.4	1.00212	1.00215	1.00218	1.00221	1.00224	1.00228	1.00231	1.00234
16.6	1.00214	1.00218	1.00221	1.00224	1.00227	1.00230	1.00233	1.00237
16.8	1.00217	1.00220	1.00224	1.00227	1.00230	1.00233	1.00236	1.00239
17.0	1.00220	1.00223	1.00226	1.00230	1.00233	1.00236	1.00239	1.00242
17.2	1.00223	1.00226	1.00229	1.00233	1.00236	1.00239	1.00242	1.00245
17.4	1.00226	1.00229	1.00232	1.00235	1.00239	1.00242	1.00245	1.00248
17.6	1.00229	1.00232	1.00235	1.00238	1.00242	1.00245	1.00248	1.00251
17.8	1.00232	1.00235	1.00238	1.00241	1.00245	1.00248	1.00251	1.00254
18.0	1.00235	1.00238	1.00241	1.00245	1.00248	1.00251	1.00254	1.00257
18.2	1.00238	1.00241	1.00245	1.00248	1.00251	1.00254	1.00257	1.00260
18.4	1.00241	1.00245	1.00248	1.00251	1.00254	1.00257	1.00260	1.00263
18.6	1.00245	1.00248	1.00251	1.00254	1.00257	1.00260	1.00264	1.00267
18.8	1.00248	1.00251	1.00254	1.00257	1.00261	1.00264	1.00267	1.00270
19.0	1.00251	1.00254	1.00258	1.00261	1.00264	1.00267	1.00270	1.00273
19.2	1.00255	1.00258	1.00261	1.00264	1.00267	1.00270	1.00273	1.00277
19.4	1.00258	1.00261	1.00264	1.00267	1.00271	1.00274	1.00277	1.00280
19.6	1.00261	1.00265	1.00268	1.00271	1.00274	1.00277	1.00280	1.00283
19.8	1.00265	1.00268	1.00271	1.00274	1.00277	1.00281	1.00284	1.00287
20.0	1.00268	1.00272	1.00275	1.00278	1.00281	1.00284	1.00287	1.00290
20.2	1.00272	1.00275	1.00278	1.00281	1.00285	1.00288	1.00291	1.00294
20.4	1.00276	1.00279	1.00282	1.00285	1.00288	1.00291	1.00294	1.00298
20.6	1.00279	1.00282	1.00285	1.00289	1.00292	1.00295	1.00298	1.00301
20.8	1.00283	1.00286	1.00289	1.00292	1.00295	1.00299	1.00302	1.00305
21.0	1.00287	1.00290	1.00293	1.00296	1.00299	1.00302	1.00305	1.00309
21.2	1.00290	1.00294	1.00297	1.00300	1.00303	1.00306	1.00309	1.00312
21.4	1.00294	1.00297	1.00301	1.00304	1.00307	1.00310	1.00313	1.00316
21.6	1.00298	1.00301	1.00304	1.00307	1.00311	1.00314	1.00317	1.00320
21.8	1.00302	1.00305	1.00308	1.00311	1.00315	1.00318	1.00321	1.00324
22.0	1.00306	1.00309	1.00312	1.00315	1.00318	1.00322	1.00325	1.00328
22.2	1.00310	1.00313	1.00316	1.00319	1.00322	1.00326	1.00329	1.00332
22.4	1.00314	1.00317	1.00320	1.00323	1.00326	1.00330	1.00333	1.00336

APPENDICE-3 *(Continued)*

Temperature	Atmospheric Pressure or Barometric Pressure (hPa or mbar)							
(°C)	850	880	910	940	970	1,000	1,030	1,060
22.6	1.00318	1.00321	1.00324	1.00327	1.00330	1.00334	1.00337	1.00340
22.8	1.00322	1.00325	1.00328	1.00331	1.00335	1.00338	1.00341	1.00344
23.0	1.00326	1.00329	1.00333	1.00336	1.00339	1.00342	1.00345	1.00348
23.2	1.00330	1.00334	1.00337	1.00340	1.00343	1.00346	1.00349	1.00352
23.4	1.00335	1.00338	1.00341	1.00344	1.00347	1.00350	1.00353	1.00356
23.6	1.00339	1.00342	1.00345	1.00348	1.00351	1.00355	1.00358	1.00361
23.8	1.00343	1.00346	1.00349	1.00353	1.00356	1.00359	1.00362	1.00365
24.0	1.00348	1.00351	1.00354	1.00357	1.00360	1.00363	1.00366	1.00369
24.2	1.00352	1.00355	1.00358	1.00361	1.00364	1.00368	1.00371	1.00374
24.4	1.00356	1.00360	1.00363	1.00366	1.00369	1.00372	1.00375	1.00378
24.6	1.00361	1.00364	1.00367	1.00370	1.00373	1.00376	1.00379	1.00383
24.8	1.00365	1.00369	1.00372	1.00375	1.00378	1.00381	1.00384	1.00387
25.0	1.00370	1.00373	1.00376	1.00379	1.00382	1.00385	1.00389	1.00392
25.2	1.00375	1.00378	1.00381	1.00384	1.00387	1.00390	1.00393	1.00396
25.4	1.00379	1.00382	1.00385	1.00388	1.00392	1.00395	1.00398	1.00401
25.6	1.00384	1.00387	1.00390	1.00393	1.00396	1.00399	1.00402	1.00405
25.8	1.00389	1.00392	1.00395	1.00398	1.00401	1.00404	1.00407	1.00410
26.0	1.00393	1.00396	1.00399	1.00403	1.00406	1.00409	1.00412	1.00415
26.2	1.00398	1.00401	1.00404	1.00407	1.00410	1.00413	1.00417	1.00420
26.4	1.00403	1.00406	1.00409	1.00412	1.00415	1.00418	1.00421	1.00424
26.6	1.00408	1.00411	1.00414	1.00417	1.00420	1.00423	1.00426	1.00429
26.8	1.00413	1.00416	1.00419	1.00422	1.00425	1.00428	1.00431	1.00434
27.0	1.00418	1.00421	1.00424	1.00427	1.00430	1.00433	1.00436	1.00439
27.2	1.00422	1.00426	1.00429	1.00432	1.00435	1.00438	1.00441	1.00444
27.4	1.00427	1.00431	1.00434	1.00437	1.00440	1.00443	1.00446	1.00449
27.6	1.00433	1.00436	1.00439	1.00442	1.00445	1.00448	1.00451	1.00454
27.8	1.00438	1.00441	1.00444	1.00447	1.00450	1.00453	1.00456	1.00459
28.0	1.00443	1.00446	1.00449	1.00452	1.00455	1.00458	1.00461	1.00464
28.2	1.00448	1.00451	1.00454	1.00457	1.00460	1.00463	1.00466	1.00469
28.4	1.00453	1.00456	1.00459	1.00462	1.00465	1.00468	1.00471	1.00474
28.6	1.00458	1.00461	1.00464	1.00467	1.00471	1.00474	1.00477	1.00480
28.8	1.00464	1.00467	1.00470	1.00473	1.00476	1.00479	1.00482	1.00485
29.0	1.00469	1.00472	1.00475	1.00478	1.00481	1.00484	1.00487	1.00490
29.2	1.00474	1.00477	1.00480	1.00483	1.00486	1.00489	1.00492	1.00496
29.4	1.00480	1.00483	1.00486	1.00489	1.00492	1.00495	1.00498	1.00501
29.6	1.00485	1.00488	1.00491	1.00494	1.00497	1.00500	1.00503	1.00506
29.8	1.00490	1.00493	1.00496	1.00500	1.00503	1.00506	1.00509	1.00512
30.0	1.00496	1.00499	1.00502	1.00505	1.00508	1.00511	1.00514	1.00517

1.2 CALIBRATION OF BUTYROMETER

1.2.1 INTRODUCTION

Gerber method is the method invented by Swiss chemist Niklaus Gerber for estimation of fat content in milk and milk products. Butyrometers are the graduated glass tubes used in the Gerber method for estimation of fat content. Butyrometers (Table 1.8) are divided into 11 types depending upon the scale and their end uses [15].

TABLE 1.8 Types of Butyrometers [1]

Sl. No.	Scale	Used for
1.	0–10%	Whole and evaporated (unsweetened) milk.
2.	0–8%	
3.	0–6%	
4.	0–0.5%	Skim milk.
5.	0–4%	Separated milk, partly skim milk and butter milk.
6.	0–70%	Cream.
7.	0–40%	Cheese.
8.	0–20%	Dry milk powder.
9.	0–20%	Ice cream and condensed milk.
10.	0–90%	Butter.
11.	0–40%	Dry whole milk powder.

Butyrometer is made up of borosilicate glass 3.3 (Type I) having a coefficient of linear thermal expansion of 3.3 x 10^{-6}°C^{-1}. It shall be resistant to chemicals and thermal shock. It shall be well annealed and free from the presence of air bubbles, cracks, blisters, and any other visible defects. Graduation marks on the scale shall be of uniform thickness (not less than 0.1 mm and not more than 0.2 mm), clear, fine, and permanent line. Space between the different graduation marks should be equal and there should not be any irregularities. Graduation marks shall be lying in planes perpendicular to the longitudinal axis of the graduated tube [15]. Table 1.9 indicates permissible error for different types of butyrometers.

1.2.2 PRINCIPLE

The graduation on the milk butyrometer is such that each 1% corresponds to an internal volume of 0.125 ml (while in the case of cream and cheese

TABLE 1.9 Permissible Error for Different Types of Butyrometers

Graduation Scale	Quantity of Material Added	Internal Volume of Graduation Scale (Corresponding to 1% Fat)	Numbered at Each %	Error, g of Fat for 100 g of Sample, Max.
%	ml	ml	–	%
0–6	10.75 ml	0.125	1	0.05
0–8	10.75 ml	0.125	1	0.05
0–10	10.75 ml	0.125	1	0.05
0–4	10.75 ml	0.125	1	0.025
0–0.5	21.5 ml	0.125	0.1	0.01
0–40	3 g	0.0338	10	0.50
0–70	5 g	0.0568	10	0.50

the internal volumes are 0.0568 and 0.0338 ml, respectively). Therefore, to check the accuracy of milk butyrometer, internal volume of each 1% mark is determined, which shall be 0.125±0.001 ml. At least, three points on the scale should be checked [15].

1.2.3 MATERIAL REQUIRED

- Mercury/micro-pipette (1 or 2 ml, capable of measuring 0.125/0.250 ml fraction);
- Burette stand;
- Butyrometer;
- Cotton.

1.2.4 PROCEDURE

1.2.4.1 METHOD I

It is required to test four points, on the scale of 10% milk butyrometer. The butyrometer previously cleaned and dried, is clamped vertically with the small bulb pointing downwards. Mercury is run into the butyrometer care being taken to avoid trapping air bubbles, until the top of the mercury meniscus is in horizontal plane passing through the bottom edge of the 10% graduation mark. Before making the final adjustment to the mark, the butyrometer is getting tapped to obtain a well-formed meniscus. The mercury pipette is then filled and the mercury delivered into the butyrometer. The butyrometer is again tapped and the position of the top of the mercury meniscus relative to the scale is noted. Since 0.25 ml of mercury has been transferred to the butyrometer, the scale reading should be 2.0% and any deviation from this reading represents the error on the scale. Successions 0.25 ml volume of mercury is transferred from the pipette to the butyrometer and scale reading after each delivery is noted [15].

1.2.4.2 METHOD II

Take previously cleaned and properly dried milk butyrometer. Place the butyrometer in a flask in the vertical position with the bulb downside. Fill the mercury up to the mark of 10% in milk butyrometer with the help of micro-burette. The meniscus of the mercury should be adjusted exactly at 10% with the help of magnifying lens. Note down the weight of the butyrometer (W_1). It is then filled to the next graduation mark (as per the type of butyrometer) and the weight of mercury filling 1% graduation (10 to 9%, 9 to 8%, and so on) is recorded. The weight of the mercury (W_2) filled in each 1% (depending upon the type of butyrometer) graduation mark is calculated. Volume of mercury in 1% graduation is calculated using the weight and density of mercury at room temperature. In case of milk butyrometer, each 1% graduation mark should have an internal volume of 0.125±0.001 ml. Cream butyrometers are checked at each 10% graduation marks (at three different points). Internal volume for 10% graduation mark should be 0.568±0.004 ml. Cheese butyrometers are checked at each 5% graduation mark (at three different points). Internal volume for 5% graduation mark should be 0.169±0.002 ml [15].

1.2.5 CALCULATION

$$V = (W_2 - W_1)/D \tag{3}$$

where; V is the internal volume (ml) of graduation scale under test; W_1 is the weight of butyrometer with mercury at point 1; W_2 is the weight of butyrometer with mercury at point 2; and D is the density of mercury at laboratory temperature.

1.2.6 DATA COLLECTION SHEET

Title: Calibration of butyrometer.

Location and Organization:________________________________

Participants: ____

Date (Month/Day/Year): ________________________________

Observations:

Parameter	Repeat 1	Repeat 2	Repeat 3	Average
Weight of butyrometer with mercury at point 1 (W_1 g)				
Weight of butyrometer with mercury at point 2 (W_2 g)				
Temperature of laboratory (tC)				
Density of mercury at t°C (D)				
Internal volume (V ml)				

Comments:____.

1.3 CALIBRATION OF THERMOMETER

1.3.1 INTRODUCTION

Thermometer is used to measure the temperature or temperature gradient [19]. In 1624, J. Leurechon coined the word "thermometer" (French form) which is the combination of two Greek words, "thermos (meaning hot)" and "metron (meaning measured)" [16]. Thermometer is generally sold with the calibration certificate; however, if not then it is calibrated against the reference thermometer. Due to the thermal cycling, from room temperature to lower or higher temperature characteristic of the thermometer changes with time, hence there is the need from time-to-time calibration of thermometer usually after every 12 months [17, 19]. Higher quality thermometers are more sensitive to mishandling, i.e., mechanical stresses or shocks [19]. Tolerance values for different types of thermometers are given in Table 1.10 [18].

Temperature drift in the thermometers occurs due to both reversible and irreversible changes. Reversible changes include gap or presence of gas bubble in the liquid column and temporary change in the volume of bulb due to the high temperature exposure. Irreversible changes include permanent change in bulb volume, microscopic alteration and cracking at higher temperature, thermometer liquid degradation and physical damage [17].

TABLE 1.10 Tolerance Values for Different Types of Thermometers

Devices	Maximum Division Type I (Liquid-in-Glass)		Maximum Division Type II (Digital)		Typical Clearance		Tolerance	
	(°C)	(°F)	(°C)	(°F)	(°C)	(°F)	(°C)	(°F)
Petroleum Products								
Wholesale petroleum meters	1	0.5	0.1	0.1	0 to 120	0 to 50	1	0.5
Liquefied petroleum gas meters	1	0.5	0.1	0.1	–20 to 120	0 to 50	1	0.5
Hydrocarbon gas vapor meters	1	0.5	0.1	0.1	60 to 120	15 to 50	1	0.5
Grain Moisture Testing								
Accessory thermometer	1	0.5	0.1	0.1	32 to 104	0 to 40	1	0.5
Standard thermometer	0.2	0.1	0.01	0.01	32 to 102	0 to 40	0.2	0.1
Quality Control								
Alcoholic beverages	1	0.5	0.1	0.1	–20 to 120	–30 to 50	1	0.5
Compressed gas in cylinders	1	0.5	0.1	0.1	20 to 100	–10 to 40	1	0.5
Petroleum products	1	0.5	0.1	0.1	–20 to 120	–30 to 50	1	0.5
Aerosol commodities	1	0.5	0.1	0.1	20 to 100	–10 to 40	1	0.5
Frozen foods	1	0.5	0.1	0.1	–30 to 120	0 to 50	1	0.5

Liquid-in-glass thermometer is the type of thermometer in which liquid is filled in the glass tube (Figure 1.5). Liquid-in-glass thermometer is generally filled with mercury or organic liquid. Mercury thermometer can be used up to 200°C temperature and recalibrated at ice point (0°C). Full recalibration of mercury thermometer is not required. Mercury thermometer is not affected by aging, however, results of organic liquid filled thermometer varies with time. This is due to the lower ratio of bulb to capillary volume, less stability of organic liquid as compared to mercury and liquid interaction with capillary wall.

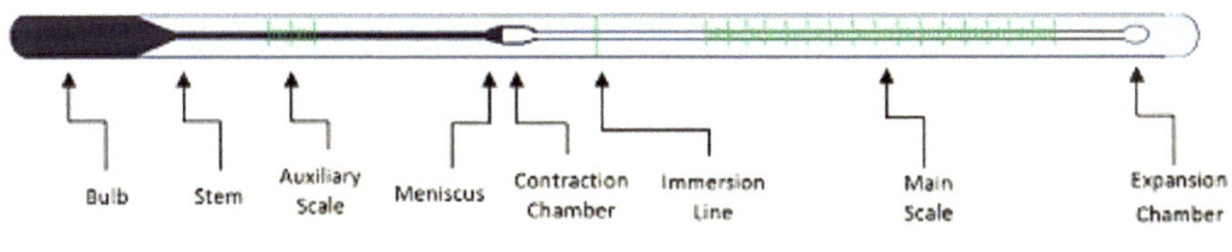

FIGURE 1.5 Liquid-in-glass thermometer.

Thermometer can be further divided into two groups, i.e., total immersion thermometer and partial immersion thermometer (Figure 1.6). Partial immersion thermometer is provided with the immersion line and thermometer needs to be immersed up to that line prior to read the temperature on thermometer. Total immersion thermometer is not provided with any line, these needs to be fully immersed except last 1 cm of thermometer, in liquid for which temperature needs to be checked [17].

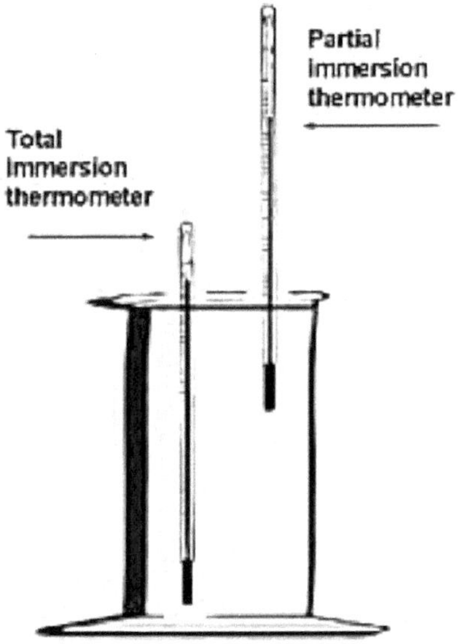

FIGURE 1.6 Partial and total immersion thermometers.

1.3.2 PRINCIPLE

Liquid-in-glass thermometer is the type of thermometer in which liquid is filled in the glass tube. Glass tube is provided with the bulb and stem. Bulb is filled with the liquid material and connected with narrow graduated capillary. Liquid filled in thermometer has higher thermal expansion as compared to glass either mercury or organic liquid. During the measurement of temperature, liquid will expand or contract demanding upon the temperature of solution for which temperature is being measured. Liquid expansion forces the liquid in the capillary and based upon the liquid meniscus position on the graduated capillary, temperature value is measured. If the thermometer is calibrated, then additional correction factor (either positive or negative) will be added to the measured temperature. Positive correction factor, if the measured temperature is less than the actual value and negative if it more [17].

Partial immersion thermometer has greater sensitivity (for variation in ambient temperature) and lower reproducibility than total immersion thermometer, as the stem is not maintained at controlled temperature. The sensitivity to change in ambient temperature is also higher for organic liquid filled thermometer than mercury thermometer. Incorrect immersion of both the thermometers can lead to large measurement errors.

1.3.3 MATERIALS REQUIRED

- Thermometer.
- Ice point: Ice point is prepared using the crushed ice, water, and insulated flask. It is a secondary fixed point which can be easily prepared in any laboratory with the uncertainty of 0.002°C.

1.3.4 PROCEDURE

1.3.4.1 PRECONDITIONING OF THERMOMETER

Liquid thermometer, which has not been recently used, and has maximum temperature 150°C or more, needs to be preconditioned (or thermally recycled) prior to recalibration. Heat the thermometer up to the maximum temperature of its use and hold it at that temperature until the bulb is heated within 1°C of its equilibrium temperature, and then cool it to room temperature. After thermal cycling, wait for 72 h prior to recalibration.

If the thermometer is used above 300°C, then ice point measurement also needs to be done, before and after thermal cycling. If ice point measurement is not correct within one scale division then reject it for calibration and use.

1.3.4.2 ICE POINT PREPARATION

Ice used for the preparation of ice point is made up of deionized or distilled water. Crushed ice with a diameter of 2 to 5 mm is placed in the insulated vessel with distilled water and firmly packed. Completely fill the vessel with ice and crevices with distilled water. Wait for 15–30 min and afterwards all the melted water has been siphoned. Space created by the removal of water is again filled with the crushed ice. The process is repeated until the ice sets for 15–30 min. Ice should become saturated with the distilled water and it should not float in the vessel. Excess water on the surface should also be avoided. Excess water produced during the process is removed from the bottom of the vessel through siphon tube. During the whole duration of the experiment, siphoned water is replaced with ice [17]. Expected temperature for ice point calibration is 0°C.

1.3.4.3 RECALIBRATION OF THERMOMETER

Prior to the calibration, the thermometer needs to be checked for any visual separation of the liquid column, filling liquid degradation, cracks, and degradation of glass and loss of scale legibility. Liquid column of thermometer is observed under the microscope (magnification 10X) or using hand magnifier for better accuracy. Ice point should be free from any type of contaminants. Person who is calibrating the thermometer should also wash their hands just before starting the calibration or worn the plastic gloves. The thermometer should be properly cleaned prior to reading the ice point reading. It should be washed with the distilled water at room temperature or even cold water. Warm water expands the bulb volume and resulted in erroneous reading. During ice point reading partial immersion thermometer is immersed up to the immersion line and total immersion thermometer is immersed completely except last 1 cm of the thermometer. Hold the thermometer for two minutes for temperature equilibration and then take out the thermometer from the insulated vessel and read the reading on the thermometer graduated scale. If it reads zero, then the thermometer is properly calibrated and if it is higher and lower than zero, then there is a need to introduce the correction factor [17].

1.3.5 CALCULATION

$$\text{Correction factor for the thermometer} = (\text{Expected temperature}) - (\text{Actual temperature}) \quad (4)$$

1.3.6 DATA COLLECTION SHEET

Title: Calibration of thermometer.

Location and Organization: ______________________________

Participants: ______________________________

Date (Month/Day/Year): ______________________________

Observations:

Parameter	**Repeat 1**	**Repeat 2**	**Repeat 3**	**Average**
Expected temperature				
Actual temperature measured by thermometer at ice point				
Correction factor				

Comments: ______________________________.

1.4 CALIBRATION OF HYDROMETER

1.4.1 INTRODUCTION

Hydrometer is made up of glass consisting of one bulb attached with stem (Figure 1.7). Specific weight is present in the lower portion of the bulb and graduated scale is being attached to the stem. Hydrometer is calibrated to be used at the specific temperature. Variation in temperature of use can be resolved using the appropriate correction factors. Lactometer is a type of hydrometer which is used to measure the density of milk [21].

Lactometer is a type of specific gravity hydrometer to measure the density and specific gravity of milk [22]. Three types of lactometers, i.e.: (i) ISI lactometer; (ii) Quevenne lactometer and iii) Zeal lactometer is being used for the estimation of density and specific gravity of milk. These lactometers are used at different temperature, i.e., ISI at 27°C, Quevenne at 15.5°C and Zeal at 29°C. The temperature at which these are being used, is mentioned on the lactometer [25]. In the USA, Quevenne lactometers are more common; and in India, ISI lactometers are generally used based upon the ambient temperature.

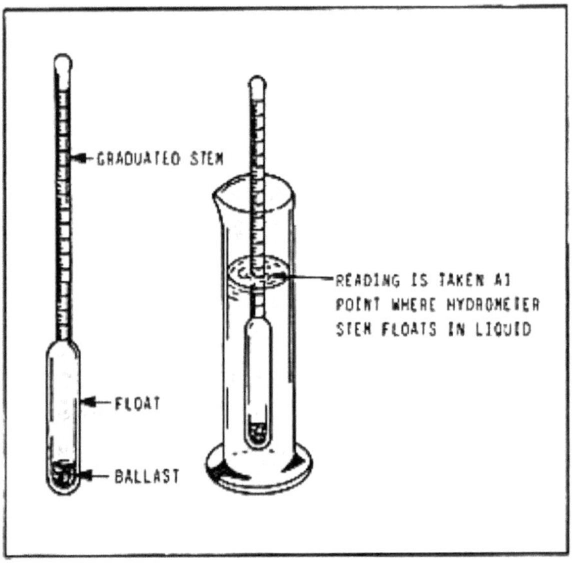

FIGURE 1.7 Hydrometer.

Primarily hydrometer is calibrated by two methods. In first method, hydrometer is floated in different known density fluids at specific temperature and density is measured. In the second method (also known as Cuckow's method), hydrometer is hydrostatically weighed when it is immersed in a single known density liquid to different levels. Cuckow's method has several advantages, i.e., different density fluid need not to be stored (i.e., different acids and various volatile hydrocarbons), calibration can be done at specific scale marks without preparation of specific density fluid, and method can be easily automated, however, it requires a specific step for hydrostatic weighing of lactometer. Hence, for regular laboratory activity first method is generally used [26].

Different fluids based upon surface tension (ST) and density can be used for calibration of hydrometers. ST of the fluid also affects the calibration of the hydrometer. Hence, prior to calibration, it is necessary to know the application of hydrometer. Hydrometer is calibrated in a liquid which has the similar ST to the fluid for which it is being used. ST of milk is 50 mN/m (millinewtons per meter); hence, lactometer is calibrated in the liquid for which ST should be 50 mN/m [23]. List of different fluids used for the preparation of different density fluid is represented in Table 1.11 [20].

TABLE 1.11 Different Density Fluids Used for Calibration of Hydrometer

Liquid Used for Hydrometer Calibration	Relative Density at 60°F
Hydrometer Used for Low Density Fluid Other Than Alcohol	
2,2-dimethylbutane	0.6540
3-methylsulfolane	1.070
Cyclopentane	0.7504
Dimethylpropane	0.5967
Isooctane	0.6962
Isopentane	0.6247
Mineral spirits (relative density 0.775) (density can be decreased using petroleum ether and increased using tetrabromoethane)	0.657–1.100
n-Pentane	0.6310
Petroleum ether	0.6250
Pure grade n-butane	0.5844
Pure grade propane	0.5077
Safrole	1.100
Hydrometer Used for Alcohol	
Mixture of water and ethanol	0.750–1.000
Hydrometer Used for Heavy Liquid	
Alcohol and tetrabromoethane (very heavy solution)	1.800 to 2.000 and above
Mixture of sulfuric acid and water	1.000–1.830
Thoulet solution (adding potassium iodide and mercuric iodide to water)	1.830–2.000

1.4.2 PRINCIPLE

Hydrometer (lactometer) works on the Archimedes principle. According to this principle, if any object is immersed in a liquid, it will be in equilibrium if it displaces the weight of the liquid equal to the weight of the object. If it displaces the less weight of the liquid as compared to the weight of the object, then it will be sink. If an object is immersed in a liquid, two types of forces act on the same. First one is buoyancy force acting in upward direction due to volume of bulb and stem below the surface of displacing fluid and portion of stem displacing the air and second force is gravitational force (due to the weight of hydrometer) acting in downward direction. Object will come to the rest in the liquid when both the forces are in equilibrium.

Specific weight is present inside the hydrometer, and it will be resulted in displacement of similar weight of the liquid. If density of liquid is high then lower volume of liquid will be displaced and if density of liquid is low then higher volume will be displaced. Graduated scale in the hydrometer has lower values on the top portion of the stem and higher values at the lower end of the stem. Calibration of hydrometer is done at specific temperature using the different density liquids having the similar ST of the fluids for which it is being used [24, 26].

1.4.3 MATERIALS REQUIRED

- Hydrometer (lactometer).
- Hydrometer jar.
- Thermometer.
- Comparator: Depth of the comparator should be in a way that hydrometer should float freely around 25 mm above the inside bottom, and it should also maintain 12.5 mm from the inner wall of the comparator [20].

1.4.4 PROCEDURE

Prior to calibration hydrometer is visually inspected for the presence of any cracks, air bubbles, rough area, deep scratches, or any other damage on the glass. There should not be any presence of foreign particles within the hydrometer. Scale inside the stem of the hydrometer should not be folded or twisted. Check the lactometer for the presence of scale slippage indicator (thin red glass strand) or line on the hydrometer which should match with the reference line on the scale. If any one of the above-mentioned defects is being observed in the lactometer, then it will be rejected for further calibration. Baume hydrometer scale has graduation lines at equal interval, however, specific gravity or density lactometer has smaller distance between the graduation lines near to the bottom of the scale.

ST of the liquid in which the hydrometer is dipped affects the reading of the hydrometer. Therefore, the intended use of hydrometer needs to be mentioned on the hydrometer. It should be calibrated in the fluid of similar ST for which it is being designed to be used.

Density of different fluids is provided in Table 1.11. Depending upon the ST of the liquid for which hydrometer is designed to be used, one solution or combination of different solutions can be used for the preparation of different density liquid needed for the calibration of hydrometer.

Standard (reference) hydrometer and test hydrometer bulb and stem should be properly cleaned and dried, so that the liquid forms a uniform layer around the bulb and stem. These should be at the temperature of the liquid in which it will be immersed for the calibration. Liquid should be uniform and at the temperature as specified on the hydrometer. Laboratory temperature should also be maintained and there should not be much difference in liquid and surrounding temperature.

Density of testing liquid is being confirmed by the standard lactometer. Pour the testing liquid in the comparator from the sides of the comparator wall to prevent the formation of air bubbles. Mix the testing liquid in the comparator and wait for 2–3 minutes. Take the necessary precautions for avoiding the formation of air bubbles. Smaller ST compounds and air bubbles used to come on the surface of the comparator. These can be removed by removing the upper layer. Thoroughly rinse the standard hydrometer with the testing liquid and slowly immerse it into the testing fluid slightly beyond the test point and left it to float freely. Observe the reading on the hydrometer scale. Graduation line just below the surface of the fluid will be considered as hydrometer reading. Raise the hydrometer in the testing fluid until the graduation line has been observed and record it. Add the correction factor (if any) in the hydrometer reading and note it down on the observation sheet as corrected hydrometer reading (CHR) for standard hydrometer.

Take out the standard hydrometer from the testing fluid and immerse the pre-rinsed test hydrometer in the testing fluid. Follow the similar process as described above with the test hydrometer. If the size of the comparator is sufficiently big to immerse both the standard and test hydrometer at the same time, then both the hydrometers are immersed simultaneously. Record three successive readings for both standard and test hydrometer; and find out the average. Calculate the correction factor for the test hydrometer by subtracting the standard hydrometer reading from the test hydrometer reading. Caution: Temperature variation during the analysis needs to be controlled. Each hydrometer needs to be calibrated at minimum three points (generally: low, medium, and high scale values). Testing fluids need to be prepared for calibration of hydrometer at all three scale values. Correction value needs to be calculated at all three points, separately [20].

1.4.5 CALCULATION

Correction factor for hydrometer at specified temperature = Standard hydrometer reading – Test hydrometer reading

$$\text{Corrected hydrometer reading (CHR)} = \text{Hydrometer reading} + \text{Correction factor}$$

$$\text{Specific gravity of fluid} = [1 + (\text{CHR}/1{,}000)] \quad (5)$$

1.4.6 DATA COLLECTION SHEET

Title: Calibration of hydrometer

Location and Organization: ________________________________

Participants: ________________________________

Date (Month/Day/Year):________________________________

Observations:

Parameter	**Repeat 1**	**Repeat 2**	**Repeat 3**	**Average**
Standard hydrometer reading				
Test hydrometer reading				
Correction factor				
Corrected hydrometer reading				
Specific gravity				

Comments: ________________________________.

KEYWORDS

- **burette**
- **butyrometer**
- **calibration**
- **lactometer**
- **measuring cylinder**
- **pipette**
- **thermometer**
- **volumetric flask**

REFERENCES

1. ASTM, (2018). *ASTM E694: Standard Specification for Laboratory Glass Volumetric Apparatus* (pp. 1–3). Philadelphia – USA: American Society for Testing and Materials (ASTM).
2. Carmignato, S., (2016). Calibration. In: Chatti, S., Laperrière, L., Reinhart, G., & Tolio, T., (eds.), *CIRP Encyclopedia of Production Engineering*. Berlin – Germany: Springer Publishing Company. doi: 10.1007/978-3-642-35950-7_16754-1.
3. Harris, G. L., & Miller, V. R., (2019). GLP-10: Good laboratory practices (GLP) for purity of water. In: *Selected Procedure for Volumetric Calibration* (pp. 1–6). Gaithersburg, MD – USA: National Institute of Standards and Technology.
4. Harris, G. L., & Miller, V. R., (2019). SOP-14: Recommended standard operating procedure (SOP) for gravimetric calibration of volumetric standards using an electronic balance. In: *Selected Procedure for Volumetric Calibration* (pp. 1–20). Gaithersburg, MD – USA: National Institute of Standards and Technology (NIST).
5. IS. *IS 1117-1975: One-Mark Pipettes (CHD 10; Glassware)* (pp. 1–14). New Delhi: Bureau of Indian Standards (BIS); (1975), (R2017).
6. IS. *IS 1997: Laboratory Glassware-Burettes* (3rd revision, pp. 1–13). New Delhi: Bureau of Indian Standards (BIS); (2008).
7. IS. *IS 878: Laboratory Glassware-Graduated Measuring Cylinders* (2nd revision, pp. 1–7). New Delhi: Bureau of Indian Standards (BIS); (2008).
8. IS. *IS 4787: Laboratory Glassware — Volumetric Instruments — Methods for Testing of Capacity and for Use* (pp. 1–21). New Delhi: Bureau of Indian Standards (BIS); (2010).
9. IS. *IS 915: Laboratory Glassware – One Mark Volumetric Flasks* (3rd revision, pp. 1–8). New Delhi: Bureau of Indian Standards (BIS); (2012).
10. ISO. *ISO-1042: Laboratory Glassware – One Mark Volumetric Flasks* (4th edn., pp. 1–7). Geneva – Switzerland: International Organization for Standardization (ISO); (1998).
11. ISO. *ISO-385: Laboratory Glassware – Burettes* (1st edn., pp. 1–12). Geneva – Switzerland: International Organization for Standardization (ISO); (2005).
12. ISO. *ISO-4788: Laboratory Glassware- Graduated Measuring Cylinders* (2nd edn., pp. 1–6). Geneva: International Organization for Standardization (ISO); (2005).
13. ISO. *ISO-648: Laboratory Glassware- Single Volume Pipettes* (2nd edn., pp. 1–8). Geneva – Switzerland: International Organization for Standardization (ISO); (2008).
14. NABL (National Accreditation Board for Testing and Calibration Laboratories), (2019). Mechanical calibration. In: *Specific Criteria for Accreditation of Calibration Laboratories (Mechanical, Fluid Flow, Radiological, Electro-Technical & Thermal Calibration)* (pp. 41–47). Gurugram – Haryana: NABL House.
15. IS. *IS 1223: Apparatus for Determination of Milk Fat by Gerber Method-Specification* (3rd revision, pp. 1–19). New Delhi: Bureau of Indian Standards (BIS); (2001).
16. Benedict, R. P., (1984). *Fundamentals of Temperature, Pressure, and Flow Measurements* (3rd edn., p. 4). New York – USA: John Wiley & Sons Inc.
17. Cross, C. D., Miller, W. W., Ripple, D. C., & Strouse, G. F., (2009). *NIST Special Publication 1088: NIST Maintenance, Validation, and Recalibration of Liquid-in-Glass Thermometer* (pp. 1–26). Gaithersburg, MD – USA: National Institute of Standards and Technology (NIST).

18. NIST, (1997). NIST Handbook 105-6: Specifications and tolerances for thermometers. In: *Specifications and Tolerances for Reference Standards and Field Standard Weights and Measures* (pp. 1–8). Gaithersburg, MD – USA: National Institute of Standards and Technology (NIST).
19. Pavese, F., (1999). Temperature measurements, scales and calibrations. Chapter 65. In: Webster, J., & Eren, H., (eds.), *The Measurement, Instrumentation, and Sensors Handbook (*pp. 65.1–65.15*)*. Boca Raton – Florida: CRC Press.
20. *ASTM E126-19: Standard Test Method for Inspection, Calibration, and Verification of ASTM Hydrometers* (pp. 1–9). West Conshohocken, PA – USA: ASTM International; (2019).
21. Bowman, H. A., & Gallagher, W. H., (1969). An improved high-precision calibration procedure for reference standard hydrometers. *Journal of Research of the National Bureau of Standards – C: Engineering and Instrumentation, 73C*(3, 4), 57–65.
22. Hughes, J. C., (1954). *NBS Circular 555: Testing of Hydrometers* (pp. 1–10). Washington, DC – USA: National Bureau of Standards (NBS).
23. IS. *IS 9585: Lactometers* (pp. 1–14). New Delhi: Bureau of Indian Standards (BIS); (1980).
24. Lorefice, S., & Malengo, A., (2006). Calibration of hydrometer. *Measurement Science and Technology, 17*, 2560–2566.
25. Mathur, M. P., Datta, R. D., & Dinakar, P., (2005). *Textbook of Dairy Chemistry* (pp. 19–21). New Delhi: Indian Council of Agricultural Research (IARI).
26. Wright, J. D., Bean, V. E., & Aguilera, J., (2020). *NIST Special Publication 250-78-RL: NIST Calibration Services for Hydrometers* (pp. 1–10). Gaithersburg, MD – USA: National Institute of Standards and Technology (NIST).

CHAPTER 2

SAMPLING METHODS OF MILK AND MILK PRODUCTS

PARTHA P. DEBNATH, KUNTAL ROY, and N. VEENA

ABSTRACT

Sampling enables comprehensive analysis of any given product for assurance of quality. Sampling is a procedure that needs more careful attention. Analysis of the laboratory samples is to be valid only when collection, transport, and storage of samples are critically done. The sample must be a true representative of the bulk, and hence, it is important to carry out a proper sampling to obtain accurate results. Samples obtained from inadequately mixed dairy products may not be representative of the entire batch, even if collection and transport are conducted under ideal circumstances. The sampling personnel must take care that no contamination (chemical or microbiological) is introduced into the sample during the collection and transportation of the sample. This chapter covers the sampling method for the chemical analysis of milk and milk products.

2.1 GENERAL INSTRUCTIONS FOR SAMPLING

2.1.1 INTRODUCTION

Assessing the quality of dairy products is done through carrying out quality assurance tests, which begin by sampling of the product. It is significant for safety and economic reasons. Sampling enables comprehensive analysis of any given milk product for assurance of quality and safety. Sampling is essential because it is practically not possible to test in the entire batch.

Analytical Methods for Milk and Milk Products, Volume 1: Sampling Methods and Chemical and Compositional Analysis. Megh R. Goyal, N. Veena, and Santosh Kumar Mishra (Eds.)

2.1.2 GENERAL ARRANGEMENTS

2.1.2.1 SAMPLING PERSONNEL

Authorized agent or sampling personnel, adequately trained in suitable techniques and is well acquainted with knowledge of the subject should perform the sampling. He must be free from any contagious diseases.

2.1.2.2 SAMPLE SEALING AND LABELLING

Sample must be sealed and labeled to give identification number, product type, batch number of the milk product, sampling date, name, and signature of the person, who was involved in sampling. If required, further details may be included, for instance, sample weight/volume, purpose of sampling, storage conditions, and condition of milk product during sampling and unit from which sample was taken.

2.1.2.3 SAMPLE REPLICATES

All samples must be taken in duplicate or in larger quantities, for legal purposes or if agreed between parties concerned. When agreed between parties, an additional number of samples must be collected and retained for independent arbitration, if needed.

2.1.2.4 SAMPLING REPORT

The report should include the following information:

- Name, designations, and signature of the person responsible for taking the sample;
- Time, date, and place of sampling;
- Purpose and method of sampling including preparation of sample;
- Nature and number of units constituting the consignment;
- Any code marking or identification number of the batch from which samples are drawn;
- Number of samples to be drawn from the particular batches as to be identified;
- Place to where samples will be sent, if required; and
- Manufacturer or supplier's name and address, etc.

When suitable, the report must also comprise pertinent conditions, such as: humidity and temperature of the atmosphere, condition of containers and their surroundings, product age, sampling equipment, sterilization methods, whether the samples were added with preservative, and any other specific information related to the milk and milk product being sampled. Test portion size and handling varies according to the purpose of analysis. Preparation of sample is also included in the sampling method. Therefore, laboratory sample preparation is clearly included in the sample report or a separate laboratory report.

2.1.3 APPARATUS

2.1.3.1 EQUIPMENTS FOR SAMPLING

Use grease- and water-proof material (stainless steel, glass, plastic, etc.) of suitable shape and capacity and the quality suitable for sterilization, if needed to be sampled. Such equipment must safeguard the sample and there should be no effect on flavor, odor, and composition of the milk product. Surfaces of the equipment must be smooth with no cracks.

For physical and chemical analysis, equipment used for sampling must be dry and clean and should not impact on the physicochemical properties and composition of the milk product. Odorless permanent ink or felt-tip pens are used for marking the samples that should not influence the product composition [1].

2.1.3.2 CONTAINERS

Containers and closures must be of suitable material (glass, stainless steel and some plastics, such as polypropylene) and construction that sufficiently safeguard the sample with no effect on flavor, odor, and composition of the milk product. The containers should be usually opaque. If containers in use are transparent, then these have to be kept in a dark place. Containers or closures must be clean, dry, and either sterile or appropriate for sterilization. Avoid the use of glass containers inside the production area for sampling.

Sample containers with suitable shape and capacity should be used for the specific requirements for the product to be sampled. One can use aluminum foil of adequate strength as well as plastic containers and appropriate plastic bags with closure.

Securely close the containers with plastic or a screw cap made of plastic or metal. The screw cap should have, if needed, a liquid-tight plastic liner that is non-absorbable, insoluble, and grease-proof and which shall not affect odor, flavor, chemical composition as well as properties of the milk product being sampled. If stoppers are used, they should be made from flavorless and non-adsorbent material. Airtight containers must be used for avoiding the contamination and air way in and should be wide mouthed for solid, semi-solid or viscous milk products.

Small retail containers are taken as sample containers. Contents of intact and unopened containers are used for the samples.

2.1.4 SAMPLING

Sampling should be true representative of the bulk. If samples are taken in a laboratory separately for physical, chemical, and microbiological analysis as well as organoleptic evaluation; then those for bacteriological examination should be taken first aseptically using sterilized equipment and containers. It is impossible to depend on single sampling method, which will be relevant in all cases. The sampling technique, therefore, differs depending on the type of the product and the purpose for which it is needed. It may be required to increase the size of the sample, if coarse particles are present in the milk product. After sampling, sample containers should be closed immediately to prevent contamination of the sample. For retail units, the sample consists of one or more packed containers shall be selected [1].

2.1.5 PRESERVATION OF SAMPLES

Normally preservatives are added to the cheese or liquid milk products proposed for chemical analysis but not for sensory or microbiological analysis. No preservatives should be added to solid (excluding cheese), or semisolid or dried milk samples envisioned for chemical analysis. Sample must be cooled and stored quickly at 0 to 5°C in the refrigerator. Dry milk or milk powders may be stored at ambient temperature.

The preservatives added in milk products should not impede with the subsequent analysis (a suitable correction must be used if the preservative interferes with the analyte). Indicate the nature and amount of preservatives addition on the label as well as in the sampling report. Follow the safety instructions of the preservative in use.

2.1.6 SAMPLES STORAGE AND TRANSPORT

During the storage and transit of samples, the state of the samples at the time of sampling should not be altered until the time of initiating the test procedure. After sampling, samples should reach as quickly as possible to the lab. If possible, the dispatch time of the samples to the testing lab should be short, preferably <24 hours. Take precautions during transit to avoid exposure to sunlight, off-flavors, and other adverse conditions. The minimum temperature requirements to be met for cooling the samples is based on manufacturer instructions or legally requested. The temperature of storage after sampling should be accomplished as rapidly as possible [1]. Storage temperature for different dairy products is represented in Table 2.1. Sample should be analyzed immediately after the preparation of the test portion.

TABLE 2.1 Preservation, Temperature of Storage and Sample Size for Milk Products

Type of Milk Product	Preservation of Samples Considered for Physical and Chemical Analysis	Storage Temperature (Before and During Transit), °C	Size of Sample (Minimum)
Butter and other related products.	No	1–5	50 g
Butter oil and other related products.	No	1–5	50 g
Edible ices and semi-finished ice products.	No	≤ -18	100 g
Fresh, ripened, processed, and other cheeses.	No	1–5	100 g
Liquid and non-sterilized milk.	Yes	1 to 5	100 g or 100 ml
Milk powder and dried milk.	No	Room temperature (< 30)	100 g
Semi-solid or solid dairy products other than cheese and butter.	No	1–5	100 g
Sweetened condensed milk (SCM), evaporated milk, and concentrated milk.	No	Room temperature (< 30)	100 g
UHT milk, sterilized milk, and sterilized liquid milk products after sampling from the production line or from several unopened packs.	Yes	1 to 5	100 g or 100 ml
Unopened, original UHT milk, sterilized milk, and sterilized liquid milk products in containers.	No	Room temperature (< 30)	100 g or 100 ml

2.1.7 DATA COLLECTION SHEET

Title: General instructions for sampling of milk and milk products.

Location and Organization: ______________________________

Participants: ______________________________

Date (Date/Month/Year): ______________________________

Observations: ______________________________

Comments: ______________________________

2.2 SAMPLING OF MILK

2.2.1 INTRODUCTION

In India, dairy farmers are generally paid on the fat and SNF content of milk. Therefore, it can be inferred that milk being rich in fat is an expensive ingredient, which makes it prone to adulteration. To ensure that the customer receives superior quality milk, free of adulterants and preservatives, various chemical and microbiological analysis have to be performed. For ensuring each one of the samples in a batch or lot is free from any adulterant or any harmful additive, proper sampling becomes necessary. Here, the sampling methodology of milk provided by the BIS (IS 1479 (Part-1) – 1960, reaffirmed 2003) [3], AOAC (Association of Official Analytical Chemists) official method 970.26 [2] and International Organization for Standardization (ISO) (707:2008) [1] are discussed. As per ISO, minimum 100 g or 100 ml of milk sample must be taken for analysis [1].

2.2.2 MATERIALS

2.2.2.1 APPARATUS

2.2.2.1.1 Appliances Required for Agitating the Milk Manually

1. **Plungers:** For facilitating the uniform mixing of milk, plungers should have appropriate dimensions, as recommended by BIS. According to BIS, the plunger should have a disc having a diameter of 15 cm. The disc should be pierced by six holes, each having a

diameter of 1.25 cm and the diameter of the disc's pitch circle should be 10 cm. The disc is joined with a metal rod and the metal rod is attached to a handle. Total length of the metal rod along with the handle must be about 100 cm.

2. **Dippers:** The minimum capacity of the dipper (Figure 2.1) should be 80 ml. It should be attached to a solid handle having a minimum 15 cm length.
3. **Sampling Tube:** It should be made of metal, and it should not have any welding seam. It should have a length of 60 cm, with an interior diameter and thickness of about 0.6 and 0.16 cm, respectively.

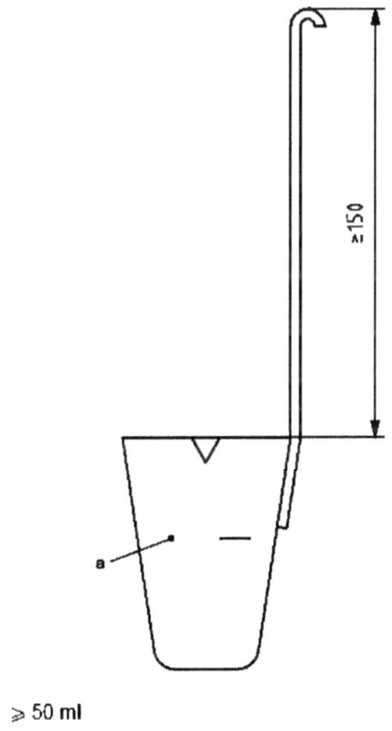

FIGURE 2.1 Dipper.

Source: Adapted from: ISO 707-2008, with permission from the ANSI on behalf of ISO.

All the above appliances should be made of stainless steel or any other appropriate metal and should possess sufficient strength so that they should not get distorted during their usage. Although they should not be too heavy so that the sampling person can easily stir the milk with these appliances. Sometimes, soldering is done on the above appliances. The soldering joints should be capable to withstand about 180°C temperature. There should not be any unevenness on the equipment's surface and round edges are preferred, particularly at the corners of the appliance. Only clean and dried appliances should be used and they should not transmit any bad aroma.

2.2.2.1.2 Appliances Required for Mechanical Agitation of the Milk

1. **Built-in-Agitators:** The design specifications of built-in agitators are primarily dependent on the type of product to be handled. Even though many types of agitators are in use, yet none of them are described or recommended in ISO and BIS.
2. **Removable Agitators:** These are sometimes used in road and rail tankers. These types of agitators generally have a propeller and are being inserted via tanker's manhole. While agitating the sample, the stirrer should be held at an inclined angle of 5–20°, this facilitates both vertical and horizontal blending of the sample.

2.2.2.1.3 Containers Required for Keeping, Storing, and Transporting the Samples

- Bottles of wide mouth, having a capacity of about 100 ml, 150 ml or 250 ml can be used. If the flavor, odor, and taste of the sample have to be determined, then the containers should be closed with a screw cap that is lined with butter paper.
  Only clean and dried containers should be used and they should not transmit any bad aroma.
- The container's design should not hinder the filling of the sample in the container up to the brim and its uniform blending before analysis. During transportation of the sample, containers should be capable to prevent churning of the fat in the sample.
- Thermally insulated containers can be used for the transportation of the sample. During sample transit (at 30±1°C), the thermally insulated container should be capable to maintain the temperature of the sample between 0°C and 5°C. Crushed ice or dry ice (solid CO_2) can be used in conjunction, to strictly control the temperature change. However, while using these adjuncts, there should not be any negative impact on the sample. There should also be a provision of continuous temperature monitor, during sample transit, by using appropriate instruments, such as, Pt 100. Apart from restricting the temperature

change, the containers also have to fulfill various other criteria. It should be capable of enduring all the stresses during transfer. There should not be any odor migration from the container to the sample, and also it should not allow any aroma migration from the sample to the surroundings. The container should not have a porous, uneven, and cracked surface.

2.2.3 SAMPLING PROCEDURE

2.2.3.1 FROM INDIVIDUAL CONTAINER

Mix the milk in the container with the help of a plunger (Figure 2.2). In absence of plunger, pour three to four times the milk from one container to another. The plunger should move in the entire container at least 10 times. It should reach the bottom and come to the top as quickly as possible so as to mix the upper and bottom layers uniformly. Do not miss to mix the fat adhere to the neck and under the shoulder of the can with the whole milk. Draw the sample immediately after thorough mixing.

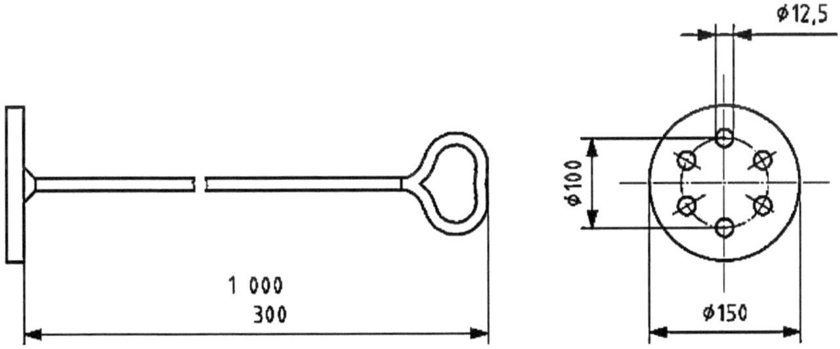

FIGURE 2.2 Plunger for cans (dimensions in mm).

Source: Adapted from: ISO 707:2008, with permission from the ANSI on behalf of ISO.

2.2.3.2 FROM MANY CONTAINERS

Normally, all the milk in cans are dumped in a vat and mixed with a mechanical agitator and then the sample is drawn from there. However, in the absence of mechanical agitator or vat, every container is agitated. From each container, an aliquot amount of milk is taken from numbers of containers. The containers are not filled full, then an equal amount of milk should be filled in each container. Mix each container with a plunger and transfer an equal amount of milk to another can. Agitate the milk in the can with a plunger (Figure 2.2) and draw the sample immediately after thorough mixing.

2.2.3.3 FROM BULK UNITS

When the bulk volume of milk is supplied from the same individual, having no variations from the qualitative point of view, then use Table 2.2 provided by BIS as a guide for sampling:

TABLE 2.2 Sampling of Bulk Units of Milk [3]

Number of Containers in the Lot (M)	Number of Containers to be Chosen for Sampling (m)
1	1
2 to 5	2
6 to 20	3
21 to 60	4
61 to 100	5
Above 100	Take five samples for initial 200 units and add one more sample (6) for further 200 units. Continue this process of addition for sample selection of the entire lot.

In case of different individual units, each bulk unit shall be separately sampled like the above-mentioned guide (Table 2.1).

2.2.3.4 FROM ROAD AND RAIL MILK TANKERS OR STORAGE TANKS

The milk tanker is thoroughly mixed by a plunger of sufficient length, by compressed air or by a mechanical agitator (Figure 2.3). In the case of plunger used for mixing the milk, it is inserted in the manhole by the operator standing on top of the tanker. The repeated cycle of thrust forward and pulled back, thrust downwards and pulled back and thrust backward and pulled back is done. This operation is continued for at least 15 minutes.

Check the samples collected at the manhole and the outlet cock for determining the uniformity of the samples with respect to milk solids and fat content. The sample is then drawn from the tank with the help of a dipper or from the valve in the discharge line from the tank.

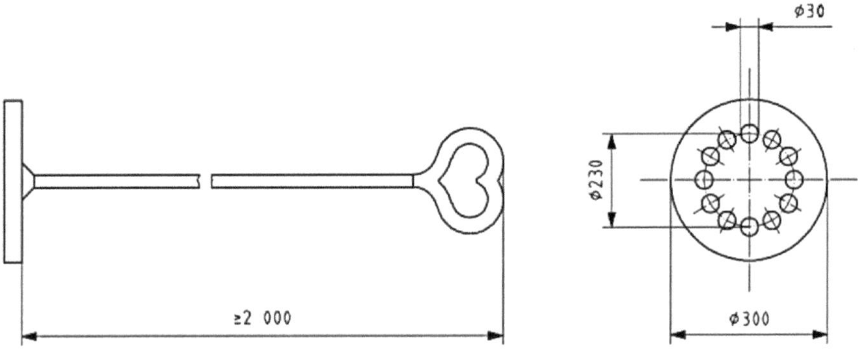

FIGURE 2.3 Plunger for storage tanks or rail and road milk tankers (dimensions in mm).
Adapted from: ISO 707:2008, with permission from the ANSI on behalf of ISO.

2.2.3.5 COMPOSITE MILK SAMPLES

Take a minimum of 175 ml milk into the composite sample bottle over the period of collection. Preserve the composite milk for a period of time till analysis by adding 0.4 ml formaldehyde (36%) in 100 ml of milk sample. During the addition of fresh milk in the sample bottle, each time the bottle is rotated to uniformly mix the content and to prevent plug formation by cream. To prevent evaporation of milk from the sample, the bottle is tightly capped and placed away from light in a cool place. The composite sample is analyzed on the last day of sample collection.

2.2.3.6 PRESERVING, TRANSPORTING, AND STORING THE MILK SAMPLES

- Add an appropriate permitted preservative if required or instructed by the laboratory. Indicate the preservative's type and quantity on the container's label. Cool the sample and store it at 0–5°C; in case there is a delay in analysis or if any preservative is not incorporated. Don't incorporate any preservative in samples that are set to undergo microbial and organoleptic analysis.
- At the earliest, deliver the samples to the testing unit. Care must be taken during transportation. Use a thermally insulated container for

transportation to maintain 0–5°C temperature. Also, keep ice in the aforesaid container, for maintaining the low temperature. Although, excessive ice usage is discouraged to prevent freezing of samples.

2.2.4 DATA COLLECTION SHEET

Title: Sampling method of milk.

Location and Organization: ________________________________

Participants: ____

Date (Date/Month/Year): ______________________________

Observations:

Identification number of sample	
Description of sample	
Batch/marking/code of sample	
Expiry date	
Number of samples	
Sample weight	
Packaging	
Place of origin	
Location (manufacturer address/packaging)	
Date of sampling	
Sampling time	
Place or laboratory to which sample to be sent	
Sampling person name	
Designation of sampler	
Signature	
Witness name	
Designation of witness	
Counter signature	
Relevant conditions (temperature and humidity) for sampling	
Preservative (type and quantity)	
Sampling method	
Preparation of test portion	

Comments: ____.

2.3 SAMPLING OF CREAM

2.3.1 Introduction

Cream is the part of milk, which is rich in fat. Cream can be separated from milk on the basis of density by using centrifugal force in a separator. The fat content of cream may range from 10–80%. According to FSSAI (2011), cream is defined as cow or buffalo milk's product or its combination. The regulatory body, FSSAI, has classified cream into three broad groups, i.e., low fat cream having minimum 25% milk fat on weight basis; medium fat cream having at least 40% milk fat (w/w); and high fat cream comprising minimum 60% milk fat (w/w). Thus, cream being rich in milk fat, possesses a pleasant and delicate flavor, which makes it suitable for varied applications, e.g., cakes, confectionaries, etc. The sampling methodology of cream as given by the BIS (IS 3509-1966) [5] and ISO [1] is discussed in this section.

2.3.2 MATERIALS

2.3.2.1 APPARATUS

2.3.2.1.1 Appliances Required for Agitating the Cream Manually

1. **Plungers:** Details are discussed in Section 2.2.1.1
2. **Dippers:** Details are discussed in Section 2.2.1.1.
3. **Sampling Tubes:** Details are discussed in Section 2.2.1.1.

2.3.2.1.2 Appliances Required for Agitating the Cream Mechanically

1. **Built-in-Agitators:** Details are discussed in Section 2.2.1.2.
2. **Removable Agitators:** Details are discussed in Section 2.2.1.2.

2.3.2.1.3 Containers Required for Keeping, Storing, and Transporting the Samples

Jars and bottles of wide mouth, having a capacity of about 50 or 100 ml can be used. If flavor, odor, and taste of the sample have to be determined, then the containers should be closed with a screw cap which is lined with butter paper. Usage of stoppers made of glass is mandatory only for microbial analysis. Storage and transport of the cream sample in container is same as that of milk (as discussed in Section 2.2.1.3).

2.3.3 PROCEDURE

2.3.3.1 UNITS SELECTION FOR SAMPLING WHILE SECURING RANDOM SELECTION

- Choose the units randomly from the lot. A lot comprises of all the units manufactured in the same batch. If the units are from various batches, then the units in an individual single batch will form a distinct lot.
- Carry out the random selection of sample as per the agreement of purchaser and supplier. Both the parties form a random number table, for facilitating the random selection.
- However, if no such agreement exists, then for securing the random selection, count the number of units as 1, 2, 3…….till m; where, m = A/B; A = Number of units present in the lot; and B = Number of units to be chosen.
- Continue the counting as above for the entire lot. All the units counted and regarded as m^{th} unit should be considered for analysis. This will provide the necessary number of units required for analysis of the sample, while maintaining the randomness.

2.3.3.2 NUMBER OF CONTAINERS TO BE CHOSEN FROM BULK UNITS FOR ANALYSIS

When bulk volume of cream is supplied in churns [6], having no variations in the quality of cream, then use Table 2.3 provided by BIS as a guide for sampling of cream:

However, when bulk volume of cream supplied in churns, varies widely from a qualitative point of view, then take sample from each and every unit.

2.3.3.3 NUMBER OF CONTAINERS TO BE CHOSEN FROM SMALL UNITS FOR ANALYSIS

When cream is supplied in small amount in bottles and small tins and has no variations in the quality [6], then use Table 2.4 provided by BIS as a guide for sampling:

TABLE 2.3 Sampling of Bulk Units of Cream

Number of Containers in the Lot (M)	Number of Containers to be Chosen for Sampling (m)
1	1
2 to 9	2
10 to 49	3
50 to 99	4
100 to 199	5
200 to 999	Take five samples for initial 200 units, and add one more sample (6th) for further 200 units. Continue this process of addition for sample selection till 999 units.
More than 1,000	Take nine samples for initial 1,000 units and add one more sample (10th) for a further 1,000 units. Continue this process of addition for sample selection for the entire lot.

TABLE 2.4 Sampling of Retail Units of Cream

Number of Containers in the Lot (M)	Number of Containers to be Chosen for Sampling (m)
1–100	1
101–1,000	2
1,001–10,000	4
More than 10,000	Take four samples for initial 10,000 units and add one more sample (5) for further 2,500 units. Continue this process of addition for sample selection for the entire lot.

However, when cream is supplied in small amount in bottles or in small tins and have wide variations from a qualitative point of view, then the sampling procedure can vary depending on the circumstances. The cream undergoing analysis is manufactured in how many batches, will also be one of the determining factors. As per BIS recommendation, the testing laboratory can decide the methodology of sample selection and the number of samples to be considered for analysis.

2.3.3.4 MIXING OF CREAM IN VARIOUS APPLIANCES FOR TAKING VARIOUS TYPES OF SAMPLES

1. **Small Containers:** Blend the cream having low fat content and less viscosity by transferring it from one container to another for six

times. The transferring of cream discussed above can be avoided by plunging the cream for minimum 10 times.

2. **Bulk Containers:** Blend the cream having medium to high fat content and high viscosity with the help of a plunger. For ensuring, uniform blending of cream lying at the bottom portion with the top layer cream, rotate the plunger throughout the container (from one place to other). Keep the plunger's disc below the cream's surface so as to prevent churning and whipping of cream. Follow similar method of plunging in the small container, which is discussed above.
3. **Sour Cream:** Care must be taken while handling sour cream. Initially, heat it to 30–40°C and then cool it to ambient temperature. Prior to sampling, gently agitate or shake the contents.
4. **Composite Sampling:** Take 300 g cream in total; equally from individually selected samples (for instance, if there are three containers, take 100 g cream from each). Blend the sample taken and sub-divide it into three equal fractions for purchaser, vendor, and referee. As per BIS recommendation, the final sample size must be 100 g, if no other requirements by the examining laboratory have been made. After proper blending of cream is done in all the above scenarios, take the sample promptly.

2.3.3.5 PRESERVING, TRANSPORTING, AND STORING THE CREAM SAMPLES

Follow the procedure as described for milk in Section 2.3.6.

2.3.4 DATA COLLECTION SHEET

Title: Sampling method of cream.

Location and Organization: ______________________________

Participants: ______________________________

Date (Date/Month/Year): ______________________________

Observations:

Identification number of sample	
Description of sample	
Batch/marking/code of sample	
Expiry date	
Number of samples	
Sample weight	
Packaging	
Place of origin	
Location (manufacturer address/packaging)	
Date of sampling	
Sampling time	
Place or laboratory to which sample to be sent	
Sampling person name	
Designation of sampler	
Signature	
Witness name	
Designation of witness	
Counter signature	
Relevant conditions (temperature and humidity) for sampling	
Preservative (type and quantity)	
Sampling method	
Preparation of test portion	

Comments:____.

2.4 SAMPLING OF BUTTER

2.4.1 INTRODUCTION

Butter is manufactured by means of churning and working of the cream. Butter is a fat rich product having almost 80% fat. According to FSSAI (2011), butter has been defined as a high fat product obtained primarily from products of both or either cow or buffalo milk. Butter has been subdivided into two variants, i.e., table butter (which consists of minimum 80% fat and maximum 3% salt); and desi cooking butter having minimum 76% fat and no salt. The role of salt is not only as a preservative, but also as a flavor enhancer. The pleasant flavor of butter, mainly because of diacetyl, makes it suitable for various applications such as in the form of spread and as a condiment. Butter is very prone to adulteration, which can be prevented by various chemical analysis of butter. Although prior to chemical analysis, proper sampling of butter is imperative. Here, the sampling methodology of butter provided by BIS (IS 3507-1966) [7] and ISO (ISO 707:2008) [1] is discussed. This methodology should be followed in a dairy industry to certify the quality of butter. As per ISO and BIS, minimum 50 g of butter sample must be taken for analysis. The butter samples must be protected from damp air, bright light, dust or soot. While drawing samples from butter, it must be at a temperature of 0–15°C and after the samples are drawn, it must be stored at 0–5°C temperature. Samples must be kept in clean and dried glass containers, having no odors. Preservatives must not be incorporated in butter during sampling.

2.4.2 *MATERIALS* (FIGURE 2.4)

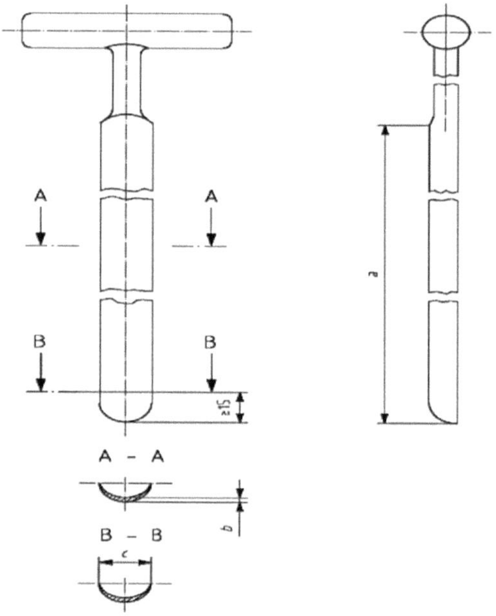

FIGURE 2.4 Butter trier (dimensions in mm; *refer* Table 2.5).

Source: Adapted from: ISO 707:2008, with permission from the ANSI on behalf of ISO.

2.4.2.1 APPARATUS

2.4.2.1.1 Appliances Required for Sampling of Butter

1. **Butter Triers:** It consists of a blade, stem, and a handle/grip. As a butter trier has to undergo sterilization, stainless steel having suitable hardness is the ideal material for manufacturing butter triers [1]. The blade part may be grooved and should be sharp, but the other components must have a smooth surface. The butter trier must have a minimum diameter of about 3 cm and adequate length such that it can get across diagonally to the container's base (Table 2.5).

TABLE 2.5 Types of Triers

Dimension	Long (Type A)	Medium (Type B)	Short (Type C)
Blade length (*a*, in mm).	540	220–260	125
From the end of blade, the frontal breadth at 15 mm (minimum) (*c*, in mm).	17	17	11
Thickness of middle of blade (minimum) (*b*, in mm).	1.8	1.5	1

2. **Spatulas and Knives:** They should be of stainless steel. These are primarily utilized to remove bit of samples from butter triers.

2.4.2.1.2 Containers Required for Keeping, Storing, and Transporting the Samples

1. **Jars and Bottles:** This of wide mouth, having a capacity of about 50, 100, 200, and 250 ml can be used. Opaque containers are considered more suitable for analysis. The containers may also be wrapped with aluminum foil, if required. If flavor, odor, and taste of the sample have to be determined, then the containers should be closed with a screw cap which is lined with butter paper. Usage of stoppers made of glass is mandatory only for microbial analysis. Only clean and dried containers should be used, and they should not transmit any bad aroma.
2. **Carton Boxes:** This can be used for 2 kg samples. If carton boxes are used for organoleptic analysis, then samples must completely fill the boxes. Samples should also be wrapped with aluminum foil or vegetable parchment paper.

3. For analysis such as fat indices, containers must be filled up to brim or be filled up with inert gas. Airtight closures are required in this case.
4. **Thermally Insulated Containers:** Refer the details discussed in Section 2.2.1.3 of this chapter.

2.4.3 PROCEDURE

2.4.3.1 UNITS SELECTION FOR SAMPLING WHILE SECURING RANDOM SELECTION

Please refer to Section 3.3.1.

2.4.3.2 NUMBER OF CONTAINERS TO BE CHOSEN FROM BULK QUANTITY FOR ANALYSIS

When bulk volume of butter is supplied in casks or boxes [7], having no variations in the quality, then use Table 2.6 provided by BIS as a guide for sampling:

TABLE 2.6 Sampling of Bulk Units of Butter [7]

Number of Containers in the Lot (M)	Number of Containers to be Chosen for Sampling (m)
1	1
2–9	2
10–49	3
50–99	4
100–199	5
Above 200	Take five samples for initial 200 units and add one more sample (6th) for further 200 units. Continue this process of addition for sample selection for the entire lot.

However, when bulk volume of butter supplied in casks or boxes, varies widely from a qualitative point of view and then takes sample from each and every unit.

2.4.3.3 NUMBER OF CONTAINERS TO BE CHOSEN FROM SMALL QUANTITY FOR ANALYSIS

When butter is supplied in small amount in packets and small tins and has no variations in the quality, then use Table 2.7 provided by BIS as a guide for sampling:

TABLE 2.7 Sampling of Retail Units of Butter

Number of Containers in the Lot (M)	Number of Containers to be Chosen for Sampling (m)
1–100	1
101–1,000	2
1,001–10,000	4
More than 10,000	Take four samples for initial 10,000 units and add one more sample (5th) for further 2,500 units. Continue this process of addition for sample selection for the entire lot.

However, when small volume of butter supplied in packets and tins, varies widely from qualitative point of view, then take sample from each and every unit.

2.4.3.4 ANALYSIS OF HARD/SEMI-HARD BUTTER WHICH IS STORED IN COLD CONDITIONS

1. **Sampling from Churns:** Insert butter triers at the core part of the butter sample. Draw four such samples. All the four places from where the sample is drawn, should be at equal distance from each other. However, remember to draw two such samples close to churn's center part.
2. **Sampling from Trollies:** Insert butter triers at two ends of the trolly and draw two samples from the core part of the butter sample. Again, insert butter trier at two sides of the trolly and draw the remaining two samples.
3. **Sampling from Boxes:** Insert vertically butter triers at three places of butter block. Draw one sample from the center of the butter block. Draw the remaining two samples from the core part of the butter

block, at two diagonally opposite corners of the butter block at the open side.

4. **Sampling from Casks:** Draw three samples from the central part of butter with the help of butter triers. Insert butter triers at three places which are at equal distance from the border of one edge of the butter cask.
5. **Sampling from Small Packets:** Use only closed packets for analysis. Initially, draw samples for microbial analysis. Use the remaining part of the sample for chemical test.

2.4.3.5 ANALYSIS OF PASTY BUTTER WHICH IS STORED UNDER WARM CONDITIONS

1. **Sampling from Small Units:** Use a spatula, spoon or a butter trier for sampling. Take three samples from the core part of the butter block. Draw one sample from the center of the butter block. Draw the remaining two samples close to 2–3 cm of the container's end. These two samples must be drawn at equal distance from the center.
2. **Sampling from Bulk Units:** Use a spatula, spoon or a butter trier for sampling. Take three samples from the core part of the butter block, which are at equal distance from one another. Draw one sample from the center of the butter block. Draw another two samples, at equal distance from the center, close to 2–3 cm from the block's close and open end.

2.4.3.6 ANALYSIS OF COMPOSITE SAMPLES

Follow the procedure as described for composite sampling of cream in Section 3.3.4 of this chapter.

2.4.3.7 PRESERVING, TRANSPORTING, AND STORING THE CREAM SAMPLES

Follow the procedure as described for milk in Section 2.3.6.

2.4.4 DATA COLLECTION SHEET

Title: Sampling method of butter.

Location and Organization: ______________________________

Participants: ______________________________

Date (Date/Month/Year): ______________________________

Observations:

Identification number of sample	
Description of sample	
Batch/marking/code of sample	
Expiry date	
Number of samples	
Sample weight	
Packaging	
Place of origin	
Location (manufacturer address/packaging)	
Date of sampling	
Sampling time	
Place or laboratory to which sample to be sent	
Sampling person name	
Designation of sampler	
Signature	
Witness name	
Designation of witness	
Counter signature	
Relevant conditions (temperature and humidity) for sampling	
Preservative (type and quantity)	
Sampling method	
Preparation of test portion	

Comments:______________________________.

2.5 SAMPLING OF BUTTER FAT (GHEE)

2.5.1 INTRODUCTION

Ghee can be defined as a pure clarified milk fat. According to FSSR (2011), ghee must contain milk fat content (minimum) of 99.5% and moisture content (maximum) of 0.5%. Ghee is manufactured by clarifying cream or butter to high temperature (>100°C). Ghee possesses a pleasant, cooked, and slightly caramelized flavor, primarily due to the presence of carbonyls, lactones, and free fatty acids (FFAs). The pleasant flavor of ghee makes it suitable for cooking and frying medium for various sweets and as a garnishing agent for rice and chapatis. Ghee being a fat rich product is a quite expensive dairy product and thus it is more prone to adulteration by vegetable fat/vanaspati and many others. Thus, strict analysis of ghee is crucial to ensure only quality product reaches to the market. However, it is quite cumbersome to analyze each and every ghee container, thus proper sampling of ghee from a lot is essential. Here, the sampling methodology of ghee provided by BIS (IS 3508-1966) [9] will be discussed. This methodology should be followed in a dairy industry to certify the quality of ghee.

2.5.2 MATERIALS

2.5.2.1 APPARATUS

2.5.2.1.1 Appliances Required for Sampling

1. **Sampling Concentric Tubes:** It comprises of 2 concentric tubes in a single instrument such that the rotation of one tube is possible, even being placed inside another tube. The opening and closing of the instrument can be adjusted by rotating the inner tube. At open position only, ghee sample can be taken into the tube, following which the tube should be closed and withdrawn from the sample.
2. **Sampling Plain Tubes:** It is either made of glass or metal of high thickness. It has a tapered conical upper and lower end. For facilitating the handling of ghee sample, two rings are fitted at the tube's upper end. The instrument should be inserted into the ghee sample to a preferred depth, by closing the tube's upper end with thumb or stopper. The sample can be taken by opening the upper end, following which; the tube should be closed and withdrawn from the sample. This instrument finds its usage when the ghee is in liquid form and has a uniform quality.

   Polished surface and stainless-steel buildup of all the appliances is preferred. The sampling instruments must be cleaned and dried, both before and after sampling. Also, the sampling instrument must not possess any off odor and flavor. Remove adhered particles attached to the sample appliance if any, before discharging the sample from the instrument.

2.5.2.1.2 *Containers Required for Keeping the Samples*

Jars, bottles, and tin containers of wide mouth, having a capacity of about 50, 100, 200, and 250 ml can be used. The containers should be closed with a screw cap lined with butter paper. However, if flavor, odor, and taste of the sample have to be determined, then the containers must be closed with glass stoppers. Press on type lids must be used as a closure for tin containers. Only clean and dried containers should be used, and they should not transmit any bad aroma.

2.5.3 *PROCEDURE*

2.5.3.1 UNITS SELECTION FOR SAMPLING WHILE SECURING RANDOM SELECTION

Please refer to Section 3.3.1.

2.5.3.2 NUMBER OF CONTAINERS TO BE CHOSEN FOR ANALYSIS

Table 2.8, provided by BIS, can be used as a guide for choosing the number of containers for sampling [9].

TABLE 2.8 Sampling of Ghee

Number of Containers in the Lot (M)	Number of Containers to be Chosen for Sampling (m)
1	1
2 to 40	2
41 to 110	3
111 to 300	5
301 to 600	7
Above 601	10

However, when ghee sample varies widely with the quality, then take sample from each and every unit and perform analysis.

2.5.3.3 ANALYSIS OF COMPOSITE SAMPLES

Follow the procedure as described for composite sampling of cream in Section 3.3.4 of this chapter. Protect the sample equipment, container, and the sample from dust particles contamination, off odor, and light. In all the above scenarios, send the sample to the testing laboratory at earliest possible. Protect the samples from light and off odor during transportation and store it in a cool and dark place.

2.5.4 DATA COLLECTION SHEET

Title: Sampling method of ghee.

Location and Organization:______________________________

Participants: ____

Date (Date/Month/Year): _______________________________

Observation:

Identification number of sample	
Description of sample	
Batch/marking/code of sample	
Expiry date	
Number of samples	
Sample weight	
Packaging	
Place of origin	
Location (manufacturer address/packaging)	
Date of sampling	
Sampling time	
Place or laboratory to which sample to be sent	
Sampling person name	
Designation of sampler	
Signature	
Witness name	
Designation of witness	
Counter signature	
Relevant conditions (temperature and humidity) for sampling	
Preservative (type and quantity)	
Sampling method	
Preparation of test portion	

Comments:____.

2.6 SAMPLING OF CONDENSED MILK AND EVAPORATED MILK (EM)

2.6.1 *INTRODUCTION*

Evaporated milk (EM) and condensed milk are concentrated milk from which part of water is removed by heat treatment or any other means before canning. Sweetened condensed milk (SCM) is the same as EM except for the sugar content. SCM contains 40–45% sugar and it imparts rich sweet flavor and caramel color as well as thickness to the product. As per FSSR (2017), SCM contains a min. of 8% fat and 28% milk solids. Whereas EM contains min. of 7.5% fat and 25% milk solids. The sampling methodology of SCM and EM as given by the BIS (IS 1166-1986) [10], AOAC official method 970.27 [2] and ISO (707:2008) [1] is discussed in this section.

2.6.2 *MATERIALS*

2.6.2.1 *APPARATUS*

- Broad-bladed metal stirrer (Figure 2.5): The stirrer should be long enough to touch the container's bottom and at the bottom should have a wide perforated disk.

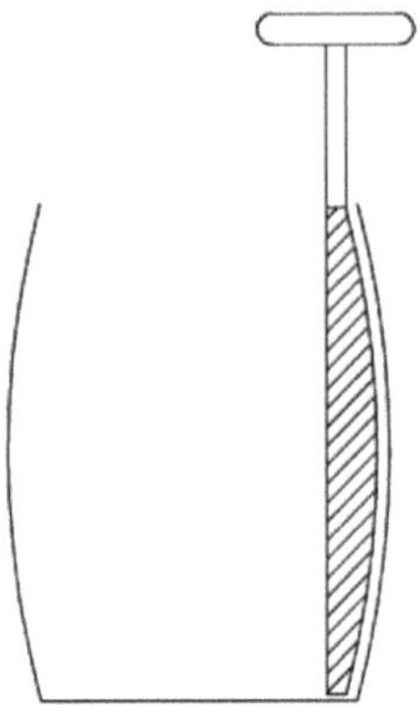

FIGURE 2.5 Stirrer for condensed milk in barrels.

Source: Adapted from: ISO 707:2008, with permission from the ANSI on behalf of ISO.

- Sample container: Jars that have a wide mouth and airtight lids.
- Sample dipper.

2.6.3 SAMPLING PROCEDURE

2.6.3.1 UNITS SELECTION FOR SAMPLING WHILE SECURING RANDOM SELECTION

Please refer to Section 3.3.1.

2.6.3.2 FOR BULK CONTAINERS

When bulk volume of condensed and EM (container between 5 kg and 20 kg) is supplied in barrels or drums having no variations in the quality, then use Table 2.9 provided by BIS as a guide for sampling:

TABLE 2.9 Sampling of Bulk Units of SCM/EM [10]

Number of Containers in the Lot	Number of Containers to be Chosen for Chemical and Sensory Analysis
1–100	2
101–300	3
301–500	4
More than 500	5

However, when the bulk volume of condensed and EM is supplied, varies widely from the qualitative point of view and then take the sample from each and every unit.

2.6.3.3 FOR RETAIL CONTAINERS

When condensed and EM is supplied in small amount in bottles and small tins (container between 400 g and 5 kg) and have no variations in the quality, then use Table 2.10 provided by BIS as a guide for sampling [10].

TABLE 2.10 Sampling of Retail Units of SCM/EM [10]

Number of Containers in the Lot	Number of Containers to be Chosen for Chemical and Sensory Analysis
1–300	3
301–500	5
501–1,000	7
More than 1,000	10

However, when condensed and EM is supplied in small amount in bottles and small tins and have wide variations from the qualitative point of view, then the sampling procedure can vary depending upon the circumstances. As per BIS recommendation, the testing laboratory can decide the methodology of sample selection and the number of samples to be considered for analysis.

2.6.3.4 BLENDING OF CONDENSED AND EVAPORATED MILK (EM) IN VARIOUS APPLIANCES FOR TAKING VARIOUS TYPES OF SAMPLES

For both the type of container, first it is heated at 40°C in a water bath. From the edge of the lid, the container has to be opened and remix the adhering of the lid to the total content of the container. Mix the whole content with a spoon or spatula. For bulk container mix with a stirrer. The mixing is done to uniformly mix the bottom layer with the upper layer. Draw the sample with a suitable sample dipper in equal quantities from different corners of the same container until 0.2 kg of sample is obtained in the airtight jar. Label the jar with code and date. For small containers, the intact container is considered as a sample.

2.6.3.5 PRESERVING, TRANSPORTING, AND STORING THE SAMPLES

- Under the supervision of the testing laboratory, add no preservative. Store the sample in ambient temperature, but it should not cross 30°C.
- At the earliest, deliver the samples to the testing unit. Care must be taken during transportation. Use a thermally insulated container for transportation to avoid a rise in temperature.

2.6.4 DATA COLLECTION SHEET

Title: Sampling method of condensed and evaporated milk.

Location and Organization: ______________________________

Participants: ______________________________

Date (Date/Month/Year): ______________________________

Observations:

Identification number of sample	
Description of sample	
Batch/marking/code of sample	
Expiry date	
Number of samples	
Sample weight	
Packaging	
Place of origin	
Location (manufacturer address/packaging)	
Date of sampling	
Sampling time	
Place or laboratory to which sample to be sent	
Sampling person name	
Designation of sampler	
Signature	
Witness name	
Designation of witness	
Counter signature	
Relevant conditions (temperature and humidity) for sampling	
Preservative (type and quantity)	
Sampling method	
Preparation of test portion	

Comments:____.

2.7 SAMPLING OF DRIED MILK

2.7.1 INTRODUCTION

Dried milk product is obtained by evaporating milk to dryness. According to FSSR (2017), milk powder contains a maximum of 5% moisture. Partially skimmed, skimmed, whole milk and cream powder contains fat content between 1.5 and 26%, < 1.5%, 26 to 42%, and > 42%, respectively. For ensuring each one of the samples in a batch or lot is free from any adulterant or any harmful additive, proper sampling becomes necessary before analysis. Here the sampling methodology of milk powder provided by the BIS (IS 1165-2002), AOAC official method 970.28 and ISO (ISO 707:2008) are discussed. This methodology should be followed in the dairy industries to certify the quality of dried milk. As per ISO, minimum 100 g of dried milk sample must be taken for analysis [1].

2.7.2 MATERIALS

2.7.2.1 APPARATUS

1. **Borer Tube or Trier (Figure 2.6 and Table 2.11):** It should be clean and dry before sampling. It should be made up of stainless steel, aluminum or an alloy of aluminum.

TABLE 2.11 Dimension of Borers

Dimension	Long (Type A)	Short (Type B)
Blade length (mm)	800	400
Blade thickness (mm)	1–2	1–2
Blade inner diameter at point (mm)	18	32
Blade inner diameter at grip (mm)	22	28
Width of slit at grip (mm)	14	14
Width of slit at point (mm)	4	20

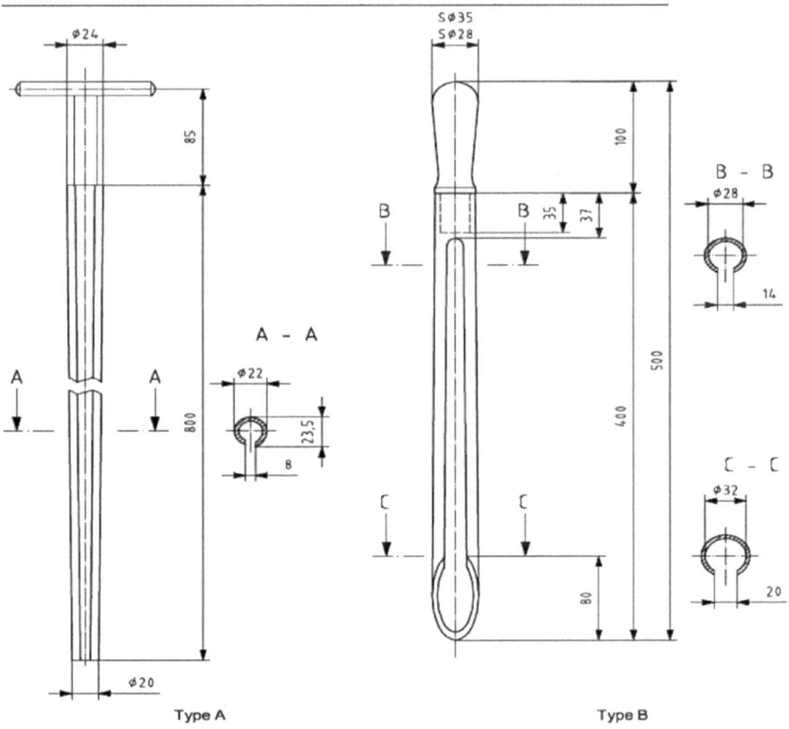

FIGURE 2.6 Dried milk borer (dimensions in mm; *refer* to Table 2.11).

Source: Adapted from: ISO 707:2008, with permission from the ANSI on behalf of ISO.

2. **Glass Bottle or Jars:** It should be clean and dry. It should have an airtight lid. The glass container should be opaque. The container should be large enough so it can be mixed by just shaking.

2.7.3 SAMPLING PROCEDURE

2.7.3.1 UNITS SELECTION FOR SAMPLING WHILE SECURING RANDOM SELECTION

Please refer to Section 3.3.1.

2.7.3.2 FOR BULK CONTAINERS

When bulk volume (container more than 5 kg) of powder milk is supplied, having no variations in the quality, then use Table 2.12 provided by BIS as a guide for sampling [11].

TABLE 2.12 Sampling of Bulk Units of Dried Milk

Number of Containers in the Lot	Number of Containers to be Chosen for Chemical and Sensory Analysis
1–50	2
51–100	3
101–300	4
More than 300	5

However, when bulk volume of powder milk is supplied, it varies widely with the quality and then take sample from each and every unit.

2.7.3.3 FOR RETAIL CONTAINERS

When powder milk is supplied in small amount in pouch packets or tin cans (container between 500 g and 5 kg) and has no variations in the quality, then use Table 2.13 provided by BIS as a guide for sampling [11].

However, when powder milk is supplied, has wide variations from the qualitative point of view, then the sampling procedure can vary depending upon the circumstances. As per BIS recommendation, the testing laboratory

TABLE 2.13 Sampling of Retail Units of Dried Milk

Number of Containers in the Lot (M)	Number of Containers to be Chosen for Chemical and Sensory Analysis (m)
1–100	3
101–300	5
301–500	7
More than 500	9

can decide the methodology of sample selection and the number of samples to be considered for analysis.

2.7.3.4 DRAWING OF SAMPLE

For both types of containers, first, pass the borer tube or trier steadily at an even rate of penetration through the powder. Withdraw the borer tube with the content, when it approaches the container's bottom and put the sample in the jar or bottle. Take more than one bore to make a 300 to 500 g sample. Seal each container air-tight after filling and provide full details of sampling on the label including manufacturer's name, code or batch no., and other important details of the consignment. Do not touch the powder or the sample with the finger. It may get contaminated and may absorb moisture. Keep the sample jar in a dry and dark place till analysis.

2.7.3.5 PRESERVING, TRANSPORTING, AND STORING THE DRIED MILK SAMPLES

Refer to Section 6.3.5 of this chapter.

2.7.4 DATA COLLECTION SHEET

Title: Sampling method of dried milk.

Location and Organization: ____________________________________

Participants: ____

Date (Date/Month/Year): _______________________________

Observations:

Identification number of sample	
Description of sample	
Batch/marking/code of sample	
Expiry date	
Number of samples	
Sample weight	
Packaging	
Place of origin	
Location (manufacturer address/packaging)	
Date of sampling	
Sampling time	
Place or laboratory to which sample to be sent	
Sampling person name	
Designation of sampler	
Signature	
Witness name	
Designation of witness	
Counter signature	
Relevant conditions (temperature and humidity) for sampling	
Preservative (type and quantity)	
Sampling method	
Preparation of test portion	

Comments: ____.

2.8 SAMPLING OF ICE CREAM

2.8.1 INTRODUCTION

Ice cream is a frozen product, which is prepared by freezing a mix. During freezing, continuous agitation is being performed for air incorporation and to maintain uniformity in all packages. This mix may comprise of milk and other milk products, water, sweeteners, and other additives (emulsifiers and stabilizers). According to FSSAI (2011), ice cream must be manufactured from a dairy source. Ice cream must have a minimum fat and total solid content of 10 and 36%, respectively. Similar to other milk products, ice cream is also prone to adulteration. Thus, strict analysis of ice cream is crucial to ensure the quality product reaches to the market. Although it is difficult to inspect each and every ice cream container, yet proper sampling of ice cream from a lot is essential. Both ISO and BIS have recommended that the sampling size of ice cream must be minimum 100 g. It has also been emphasized that the ideal temperature for sampling of ice cream is –12 to –18°C temperature. Here, the sampling methodology of ice cream provided by BIS (IS 2802-1964) [8] and ISO (ISO 707:2008) [1] will be discussed. This methodology should be followed in dairy industries to certify the quality of ice cream.

2.8.2 MATERIALS

2.8.2.1 APPARATUS

2.8.2.1.1 Appliances Required for Sampling

- Borers having adequate length such that it can reach the container's bottom.
- Spoon, knife or spatula, ice scoop.

2.8.2.1.2 Containers Required for Keeping, Storing, and Transporting the Samples

- Jars of wide mouth, having a capacity of about 100 and 200 g can be used. Containers must be glass or metal made or any other suitable sterilizable material. The container's closure must be screwing cap or any other material which is grease/fat proof. Only clean, dried, and sterilized containers should be used and they should not transmit any bad aroma. The containers used must be kept for minimum 30 minutes prior to use in a thermally insulated container, which is already refrigerated (by using, for example, solid CO_2).
- Thermally insulated containers (refer the details discussed in Section 2.2.1.3 of this chapter).

2.8.3 PROCEDURE

2.8.3.1 UNITS SELECTION FOR SAMPLING WHILE SECURING RANDOM SELECTION

Please refer to Section 3.3.1.

2.8.3.2 NUMBER OF CONTAINERS TO BE CHOSEN FROM BULK QUANTITY FOR ANALYSIS

When bulk volume of ice cream is supplied, having no variations from a qualitative point of view, then use Table 2.14 provided by BIS as a guide for sampling [8].

TABLE 2.14 Sampling of Bulk Units of Ice Cream

Number of Containers in the Lot (M)	Number of Containers to be Chosen for Sampling (m)
1	1
2 to 5	2
6 to 20	3
21 to 60	4
61 to 100	5
Above 100	Take five samples for initial 100 units and add one more sample (6th) for further 100 units. Continue this process of addition for sample selection for the entire lot.

However, when the bulk volume of ice cream varies widely from a qualitative point of view, then take sample from each and every unit for analysis.

2.8.3.3 NUMBER OF CONTAINERS TO BE CHOSEN FROM SMALL QUANTITY FOR ANALYSIS

When ice cream is supplied in retail amounts and has no variations from a qualitative point of view, then use Table 2.15 provided by BIS as a guide for sampling [8].

TABLE 2.15 Sampling of Retail Units of Ice Cream

Number of Containers in the Lot (M)	Number of Containers to be Chosen for Sampling (m)
1–100	1
101–1,000	2
1,001–10,000	4
More than 10,000	Take four samples for initial 10,000 units and add one more sample (5th) for further 2,500 units. Continue this process of addition for sample selection for the entire lot.

However, when a small volume of ice cream is supplied and it varies widely from a qualitative point of view, then take sample from each and every unit. Only closed or sealed retail packages should be used for analysis.

2.8.3.4 ANALYSIS OF ICE CREAM, WHERE SUB-DIVISION OF SAMPLE IS NECESSARY

1. **Retail Ice Cream Packages:** At first, open the ice cream packet. Use a sterile knife or spoon to divide the ice cream into equal parts as required. Keep portions from individual packages separately in individual container. Continue this process till each container has ice cream in the required amount.
2. **Ice Cream Having Multiple Layers (may be of Flavor or Color):** Keep in mind before sampling this type of ice cream, that, a sample taken must be a representative of the original sample. The proportion in which layers are present in the original ice cream, the sample also must have the identical proportion in it. Do not separate the individual layers of ice cream from the original sample. Instead, take a whole sample having different layers and place it in the container.
3. **Bulk Ice Cream Packages:** Use a sterilized spoon or knife to eliminate the ice cream's top or surface layer. Again, use another sterilized equipment to take samples from different parts of the bulk package such that a representative sample is withdrawn.

2.8.3.5 PRESERVING, TRANSPORTING, AND STORING THE ICE CREAM SAMPLES

Add a required amount of preservative if required or instructed by the laboratory. Store the ice cream sample below –15°C after sampling and transport the samples in a thermally insulated container maintained at the temperature of below –15°C.

2.8.4 DATA COLLECTION SHEET

Title: Sampling method of ice cream.

Location and Organization: ______________________________

Participants: ______________________________________

Date (Date/Month/Year): ______________________________

Observations:

Identification number of sample	
Description of sample	
Batch/marking/code of sample	
Expiry date	
Number of samples	
Sample weight	
Packaging	
Place of origin	
Location (manufacturer address/packaging)	
Date of sampling	
Sampling time	
Place or laboratory to which sample to be sent	
Sampling person name	
Designation of sampler	
Signature	
Witness name	
Designation of witness	
Counter signature	
Relevant conditions (temperature and humidity) for sampling	
Preservative (type and quantity)	
Sampling method	
Preparation of test portion	

Comments:______________________________.

2.9 SAMPLING OF CHEESE

2.9.1 INTRODUCTION

Cheese is a fermented dairy-based product, and it is a rich source of fat and protein. It is usually produced by curdling the milk by adding rennet or bacterial enzymes for coagulation of casein followed by separation of whey from solid curd and pressed into cheese. Due to number of variants, cheeses are grouped based on origin, type of microorganisms (mold-ripened, bacterial ripened), type of coagulation (acid, rennet, heat, and acid coagulation), moisture content (very soft, soft, semi-hard, hard, and extra hard), cheese characteristics (Blue, Pasta-filata), etc. The sampling methodology of cheese sample as given by the BIS (IS 2785-1979) [6], AOAC official method 970.27 [2] and ISO (707:2008) [1] is discussed in this section.

2.9.2 MATERIALS REQUIRED

2.9.2.1 APPARATUS

1. **Cheese Triers [1]:** The size and shape of cheese triers (Table 2.16 and Figure 2.7) depends on the type of cheese to be sampled.

TABLE 2.16 Dimension of Triers

Dimension	Long (Type A)	Medium (Type B)	Short (Type C)
Blade length (*a*), mm	540	150	125
Metal thickness in middle of blade (minimum) (*b*), mm	1.5	0.9	0.7
From the end of blade, the frontal breadth at 15 mm (minimum) (*c*), mm	17	14	11

2. **Knife:** It should be of stainless steel and should have a pointed blade.
3. **Sealing Material:** Mix paraffin, beeswax, and white petroleum in (1:1:2) ratio in hot condition or mix paraffin and white petroleum in (1:1) ratio.
4. **Sample Container:** It should have an airtight lid and should be large enough to put the cut piece of cheese without compression or exerting any pressure.
5. **Plate:** It should be of tin, aluminum, or any suitable metal.
6. **Cutting Wire:** It should have sufficient strength to cut the sample.
7. **Spatula:** It should be made of stainless steel.

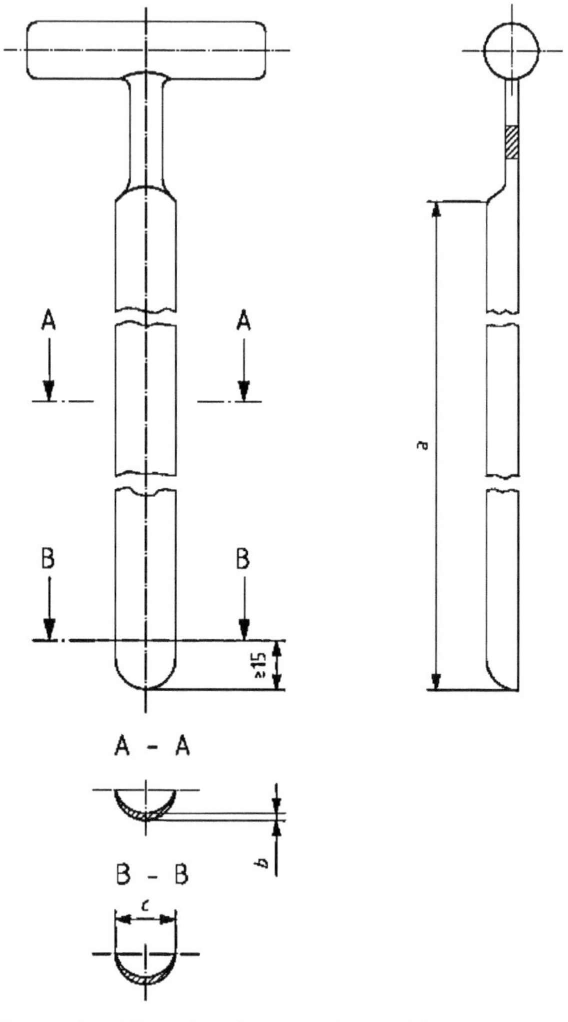

FIGURE 2.7 Cheese triers (dimensions in mm; *refer* to Table 2.16).

Source: Adapted from: ISO 707:2008, with permission from the ANSI on behalf of ISO.

2.9.3 PROCEDURE

2.9.3.1 UNITS SELECTION FOR SAMPLING WHILE SECURING RANDOM SELECTION

Please refer to Section 3.3.1.

2.9.3.2 NUMBER OF CONTAINERS TO BE CHOSEN FROM BULK QUANTITY FOR ANALYSIS

When bulk volume of cheese is supplied, having no variations from a qualitative point of view, then use Table 2.17 provided by BIS as a guide for sampling [6].

TABLE 2.17 Sampling of Bulk Units of Cheese

Number of Containers in the Lot (M)	Number of Containers to be Chosen for Sampling (m)
1	1
2–8	2
9–25	3
26–50	4
51–100	5
More than 100	8

However, when bulk volume of cheese is supplied, varies widely from a qualitative point of view and then take sample from each and every unit.

2.9.3.3 NUMBER OF CONTAINERS TO BE CHOSEN FROM SMALL QUANTITY FOR ANALYSIS

When cheese is supplied in small amount and has no variations from a qualitative point of view, then use Table 2.18 provided by BIS as a guide for sampling [6].

TABLE 2.18 Sampling of Retail Units of Cheese

Number of Containers in the Lot (M)	Number of Containers to be Chosen for Sampling (m)
1–25	3
26–100	5
101–500	8
501–1,000	10
1,001–5,000	13
More than 5,000	20

However, when cheese is supplied and has wide variations from a qualitative point of view, then the sampling procedure can vary depending upon the circumstances. The cheese undergoing analysis is manufactured in how many batches, will also be one of the determining factors. As per BIS recommendation, the testing laboratory can decide the methodology of sample selection and the number of samples to be considered for analysis.

2.9.3.4 SAMPLING TECHNIQUE

The sampling technique depends on the type, shape, size, and mass of the cheese. When collecting the samples, heterogeneity of the product is taken into account. Sample (plugs, sectors, cores, entire cheese, etc.) must be put in a suitable sample container of appropriate shape and size soon after sampling. For putting into the container, cut the sample into small pieces, but it should not be ground or compressed. To prevent formation of molds of the cheese surface during storage of cheese samples, additional measurement, such as, covering tightly with waxed paper, aluminum foil or plastic bags closable either inside or even container's outside may be used.

Unless otherwise stated and whatsoever the sampling technique is applied, the sample must be inclusive of any surface layer of the cheese (rind or crust, mold, etc.). If it is essential, specific guidelines must be followed to examine the surface layer of cheese.

2.9.3.4.1 Sampling Techniques of Cheese (Except Cheese Sold in Brine or Oil and Fresh Cheese)

1. **Sampling by Taking an Entire Cheese:** This procedure is usually used for small portion of cheese or pre-packed cheeses. In order obtain the sample, adequate number of sample portions or packets must be taken and sample must be kept in the sample container along with original packaging.
2. **Sampling by Cutting a Slices or Sectors:** First remove the outer wrapping from the cheese. Inside wrapping like plastic film or wax, should not be removed. From the center of the cheese make radially two random cuts, with a sharp knife or a cutting wire.
3. **Sampling by Taking Cores:** First remove the outer wrapping from the cheese. Inner wrapping should not be removed. If there is a delay in performing analysis after sampling, then the cores should be covered in waxed paper, aluminum foil or plastic pouches or bags (closable) before putting them in the container.

4. **Sample Including Surface Layer:** A sufficient length cheese trier must be inserted into the cheese. Rotate one complete turn of a trier and withdrawn with a core. The whole core must be moved to a sample container by using the knife. Repeat the same procedure until the sample is obtained. Suitable sealing compound is used to seal the core hole.
5. **Sample Excluding Surface Layer:** A larger diameter of cheese trier other than the sampling trier used for sampling must be inserted into the cheese to a depth of 25 mm and rotated through one complete turn. Do not withdraw this short core but keeps it separately and use later for closing the core hole. Insert a sufficient length of a smaller sampling trier through the inner surface of the cheese exposed by the above core hole. Rotate the trier through one complete turn and withdraw with the core. By using a knife, transfer the whole core to a sample container. The same procedure is repeated till the sample is obtained. Close the core hole by reinserting the first outer core.

2.9.3.4.2 Sampling of Fresh Cheeses

Intact and unopened containers should be used for sampling fresh cheese. The containers should not be opened until immediately before analysis. In order to obtain the sample, a required number of sample containers should be taken. Containers from which portions are taken should be taken as a whole.

2.9.3.4.3 Sampling of Cheese in Brine or Oil

The cheese composition will change particularly during storage in brine which mainly depends on time and temperature. The lab where the sample is testing must indicate whether the sample must be inclusive of brine or not. Usually, brine or oil is included in the sampling. If it is possible, cheese, and liquid ratio must be maintained and the liquid must immerse the cheese completely. If brine is included, an adequate quantity of brine must be taken so as to entirely immerse the cheese. If brine is excluded, using filter paper surface of the cheese is dried and then placed in sample container.

2.9.3.5 PRESERVING, TRANSPORTING, AND STORING THE CHEESE SAMPLES

Follow the same procedure as described in Section 2.3.6.

2.9.4 DATA COLLECTION SHEET

Title: Sampling method of cheese.

Location and Organization:______________________________

Participants: ______________________________________

Date (Date/Month/Year): ______________________________

Observations:

Sample identification no.	
Description of sample	
Batch/code/marking of sample	
Expiry date	
Type of cheese	
Cheese surface (rindless/rind/rind with smear/coating)	
Number of samples	
Sample weight	
Packaging (prepacked/aluminum/plastic/modified atmosphere/vacuum)	
Place of origin	
Location (manufacturer address/packaging)	
Date of sampling	
Sampling time	
Place or laboratory to which sample to be sent	
Sampling person name	
Designation of sampler	
Signature	
Witness name	
Designation of witness	
Counter signature	
Relevant conditions (temperature and humidity) for sampling	
Preservative (type and quantity)	
Sampling method for cheese	
Sampling details (did the sample includes smear or rind or how many millimeters of rind cutoff?)	
Preparation of test portion	

Comments:____.

KEYWORDS

- **labeling**
- **milk**
- **milk product**
- **sample container**
- **sampling techniques**

REFERENCES

1. ISO 707|IDF 50, (2008). *Milk and Milk Products — Guidance of Sampling* (3rd edn., pp. 1–40). Geneva, Switzerland: International Organization for Standardization.
2. AOAC, (2000). AOAC official method, 33. Dairy products. In: *Official Methods of Analysis of AOAC International* (17th edn., pp. 1–75). Gaithersburg, Maryland – USA: AOAC International.
3. IS, (2003). *IS 1479-1960: Methods of Test for Dairy Industry (Part -1), Rapid Examination of Milk* (pp. 1–44). New Delhi: Bureau of Indian Standards.
4. IS, (2012). *IS 11546-2012: Milk and Milk Products-Guidance on Sampling* (p. 40). New Delhi: Bureau of Indian Standards.
5. IS, (1983). *IS 3509-1966: Method of Sampling and Test for Cream* (pp. 5–11). New Delhi: Bureau of Indian Standards.
6. IS, (1979). *IS 2785-1979: Specification for Natural Cheese (Hard Variety), Processed Cheese, Processed Cheese Spread and Soft Cheese* (pp. 12–18). New Delhi: Bureau of Indian Standards.
7. IS, (1996). *IS 3507-1966: Method of Sampling and Test for Butter* (pp. 5–11). New Delhi: Bureau of Indian Standards.
8. IS, (2009). *IS 2802-1964: Specification of Ice Cream* (pp. 24–26). New Delhi: Bureau of Indian Standards.
9. IS. *IS 3508-1966: Method of Sampling and Test for Ghee (Butter Fat)* (pp. 5–10). New Delhi: Bureau of Indian Standards; (1990).
10. IS, (1997). *IS 1166-1986: Specification for Condensed Milk, Partly Skimmed and Skimmed Condensed Milk* (pp. 11–14). New Delhi: Bureau of Indian Standards.
11. IS, (2002). *IS 1165-2002: Milk Powder – Specification* (pp. 5, 6). New Delhi: Bureau of Indian Standards.

CHAPTER 3

CHEMICAL ANALYSIS OF MILK

N. VEENA, ANURADHA KUMARI, AHESANVARISH SHAIKH, SHRIYESH PATEL, SATISHKUMAR C. PARMAR, and AMIT K. JAIN

ABSTRACT

Milk plays an indispensable role in our daily diet. In the present scenario, various types of milk are marketed. Ensuring compliance with the regulatory standards of milk is an indispensable requirement. Milk analysis is important for the economic viability of the dairy industry, effective process control as well as consumer interests. The compositional quality of milk has a direct bearing on the quality of dairy products that are prepared from it. Therefore, it is imperative to practice systematic analysis of milk to ensure uniform quality of dairy products. This chapter includes procedures for determination of contents of fat, protein, solids-not-fat (SNF), ash; calcium, citric acid (CA), vitamins, minerals, etc.

3.1 TOTAL SOLIDS CONTENT IN MILK BY GRAVIMETRIC METHOD

3.1.1 INTRODUCTION

Total solids (TS) are one of the most important parameters of milk analysis. Determination of TS is important for the economic point of view, technological suitability, nutrition; and to ascertain freedom from possible adulteration. Several methods can be used for determination of moisture/TS, such as gravimetric method, infrared method, Karl Fischer method, specific gravity method, etc. However, the gravimetric method is one of the preferred methods due to its simplicity and convenience. TS of milk vary depending

Analytical Methods for Milk and Milk Products, Volume 1: Sampling Methods and Chemical and Compositional Analysis. Megh R. Goyal, N. Veena, and Santosh Kumar Mishra (Eds.)

upon the type of milk. Generally, TS in cow milk varies from 12.3–14.7% [3] compared with 16–17.7% in buffalo milk [4].

3.1.2 PRINCIPLE

TS content of milk is estimated by weighing the test portion that is pre-dried in steam bath followed by drying at 100±1°C for 3 h in an oven; and this is followed by weighing the dried test portion. The TS (%) is expressed as a ratio of the mass of dried milk to the original mass of milk.

3.1.3 MATERIALS

3.1.3.1 APPARATUS

- Analytical balance: capable of weighing nearest to 0.0001 g.
- Desiccator.
- Dishes (flat bottom; ≥50 mm dia.) made of stainless steel, Ni or Al.
- Forced air oven (set at 100±1°C).
- Steam bath with individual openings.
- Tongs.

3.1.4 PROCEDURE

3.1.4.1 SAMPLE PREPARATION

Bring the temperature of the test sample to 20°C and mix it until it is homogenous by transferring into the clean container and back repeatedly, followed by weighing the test portion. If the sample is old and there is a difficulty in dispersing the cream, then place the milk sample in a water bath at 38°C; and water level should be above the milk level. Mix thoroughly by inversion at least 10 times until a homogenous mixture is obtained and it is cooled to 20°C before transferring the test portion [1].

3.1.4.2 METHODOLOGY

Before taking test portion in a dish, heat the dish in a forced air oven (100±1°C for ≥ 2 h) followed by placed in a desiccator to cool at room temperature. Take the weight of pre-dried dish. Transfer three grams of test portion into the dish and note down the weight. Remove the dish from a balance and prepare blank test by taking the weight of two empty pre-dried dishes. Pre-dry the test sample by placing on steam bath (≤25 min) so that bottom surface of dish is fully exposed to steam until there is little or no flow of liquid in dish. Similarly, expose blank dishes to steam. Remove the dishes and place them in a forced air oven (@ 100±1°C) for 3 h. After drying, the dishes are cooled to ambient temperature in a desiccator for minimum of 30 min. Record the weight of dried dish. The maximum difference between duplicate samples for TS content should not exceed 0.05% [2].

- **Note:** Do not keep the dish on solid trays that will not allow circulation of heat uniformly around the dish.

3.1.5 *CALCULATIONS*

The TS content of the sample (expressed as mass %) is calculated as follows:

$$TS(\%) = \frac{M_2 - M_o - B}{M_1 - M_o} \times 100 \quad (1)$$

where; $M_0$is the empty dish weight (g); M_1 is the weight of dish plus milk (g); and M_2 is the weight of the dish plus dried milk (g); and B is the average residual weight of blank.

- **Note:** If the average blank value is negative, then make sure to add the blank weight instead of deducting it.

3.1.6 *DATA COLLECTION SHEET*

Title: Determination of total solids (TS, %) in milk by oven drying method.

Location and Organization:______________________________

Participants: ______________________________

Date (Month/Day/Year):______________________________

Observations:

Parameter	**Repeat 1**	**Repeat 2**	**Repeat 3**	**Average**
Empty dish weight (g) = M_0				
Weight of dish plus milk (g) = M_1				
Weight of the dish plus dried milk (g) = M_2				
Average residual blank weight (g) = B				
TS (%)				

Comments: ______________________________.

3.2 FAT CONTENT IN MILK BY GERBER METHOD

3.2.1 INTRODUCTION

Milk fat is considered as one of the most expensive fats. Still many countries are making payments based on the fat content of milk. Fat content of buffalo milk varies between 6% and 12% [12], compared with 2.5–5.5% of cow milk [3]. Various methods for determination of fat are gravimetric (reference method), volumetric method (Gerber and Babcock), Infrared method, Ultrasound based methods, etc.

The Gerber method is a commonly used method for routine estimation of milk fat content throughout the world, and it is very similar to the Babcock method, which is used widely in the USA. It is a simple, low-cost, and a rapid method of determination of fat in raw, processed, homogenized, separated or skim milk. This method was developed and patented by Swiss Chemist Niklaus Gerber in 1891. He used the special milk bottle called butyrometer, which is made of glass of 195±2.5 mm in length and 23 mm in diameter at the cylindrical part; and has graduations from 1 to 10% on the stem of this butyrometer. The test is in widespread applications including the payment, testing, and process standardization [11], and is the basis for numerous National [6, 8] and International Standards [7, 9, 10].

Collaborative research study was published in the Journal of AOAC International to evaluate the performance of Gerber method by comparing it with Mojonnier ether extraction (used as reference method) method for fat estimation in whole milk; and results of the study indicated that the Gerber by weight (11.13 g of milk) method should be adopted as an official method for testing fat percent in the raw and pasteurized milk [5, 11]. However, Gerber method by Pipet (10.77 ml of milk) was not adopted due to significant difference in the fat content when compared with the reference method [11]. The Gerber test is used only for routine purpose; and though they provide results comparable with gravimetric methods, yet they cannot replace the gravimetric method [8].

3.2.2 PRINCIPLE

In this volumetric method, fat is separated from milk by the addition of a specified volume of concentrated sulfuric acid and amyl alcohol to butyrometer. Primarily difference between the density of milk fat and the rest of the digested content helps in the separation of fat. Sulfuric acid helps in dissolving the casein and other organic matters and allows the fat in liquid condition by heat generated through the mixing of the components. Amyl alcohol aids in the isolation of fat from the aqueous phase. This mixture is subjected to centrifugal force in a special centrifuge for isolating the fat and it is quantified in the graduated stem of butyrometer.

3.2.3 MATERIALS

3.2.3.1 REAGENTS

- **Isoamyl Alcohol ($C_5H_{12}O$):** It shall be free of water, furfural (C_4H_3O–CHO: also called 2-furaldehyde), acid and fatty materials; and with a boiling point of 128 to 133°C and sp. gr. of 0.814 to 0.816 at 15.5°C.
- **Sulfuric Acid (H_2SO_4):** 90–91% (m/m) of sulfuric acid (also known as Gerber's acid) shall have a specific gravity (sp. gr.) of 1.820–1.825 at 15.5°C. It shall be colorless or pale amber color and free from suspended particles.

3.2.3.2 APPARATUS

- Milk butyrometer (6%, 8% and 10% scales depending on the fat content) (Figure 3.1).
- Lock stoppers.
- Key/regulatory pin.

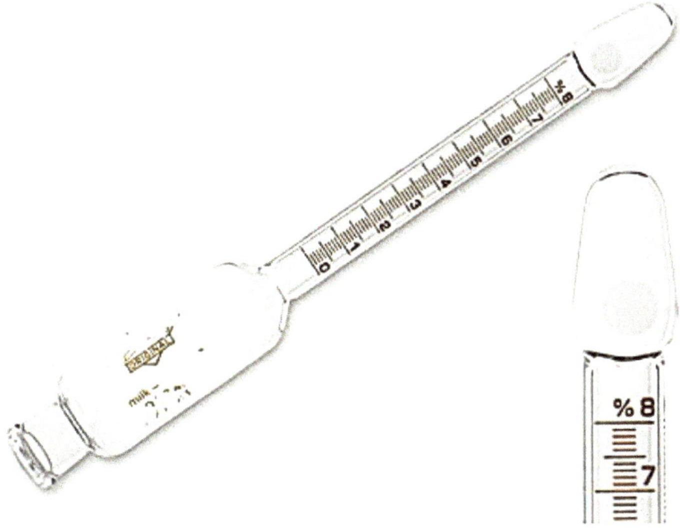

FIGURE 3.1 Gerber butyrometer (Graduated 0–10%).

- Thermostatic water bath (set at 65±2°C): It shall have adequate depth to completely immerse the butyrometer in upright position.
- Gerber's centrifuge: electric driven (1,100 rpm).
- Pipette (10.75±0.03 ml).
- Automatic tilt measure or pipette (1±0.05 ml) for amyl alcohol.
- Automatic tilt measure or pipette (10±0.25 ml) for H_2SO_4.

3.2.4 PROCEDURE

3.2.4.1 SAMPLE PREPARATION

Keep the milk sample in the water bath (39±1°C) and the water level should be above milk level. Mix thoroughly by inverting 10 times. If the fat line remains even after warming, then run hot water (50–60°C) over the outer surface of the container for 15–20 seconds. Mix thoroughly by inversion until a homogenous mixture has been obtained, and immediately take the test potion. Do not keep the milk in a water bath for >15 min. after attaining 38°C [5].

3.2.4.2 DETERMINATION

Add 10 ml of H_2SO_4 to a butyrometer by means of 10 ml of tilt measure. Then 10.75 ml of thoroughly mixed test sample is pipetted out carefully from the side of butyrometer so that two separate layers of liquid are formed. Transfer 1 ml of isoamyl alcohol ($C_5H_{12}O$) using tilt measure. Stopper the butyrometer using key/regulating pin. Thoroughly mix the contents of the butyrometer without inverting, until all the curd particles are dissolved and there is disappearance of curd particles in the mixture. Butyrometer is inverted for few times so that acid in the stem and the small bulb are thoroughly mixed with the contents of the larger bulb. The butyrometer is then placed in a water bath at 65±2°C for not less than 5 minutes.

Place it in the centrifuge, by pointing up the small bulb and diametrically opposite, and is centrifuged for four minutes at 1,100 rpm and then the butyrometer is placed, stopper down in a water bath maintained at 65±2°C for 3 to 10 minutes; and fat column shall be completely immersed in water. Reading should be taken from bringing the lower end of the fat column to coincide with the nearby whole graduation mark by adjusting with the lock stopper. Repeat the analysis, if black or white particles are found at the bottom of a fat column or if dark color or turbidity appears in the fat column [8].

Note:

- In AOAC official method 2000.18, test portion is taken on a weight basis (11.13±0.03 g) in the Gerber method [5].
- Acceptable fat column is straw yellow color and uniform with no dark or white materials. If the fluffy layer is observed at the bottom of the fat column, then the test is rejected. Correct concentration of acid should be used to avoid charring and brown specs at the interface.
- Fat is not easily separated in case of homogenized milk, and in such cases there is a need to centrifuge the butyrometer more than once. On the contrary, keeping the butyrometer for longer time at ≥65°C will result in the esterification of $C_5H_{12}O$ with a subsequent rise in the proportion of fatty layer.

3.2.5 CALCULATIONS

$$\text{Fat content (\%)} = B - A \quad (2)$$

where; B is the reading at upper meniscus of fat column; and A is the reading at lower meniscus of fat column.

3.2.6 DATA COLLECTION SHEET

Title: Determination of milk fat by Gerber method.

Location and Organization:______________________________

Participants: ______________________________

Date (Month/Day/Year): ______________________________

Observations:

Parameter	**Repeat 1**	**Repeat 2**	**Repeat 3**	**Average**
Reading at lower meniscus of fat column = A				
Reading at upper meniscus of fat column = B				
Fat (%)				

Comments:______________________________.

3.3 FAT CONTENT IN MILK BY MOJONNIER METHOD

3.3.1 INTRODUCTION

The Mojonnier method has originated from the Roese-Gottlieb method. In the Roese-Gottlieb method for fat determination, alkaline digestion is used. It is widely used for estimation of fat in all dairy products. It is considered the most accurate; and hence has been recognized as reference method by several regulatory authorities. In this section, Mojonnier method of fat estimation in milk is discussed [14].

3.3.2 PRINCIPLE

This analysis includes extraction of fat from the sample in organic solvents, evaporation of the solvent and residual moisture by heating at controlled temperature, and weighing of the fat to quantify its content in the given sample. In milk, the fat is present in the form of membrane-covered tiny droplets. Its efficient extraction requires destabilization of the emulsion system by destroying the membrane by a concentrated ammonia before the extraction by a mixture of ethers. Alcohol is also added to facilitate the destabilization of the emulsion and subsequent separation of aqueous and ethereal layers. Ether extract is evaporated, and the fat residue is then dried for the removal of residual solvent; and weight of the extracted fat (% by mass) is calculated.

3.3.3 MATERIALS

3.3.3.1 REAGENTS

- Concentrated ammonium hydroxide (NH_4OH) (sp. gr. 0.9);
- Diethyl ether (free from peroxide);
- Distilled water;
- Ethanol (C_2H_5OH, 95%);
- Petroleum ether (Boiling point ranges between 30°C and 60°C);
- Phenolphthalein indicator (0.5% solution in C_2H_5OH).

3.3.3.2 APPARATUS

- Bark cork or ground glass stopper;
- Desiccator;
- Fat collecting vessel: Metal dish (85–95 mm in diameter and 45–55 mm in length) or glass beaker (250 ml);

- Forced air oven (100±1°C);
- Glass beads;
- Measuring cylinders (25 ml);
- Mojonnier fat extraction flask (Figure 3.2);
- Tongs: For handling fat in the collecting vessel;
- Water bath or steam bath: for evaporation of solvent at ≤100°C;
- Weighing balance (Readability of 0.0001 g).

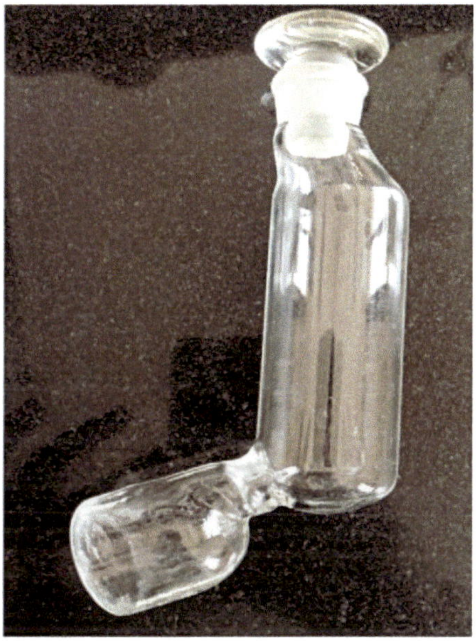

FIGURE 3.2 Mojonnier tube for fat extraction.

3.3.4 PROCEDURE

3.3.4.1 PREPARATION OF THE MILK SAMPLE [REFER TO SECTION 1.4.1]

3.3.4.2 DETERMINATION

The 10 g of prepared milk sample is accurately weighed and is transferred to a Mojonnier flask. Transfer 1.5 ml of NH_4OH into the extraction flask and mix thoroughly in the lower bulb. Ammonia solution neutralizes the acid (if it is present) and dissolves the protein. Transfer three drops of phenolphthalein solution to clearly visualize the interface between solvent and aqueous phase. Transfer C_2H_5OH (10 ml) to the flask, stopper with cork and mix thoroughly between the two bulbs by letting the contents to flow forwards and backwards for 15 s. For 1st extraction, transfer diethyl ether (25 ml) followed by closing of the flask with a cork (wetted stopper); and contents are mixed vigorously for one minute. Remove the cork and then transfer petroleum ether (25 ml). Close the flask with rewetted cork, and shake the flask again for one minute.

Centrifuge at 600 rpm for not less than 30 s. In case of non-availability of centrifuge, place the tube in a stand on position using rack for a minimum of 30 min. till the upper solvent phase is clear and has been noticeably separated from the aqueous (pink) phase. If there is a tendency to form emulsion, add few drops of alcohol to aid further separation of the layers. Take the empty weight of fat collecting vessel containing a few glass beads. Carefully, transfer the ethereal layer completely (as much as possible) into the fat collecting vessel. Evaporate the ether from the vessel at ≤100°C while conducting 2nd extraction.

To extract the remaining fat within the extraction flask, pour C_2H_5OH (5 ml) to the flask, close with cork and shake vigorously for 15 s. Followed by transfer 15 ml of diethyl ether, close with the cork and shake the flask for 1 min. Transfer 15 ml of petroleum ether, close with cork and shake again for 1 min. Centrifuge at 600 rpm for not less than 30 s. Transfer the ethereal layer carefully from the flask to the same vessel and evaporate-off completely. For 3rd extraction, follow the same procedure as used for 2nd extraction except addition of C_2H_5OH. Completely evaporate the ether from the vessel at ≤100°C.

Heat the vessel containing residual fat for ≤30 min. in a forced air oven (100±1°C), followed by transfer the vessel in a desiccator; and note the mass after cooling. Carry out the same procedure with 10 ml water instead of test sample for running blank test (run duplicate blank reagents). Record the mass of dried residue that has been collected; and it should be less than 0.0020 g for blank reagent.

- **Note:** The maximum difference between the duplicate samples for fat content should be less than 0.03%.

3.3.5 CALCULATION

$$\text{Fat, \%} = \frac{(w_1 - w_2 - B)}{w} \times 100 \qquad (3)$$

where; W_1 is the weight of fat collecting vessel with dried fat (g); W_2 is the weight of the empty vessel (g); W is the weight of test sample (g); and B is the mean residual weight of blank (g).

- **Note:** If the average of blank value is negative, then make sure to add blank weight instead of deducting it.

3.3.6 DATA COLLECTION SHEET

Title: Estimation of fat content in milk by Mojonnier method.

Location and Organization: ______________________________

Participants: ______________________________

Date (Month/Day/Year): ______________________________

Observations:

Parameter	Repeat 1	Repeat 2	Repeat 3	Average
Weight of fat collecting vessel with dried fat (g) = W_1				
Empty weight of vessel (g) = W_2				
Test sample weight (g) = W				
Mean residual mass of blank (g) = B				
Fat % (by mass)				

Comments: ______________________________.

3.4 SOLIDS-NOT-FAT (SNF) CONTENT OF MILK

3.4.1 INTRODUCTION

Solids-not-fat (SNF) of milk is an important parameter in the milk analysis. SNF is constituted of proteins, lactose, and minerals. SNF constituents have significance from the point of view of economics, technological suitability, nutrition, regulatory compliance, etc. The SNF content in milk varies depending upon the species, breed, season, feed, and several other factors.

3.4.2 PRINCIPLE

If percent fat and TS contents of milk are known, then percentage SNF can be determined by deducting the % fat from %TS content [15].

3.4.3 PROCEDURE

3.4.3.1 TOTAL SOLIDS (TS)

The TS content is determined according to the procedure described in Section 3.1 (determination of total solids in milk) of this chapter.

3.4.3.2 FAT

The fat content in milk is determined according to the procedure described in Section 3.2 (determination of fat in milk) of this chapter.

3.4.4 CALCULATION

$$\text{SNF}(\%) = \text{TS}(\%) - \text{fat}(\%) \quad (4)$$

- ➢ *Note:* Maximum difference in SNF content of duplicate samples should be $<0.06\%$.

3.4.5 DATA COLLECTION SHEET

Title: Determination of SNF content of milk.

Location and Organization: ______________________

Participants: ______________________

Date (Month/Day/Year): ______________________

Observations:

Parameter	Repeat 1	Repeat 2	Repeat 3	Average
TS (%)				
Fat (%)				
SNF (%)				

Comments: ______________________.

3.5 LACTOSE CONTENT IN MILK BY POLARIMETRIC METHOD

3.5.1 INTRODUCTION

Lactose is characteristic disaccharide that is present in milk. It is important from the point of view of several physicochemical properties of dairy products including browning in heat treated milk products, caking in milk powders and sandiness in ice-cream. Lactose is also vital from the perspective of nutrition because it is implicated in brain development, calcium absorption, lactose intolerance, etc. Although, glucose can be obtained from several natural foods, lactose is the only source for galactose (that is a crucial sugar for brain development). Lactose content in cow milk varies from 3.8 to 5.3%; while for buffalo milk, it is reported by various workers to be varying from 4.3 to 5.32% [17, 18].

3.5.2 PRINCIPLE

Lactose is an optically active compound with a chiral center. The instrument by which the optical activity of the substance is measured is called Polarimeter. When a solution of glucose or lactose is placed in the path of polarized light, it changes the plane of polarized light. The extent of optical rotation depends on nature of substance, its concentration, the length of column through which light passes, wavelength of light, and temperature of liquid. The measurement of optical rotation can be employed as a method of analysis of sugar.

3.5.3 MATERIALS

3.5.3.1 REAGENTS

- Acid-mercuric nitrate [$Hg(NO_3)_2$] solution: A known mass of Hg is dissolved in twice its mass of HNO_3. Cool and dilute with H_2O at 1:5 ratio.
- Mercuric iodide (HgI_2) solution: 13.5 g of $HgCl_2$ and 33.2 g of KI is dissolved in acetic acid (CH_3COOH) (200 ml) and water (640 ml).
- Phospho-tungstic acid (PTA) solution (5%).

3.5.3.2 APPARATUS

- Measuring cylinder (25 and 50 ml);
- Polarimeter tubes with 200 mm and 400 mm length;
- Polarimeter;
- Volumetric flasks (100 and 200 ml).

3.5.4 PROCEDURE

The 65.8 g (2 normal weight) of test milk sample is exactly measured; and it is transferred to volumetric flasks (100 and 200 ml), respectively. Transfer 30 ml of $HgCl_2$ or 20 ml of acid-$Hg(NO_3)_2$ to each flask. In a 100 ml flask, PTA solution is added up to the mark; whereas in a 200 ml flask, transfer PTA (15 ml) solution and is diluted with H_2O. Mix the contents of each flask thoroughly for 15 minutes. Use dry filter to filter the contents of the flask and polarize. Measure the solution in 100- and 200-ml flask each using 200 and 400 mm polarimeter tubes, respectively, to minimize error of reading [16].

3.5.5 CALCULATION

Calculate the % lactose in the milk sample as follows:

$$\text{Lactose (\%)} = \frac{R_2 - [(R_2 - R_1)\times 2]}{2} \quad (5)$$

where; R_1 is the reading of solution from 200 ml flask; and R_2 is the reading of solution from 100 ml flask.

3.5.6 DATA COLLECTION SHEET

Title: Determination of lactose content in milk by polarimetric method.

Location and Organization: ______________________________

Participants: ______________________________

Date (Month/Day/Year): ______________________________

Observations:

Parameter	Repeat 1	Repeat 2	Repeat 3	Average
Reading of solution from 200 ml flask (polarimeter tube of 400 mm is used) = R_1				
Reading of solution from 100 ml flask (polarimeter tube of 200 mm is used) = R_1				
% Lactose				

Comments: ______________________________.

3.6 TOTAL PROTEIN CONTENT IN MILK BY MICRO-KJELDAHL METHOD

3.6.1 INTRODUCTION

Nitrogen (Total) in milk can be determined using Micro-Kjeldahl method. Digestion of sample is carried out in block digester and distillation by automatic Nitrogen distillation system. The collected distillate is titrated using standard acid solution [19].

3.6.2 PRINCIPLE

Kjeldahl method determines the total organic Nitrogen (protein and non-protein) content of test sample by 3 major processes, such as: digestion, distillation, and titration. Addition of concentrated sulfuric acid (H_2SO_4) oxidizes the organic compounds and liberates Nitrogen (N_2) as non-volatile ammonium sulfate [$(NH_4)_2SO_4$] in the presence of catalyst (copper sulfate) and potassium sulfate, when digested at high temperature for long time depending on the product.

$$\text{Protein} \xrightarrow{catalyst+heat} \text{Ammonium sulphate} + H_2O + CO_2$$

Sulfuric acid is then neutralized with 50% NaOH solution before the distillation, where free ammonia (NH_3) and water is formed; and is collected in saturated H_3BO_3 to form ammonium borate.

$$\text{Ammonium sulfate} \xrightarrow{neutralize\ and\ distillation} \text{Ammonia} + \text{water}$$

$$\text{Free ammonia} + \text{boric acid} \xrightarrow{steam} \text{ammonium borate}$$

The compound generated is thereafter titrated against standard HCl solution that will result in the amount of ammonia generated [19].

3.6.3 MATERIALS

3.6.3.1 REAGENTS

- 0.1 M standard HCl solution.
- 5.0% copper sulfate solution: Dissolve 5 g of $CuSO_4.5H_2O$ in 100 ml distilled water.
- Ammonium sulfate [$(NH_4)_2SO_4$] (99.9%): To calculate Nitrogen loss.

- Boric acid (H_3BO_3, 4.0% w/v) solution with addition of mixed indicator (3 ml in one liter of boric acid solution). This solution gives orange color.
- Concentrated sulfuric acid (95–98%).
- Mixed indicator: Prepare 0.2% methyl red solution in C_2H_5OH (95%) and 0.2% bromocresol green (BCG) in C_2H_5OH (95%). Add one part of methyl red (0.2%) into 5 parts of BCG (0.2%) to prepare a mixed indicator.
- NaOH solution (40% w/w).
- Potassium sulfate (K_2SO_4).
- Pure tryptophan or lysine hydrochloride (99%): To analyze digestion efficiency.
- Sucrose.

➢ ***Note:*** All the reagents should be Nitrogen free.

3.6.3.2 APPARATUS

- Burette: 50 ml.
- Digestion block: Made of aluminum alloy with adjustable temperature control and connected with manifold and fume neutralization system (Figure 3.3(A)).
- Digestion tubes (250 ml capacity): for block digestion.
- Distillation unit: For steam distillation (Figure 3.3(B)).
- Titration flasks (500 ml).

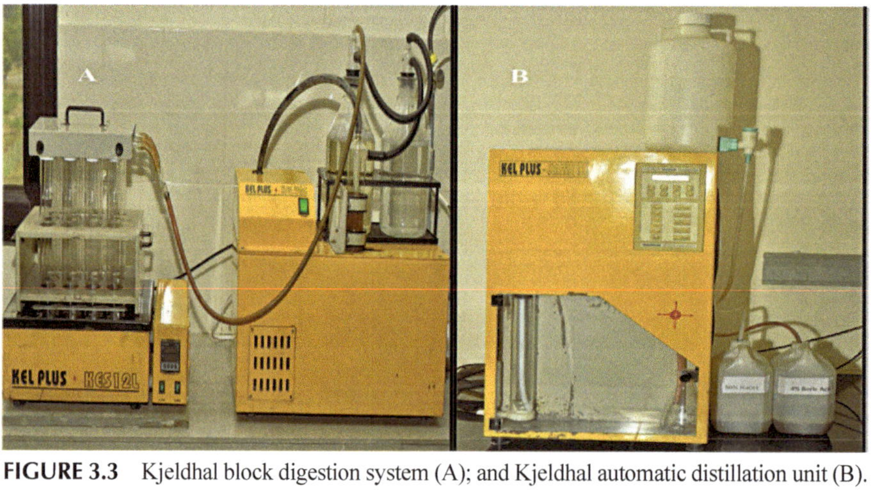

FIGURE 3.3 Kjeldhal block digestion system (A); and Kjeldhal automatic distillation unit (B).

3.6.4 PROCEDURE

3.6.4.1 PREPARATION OF THE TEST SOLUTION

Weigh 5 ml of warm milk sample (at 38±1°C and to nearest 0.0001 g) into a digestion tube containing K_2SO_4 (12 g) and $CuSO_4$ (1 ml) solutions. Thereafter, add 20 ml of H_2SO_4 and keep it for digestion. Prepare a blank using the same amount of reagents but without milk sample.

3.6.4.2 DIGESTION

Place sample tubes in a block initially at low temperature (180–230°C) to control the foaming. Tubes are connected with aspirator in the digestion block; suction shall be just sufficient to take out fumes. Samples are digested at 180–230°C for half an hour or till whitish fumes have developed. Raise the temperature (410–430°C) to heat intensively; and continue the digestion of sample until clear digest has been obtained. In order to control foaming, it may be essential to raise the temperature slowly for 20 min. Do not allow foam within tubes to rise greater than 4–5 cm from top of tube. Generally, clear digestion is indicated by bluish green color. Carry on the digestion (sulfuric acid should be boiling) for a minimum of 60 min. The total duration for digestion takes about 1.75 to 2.5 h.

At the end, the digested mixture must be clear and there are no undigested particles. Cool the digested mixture to ambient temperature for about 25 min. After cooling, the contents should appear as clear liquid. Add carefully 85 ml of water (100 ml for blank) in each tube and swirl the contents so that any developed crystals will be dissolved and allowed to cool at room temperature.

- ***Note:*** Extreme crystallization represents low amount of acid remaining at the end of digestion, and it may lead to lower results. Aspiration rate of fume is reduced to minimize the loss of acid during digestion.

3.6.4.3 DISTILLATION

Transfer 40% of NaOH in alkali tank and quantity dispensed for alkali is adjusted to 65 ml. Tube containing digested solution is attached to the distillation apparatus. Keep the titration flask (500 ml) containing H_3BO_3 (50 ml) with mixed indicator into which another outlet or delivery tube is inserted. This delivery tube from the condenser is used for redirecting the liberated ammonia and trap into the H_3BO_3 solution. Carry out steam distillation immediately, until more than 150 ml of distillate has been collected. Take out the titration flask, and let the contents to cool to ambient temperature. Similarly, run the blank using the same reagents.

3.6.4.4 *TITRATION*

Carry out the titration of collected distillate with hydrochloric acid (HCl) (0.1 M), until there is an appearance of pink color as the end point. Record the titer value.

3.6.4.5 *NITROGEN RECOVERY VERIFICATION*

1. **Nitrogen Loss:** Prepare digestion tube by adding 0.85 g of sucrose, 0.12 g of $(NH_4)_2SO_4$ and other reagents are added as mentioned in the preparation of test solution (Part 6.4.1 in this section). Run the digestion and distillation according to the procedure; and calculate the recoveries of Nitrogen (it should be >99% recovery).
2. **Digestion Efficiency:** Take either lysine hydrochloride (0.16 g) or tryptophan (0.18 g) along with sucrose (0.67 g), and all other reagents are added in digestion tube as described in the preparation of test solution (Part 6.4.1 in this section); and carry out digestion and distillation according to the procedure in this section. At least 98% recoveries should be obtained to get the accurate results [19].

3.6.5 *CALCULATIONS*

$$\%\text{ Nitrogen} = \frac{1.4007 \times (A - B) \times M}{W} \tag{6}$$

$$\%\text{ Protein} = 6.38 \times (\%\text{ Nitrogen}) \tag{7}$$

where; A is the titer value of HCl used for milk sample (ml); B is the titer value of HCl used for blank (ml); M is the molarity of HCl; and W is the weight (g) of test sample.

- Note: Maximum recommended difference in the protein content of duplicate samples is 0.03%.

3.6.6 DATA COLLECTION SHEET

Title: Estimation of total protein content in milk by Kjeldahl method.

Location and Organization:______________________________

Participants: ______________________________________

Date (Month/Day/Year):________________________________

Observations:

Parameter	**Repeat 1**	**Repeat 2**	**Repeat 3**	**Average**
Weight of test sample, g = W				
Molarity of HCl = M				
Titer value of HCl used for sample, ml = A				
Titer value of HCl used for blank, ml = B				
% Nitrogen				
% Protein = (6.38 × % Nitrogen)				

Comments: ____.

3.7 CASEIN CONTENT IN MILK

3.7.1 INTRODUCTION

Casein is the principal protein of milk that constitutes about 80% in both the major milk producing species (i.e., cow and buffalo). The typical protein in mammalian milk possesses several unique properties significant from the technological as well as nutritional point of view. The casein content of buffalo milk varies from 3.03 to 3.24% [22]; while in cow milk, varies from 1.7 to 3.5% [3].

3.7.2 PRINCIPLE

Total nitrogen (TN) and non-casein Nitrogen (NCN) contents of milk are estimated separately. Difference between these two parameters gives casein Nitrogen content of milk sample. The casein Nitrogen thus obtained can be converted to get casein percent by multiplying with the factor.

3.7.3 MATERIALS

3.7.3.1 REAGENTS

- 5.0% copper sulfate solution: Dissolve 5 g of $CuSO_4.5H_2O$ in 100 ml of distilled water.
- Acetic acid solution (10% w/v).
- Ammonium sulfate (99.9%): To calculate Nitrogen loss.
- Boric acid (H_3BO_3, 4% w/v) solution with addition of mixed indicator (3 ml in one liter of boric acid solution). This solution gives orange color.
- Concentrated sulfuric acid (95–98%).
- Mixed indicator: Prepare 0.2% methyl red solution in C_2H_5OH (95%) and 0.2% BCG in C_2H_5OH (95%). Add one part of methyl red (0.2%) into 5 parts of BCG (0.2%) to prepare a mixed indicator.
- NaOH solution (40% w/w).
- Potassium sulfate (K_2SO_4).
- Pure tryptophan or lysine hydrochloride (99%): To analyze digestion efficiency.
- Sodium Acetate (CH_3COONa; 1M) solution: Transfer 4.1 g of CH_3COONa or 6.6 g of $CH_3COONa.3H_2O$ in 50 ml volumetric flask and dilute it with H_2O (20°C).
- Standard HCl (0.1 M) solution.
- Sucrose.

➢ **Note:** All the reagents should be Nitrogen free.

3.7.3.2 APPARATUS

- Burette: 50 ml.
- Digestion block: Made of aluminum alloy with adjustable temperature control.

- Digestion tubes (250 ml capacity): for block digestion.
- Distillation unit: For steam distillation.
- Pipette (50 ml).
- Titration flasks (500 ml).
- Volumetric flask (50 and 100 ml).
- Whatman filter paper (grade 1; 15 cm; Nitrogen free).

3.7.4 PROCEDURE

3.7.4.1 DETERMINATION OF TOTAL NITROGEN (TN)

TN content of milk sample is determined as described in Section 3.6 in this chapter [19].

3.7.4.2 DETERMINATION OF NON-CASEIN NITROGEN (NCN)

Into a pre-weighed 100 ml volumetric flask, 10±1 ml of warm milk (38±1°C) is pipetted out and weight is recorded nearest to 0.0001 g. Transfer 75±0.5 ml of H_2O (38°C) and 1±0.02 ml of acetic acid (10%) solution is added, and it is mixed thoroughly; and kept in water bath at 38°C for 10 minutes. Thereafter, transfer 1±0.02 ml of CH_3COONa solution, mix thoroughly and then cool it to 20°C. Volume is made up to the mark using H_2O (20°C), mix properly and allow the precipitate to settle. Using filter paper, filter the content and complete filtrate is collected. Pipette out 50±0.05 ml of filtrate into Kjeldahl tube containing K_2SO_4 and $CuSO_4$ solution. Digestion, distillation, and titration steps are carried out as described in the Section 3.6 in this chapter (4). Similarly, a blank test is also performed (50±1 ml H_2O, 0.5±0.02 ml CH_3COOH and 0.5±0.02 ml CH_3COONa) using the same procedure but without the test sample [21].

3.7.5 CALCULATIONS

$$NCN\ (\%) = \frac{1.4007 \times (A - B) \times M \times 2 \times 0.994}{W} \tag{8}$$

$$Casein\ Nitrogen\ (\%) = TN\ (\%) - NCN\ (\%) \tag{9}$$

$$Casein\ content\ (\%)\ in\ milk = Casein\ Nitrogen\ (\%) \times 6.38 \tag{10}$$

where; A is the titer value of HCl used for test sample (ml); B is the titer value of HCl used for blank (ml); W is the weight of test portion (g); and M is the molarity of HCl solution.

3.7.6 DATA COLLECTION SHEET

Title: Determination of casein content in milk by Kjeldahl method.

Location and Organization:________________________________

Participants: ____

Date (Month/Day/Year):__________________________________

Observations:

Parameter	Repeat 1	Repeat 2	Repeat 3	Average
Weight of test sample, g = W				
Titer value of HCl used for sample, ml = A				
Titer value of HCl used for blank, ml = B				
TN (%)				
NCN (%)				
Casein nitrogen (%) = TN(%) – NCN(%)				
Casein content (%)				

Comments: ____.

3.8 NON-PROTEIN NITROGEN (NPN) CONTENT IN MILK

3.8.1 INTRODUCTION

NPN content is expressed as soluble Nitrogen in TCA (12%), which represents 5–6% of TN present in milk; and it varies from 25–30 mg/100 ml [24]. Several heterogeneous components contribute to NPN of milk, such as: urea, ammonia, amino sugars, creatinine, creatine, uric acid, nucleotides, α-amino Nitrogen, hippuric acid, etc. Urea is one of the principal components of NPN, contributing about 50% of total NPN [25]. NPN is also related with the protein metabolism of the animal and thus efficient utilization of the feed protein. Determination of NPN is an important parameter from the point of view of feed efficiency, possible adulteration of NPN substances, udder health, etc.

3.8.2 PRINCIPLE

The protein portion is precipitated out from a milk sample by treating with a solution of TCA. The filtrate thus obtained is then digested with sulfuric acid for converting Nitrogen into ammonium sulfate. Ammonia is released using sodium hydroxide (NaOH). Ammonia is subsequently distilled into H_3BO_3, and titration is done against HCl solution. The NPN content is then determined from the amount of the HCl consumed [24].

3.8.3 MATERIALS

3.8.3.1 REAGENTS

- 0.01 M standard HCl solution.
- Boric acid (H_3BO_3, 4.0% w/v) solution with addition of mixed indicator (3 ml in one liter of boric acid solution). This solution gives orange color.
- Concentrated sulfuric acid (95–98%).
- Copper sulfate solution (5%): Dissolve 5 g of $CuSO_4.5H_2O$ in 100 ml distilled water.
- Mixed indicator: Prepare 0.2% methyl red solution in C_2H_5OH (95%) and 0.2% BCG in C_2H_5OH (95%). Add one part of methyl red (0.2%) into 5 parts of BCG (0.2%) to prepare a mixed indicator.
- NaOH solution (40% w/w).
- Potassium sulfate (K_2SO_4).
- Sucrose.
- TCA (CCl_3COOH) solution (15% w/v).

➢ **Note:** All the reagents should be Nitrogen free.

3.8.3.2 APPARATUS

- Digestion tubes (250 ml capacity): for block digestion.
- Digestion block: Made of aluminum alloy with adjustable temperature control.

- Distillation unit: For steam distillation.
- Titration flasks (500 ml).
- Burette: 50 ml.
- Whatman No. 1 filter paper (15 cm in diameter; Nitrogen free).

3.8.4 PROCEDURE

3.8.4.1 PREPARATION OF TEST SOLUTION

Pipette and weigh 10±1 ml of warm milk (38+1°C) sample into a pre-weighed 125 ml Erlenmeyer flask and record the weight nearest to 0.0001 g. Transfer 40±0.5 ml of CCl_3COOH (15%) into the flask. Again, note the weight. Mix the contents thoroughly and let the precipitate to settle for five minutes. Using a filter paper, filter the contents and collect whole filtrate (repeat the preparation, if the filtrate obtained is not clear and contains particulate matter). Mix the filtrate by swirling followed by transfer 20 ml filtrate into a 50-ml beaker and record the weight. Transfer the filtrate into digestion tubes containing K_2SO_4 (12 g), and $CuSO_4$ (1 ml) and H_2SO_4 (20 ml) solutions. Immediately, note the empty weight of the beaker (50 ml). Similarly, prepare a blank test using 16±0.5 ml of TCA solution but without test sample [23].

3.8.4.2 DETERMINATION

Carry out the digestion and distillation as described in the Section 6.4 in this chapter. Titration shall be carried out using 0.01 M HCl instead of 0.1 M HCl.

3.8.5 CALCULATION

$$Nitrogen\ (\%) = \frac{[1.4007 \times (A - B) \times M]}{[(w_f \times w_m) \div \{w_t - (w_m \times 0.065)\}]} \quad (11)$$

NPN (expressed as % protein equivalent) = Nitrogen (%) × 6.38 (12)

where; A is the titer value of HCl used for sample (ml); B is the titer value of HCl used for blank (ml); W_f is the mass of the 20.0 ml filtrate (g); W_m is the mass of test sample (g); W_t is the mass of sample + 40 ml of CCl_3COOH solution (g); and M is the molarity of HCl.

- Note: 0.065 = This factor assumes that milk with ~ 3% true protein and 3.5% fat has a factor of "0.030 + 0.035 = 0.065"; and for partial skimmed milk with 2.7% fat, the factor is 0.057.

3.8.6 *DATA COLLECTION SHEET*

Title: Determination of NPN content of milk by Kjeldahl method.

Location and Organization:______________________________

Participants: ______________________________

Date (Month/Day/Year):______________________________

Observations:

Parameter	**Repeat 1**	**Repeat 2**	**Repeat 3**	**Average**
Titer value of HCl used for sample (ml) = A				
Titer value of HCl used for blank (ml) = B				
Molarity of HCl = M				
Mass of the 20.0 ml filtrate (g) = W_f				
Mass of test sample (g) = W_m				
Mass of test sample + 40 ml of TCA solution, (g) = W_t				
Nitrogen (%)				
NPN (expressed as % protein equivalent)				

Comments:______________________________.

3.9 PROTEIN NITROGEN CONTENT IN MILK

3.9.1 *INTRODUCTION*

The Nitrogen content of milk is broadly distributed into three fractions, i.e., casein, serum protein and NPN. Of the TN of milk, 95% is associated with protein Nitrogen and remaining 5% with NPN.

3.9.2 PRINCIPLE

The TN and NPN of the test sample are determined separately. The percent protein Nitrogen of milk is determined by subtracting NPN content from TN content, and results are multiplied by 6.38 [26].

3.9.3 PROCEDURE

- NPN (%) of milk is determined based on the procedure in the Section 3.8 in this chapter.
- TN (%) of milk is determined based on the procedure in the Section 3.6 in this chapter.

3.9.4 CALCULATION

$$\text{Protein Nitrogen (\%)} = \text{TN (\%)} - \text{NPN (\%)} \tag{13a}$$

$$\text{Protein (\%)} = \text{Protein Nitrogen (\%)} \times 6.38 \tag{13b}$$

3.9.5 DATA COLLECTION SHEET

Title: Estimation of protein Nitrogen content (%) of milk.

Location and Organization:______________________________

Participants: ______________________________________

Date (Month/Day/Year):________________________________

Observations:

Parameters	**Repeat 1**	**Repeat 2**	**Repeat 3**	**Average**
TN (%)				
NPN (%)				
Protein nitrogen (%)				

Comments:____

3.10 DETERMINATION OF ASH CONTENT OF MILK BY GRAVIMETRIC METHOD

3.10.1 INTRODUCTION

Ash is an inorganic matter left after the removal of organic components either by ignition or by complete oxidation of dairy products. It gives an indication of total minerals that are present in the test sample. Ash content indicates the inorganic minerals which remain after complete incineration of the food materials. In bovine milk, ash content varies from 0.7 to 0.8% which is relatively constant; however, it varies considerably with the relative content of several ions. The determination of percent ash in milk gives an idea about abnormal milk, possible adulteration, technological suitability, etc.

3.10.2 PRINCIPLE

Dry ashing method involves incinerating the milk at ≤550°C in a furnace; and weighing the residue that is left after incineration [27].

3.10.3 MATERIALS

3.10.3.1 APPARATUS

- Analytical balance.
- Desiccator;
- Marking pen for crucible: Scribed steel pin is usually used for marking;
- Muffle furnace (set at 550°C);
- Platinum or silica crucible dish: Should be stable with chemicals and withstand at higher temperature;
- Steam bath;
- Tongs (long arms): to hold the crucible.

3.10.4 PROCEDURE

3.10.4.1 PREPARATION OF THE MILK SAMPLE [REFER TO SECTION 1.4.1]

3.10.4.2 DETERMINATION

Take the weight of an empty dry silica crucible. Transfer 5 ml of test sample in a silica crucible and record the weight. Keep the sample containing crucible on steam bath and evaporate the moisture to dryness. Transfer the crucible to the muffle furnace and incinerate the sample at ≤550°C until there is a formation of white ash free from carbon. Cool the sample in desiccator. Take the weight of crucible + ash. Repeat the heating in a muffle furnace for 30 min. followed by cooling; and weigh till the differences in weight between two consecutive observation is < 1 mg. Record the lowest mass.

3.10.5 CALCULATIONS

$$\text{Ash content } (\% \text{ by mass}) = \frac{W_2 - W}{W_1 - W} \times 100 \tag{14}$$

where; W is the weight of empty crucible (g); W_1 is the weight of crucible plus test sample (g); and W_2 is the weight of crucible plus residual ash (g).

3.10.6 DATA COLLECTION SHEET

Title: Determination of ash content of milk by gravimetric method.

Location and Organization:______________________________

Participants: ______________________________

Date (Month/Day/Year):______________________________

Observations:

Parameter	**Repeat 1**	**Repeat 2**	**Repeat 3**	**Average**
Empty crucible weight (g) = W				
Crucible plus test sample weight (g) = W_1				
Crucible plus residual ash weight (g) = W_2				
Ash content (% by mass)				

Comments:______________________________.

3.11 COPPER CONTENT IN MILK BY PHOTOMETRIC METHOD

3.11.1 INTRODUCTION

Copper (Cu) is vital for absorption of Fe, and it acts as a co-factor for the enzyme that takes part in the synthesis of polar lipids, hemoglobin, and connective tissues and metabolism of glucose. Nutritional deficiency of Cu is rare, excluding in conditions of extreme malnutrition [29]. Milk is not a rich source of Cu and there is very little contribution to the total dietary intake of Cu [33]. Average content of Cu in cow milk is about 0.09 mg/liter [32]. The Cu content of milk can be improved by dietary supplementation of Cu [31] or by using Cu-metal containers and processing equipment, which are made of Cu [34]. Contamination with Cu from brass containers has been reported as the chief source of Cu in milk [28]. Spectroscopic and electro-thermal methods are available for determination of Cu in milk; however, spectroscopic methods are widely used.

3.11.2 PRINCIPLE

Digestion of the organic material in the milk sample is done with a mixture of HNO_3 and H_2SO_4 followed by neutralization of solution with ammonium hydroxide; and then complexing the Cu as a salt of diethyl-dithiocarbamic acid. Extraction of the copper (II) salt is done with amyl acetate followed by photometric measurement of the absorbance of the yellow solution [30].

3.11.3 MATERIALS

3.11.3.1 REAGENTS

Chemicals used are of AR (analytical reagents) grade (with exception of standard $CuSO_4$ (II) solution), free from copper. The water used shall be double distilled and final distillation should be done in a Cu-free distillation unit. If distillation of ethanol is required, then ether and HNO_3 should be done in a Cu-free distillation unit.

- Ammonium hydroxide solution (density 0.91 g/ml at 20°C): If required, purify it by Cu-free vacuum distillation unit.
- Amyl Acetate: Four liters of amyl acetate is dried for 24 h with 15 g anhydrous Na_2SO_4. It is distilled in a Cu-free distillation unit, and the fraction is collected, which distills between 136°C and 140°C.

➢ **Note:** In place of amyl acetate, xylene distilled in Cu-free distillation unit can be used.

- Citrate-Ethylene-Dinitrilotetra-Acetic Acid (EDTA) Solution: Dissolve 400 g of ammonium citrate [$(NH_4)_3C_6H_5O_7$] and 100 g of

EDTA di-sodium salt dihydrate ($Na_2C_{10}H_{14}N_2O_8.2H_2O$) in H_2O. Dilute to 1,000 ml in a volumetric flask. Extract the citrate-EDTA solution with 25 ml of diethyl ammonium diethyldithiocarbamate [$(C_2H_5)_2NCSSNH_2(C_2H_5)_2$] solution (1 g per liter) in carbon tetrachloride (CCl_4). Extraction is repeated with 25 ml of the diethyl ammonium diethyldithiocarbamate solution, until the carbon tetrachloride remains nearly colorless.

- Concentrated H_2SO_4 (density 1.84 g/ml at 20°C).
- Concentrated HNO_3 (density 1.42 g/ml at 20°C).
- Diethyl ether.
- Dilute H_2SO_4 (0.5 mol/liter).
- Ethyl alcohol (C_2H_5OH, 96% v/w).
- H_2O_2 solution (density 1.099–1.103 g/ml at 20°C).
- KCN solution (5%).
- NaOH solution (1 M).
- Petroleum ether (light petroleum: boiling point of 40 to 60°C):
- Phenolphthalein solution: 1% solution in ethanol (90% v/v).
- Sodium diethyldithiocarbamate solution: 400 mg sodium diethyldithiocarbamate [$(C_2H_5)_2NCSSNa$] (or 500 mg of trihydrate) is dissolved in 90 ml of water. The volume is made to 100 ml with an ammonium hydroxide solution. The solution is stored in the dark place at a temperature of 0 to 8°C. Every week, the solution is renewed.
- Standard $CuSO_4$ solution:
- Stock solution: Dissolve 196.5 mg of copper (II) sulfate pentahydrate ($CuSO_4.5H_2O$) in water. Add 5 ml H_2SO_4 solution (0.5 M) and it is diluted with H_2O to 1,000 ml.
- Working solution (1 μg Cu/ml): 10 ml of $CuSO_4$ stock solution is added to 5 ml of H_2SO_4 (0.5 M); and volume is made to 500 ml with H_2O. Freshly prepare before use.

3.11.3.2 APPARATUS

Clean all the glasswares including the glass beads with 10% (m/m) HNO_3 solution. Rinse these thrice with distilled water and then with double distilled water before use. Dry, if required, by successively rinsing with ethyl alcohol and diethyl ether.

- Analytical balance;
- Digestion flasks (Kjeldahl) (with ground glass stoppers, calibrated on the lower part of the neck at 50 ml);
- Glass beads;
- Measuring cylinders (capacity 5 and 25 ml);
- Micro-gas burner;
- Pipettes (capacity 5 and 25 ml);
- Pipettes (capacity 1, 2, 5 and 20 ml);
- Spectrophotometer, operating with cells of optical path length of 10 mm;
- Volumetric flask (100, 500 and 1,000 ml);
- Water baths (maintaining temperatures of 30 to 40°C, 40 to 60°C, 60 to 65°C and 80 to 90°C).

3.11.4 PROCEDURE

3.11.4.1 SAMPLE PREPARATION

Bring the milk sample temperature to 20°C and carefully mix it. If milk fat is not uniformly distributed, the sample is slowly heated to 40°C in the water bath, it is gently mixed by tilting and quickly cooled to 20°C. Weigh 20 g sample (milk or skim milk) to the nearest of 10 mg, in a digestion flask. The 3 ml HNO_3 and 2.5 ml H_2SO_4 are added into it.

3.11.4.2 DIGESTION

Add 3 glass beads into the digestion flask. The flask is placed in a tilted position and is heated with a micro-burner in the fume hood. Regulate the flame height to control the foam formation in the digestion flask. Foaming is allowable up to the neck, but it should not come out. The mixture is gently boiled (avoid local overheating). After the solution becomes brown, add 3–5 drops of HNO_3. Heat it vigorously. Continue heating and adding HNO_3 (5 to 20 drops at a time), remove the wall adhered material by occasionally swirling the flask, till the mixture becomes colorless. Cool to the ambient temperature. Add water (2 ml) and H_2O_2 (1 ml) solution to the flask. Again, swirl and heat till white fumes are released. Further add 0.5 ml H_2O_2, if the color of the solution turns yellow. Heating is continued for 45 min, when the white fumes start emitting. The contents are cooled to ambient temperature and the volume is made to about 25 ml by adding water. Transfer citrate-EDTA (5 ml) and phenolphthalein (0.1 ml) solution. Add NH_4OH solution with occasional swirling, till the mixtures remain red in color.

3.11.4.3 COLOR DEVELOPMENT

- ***Note:*** Unnecessary light exposure more than 150 lx illuminance (subdued daylight) is avoided at the time of color development and photometric measurement.

Sodium diethyldithiocarbamate (5 ml) is added to a flask with frequent swirling. The volume is made up to 50 ml with water. Pipette out amyl acetate (4 ml) into the solution and the flask is closed with a stopper. The closed flask is heated in a water bath (60–65°C for 10 min). Vigorously shake the flask for 1 minute with frequent inversion, ensuring that stopper remains in position (***Note:*** If xylene is used instead of amyl acetate, the time of shaking shall be 2 min).

The closed flask is quickly cooled under running water for a minimum of 10 min. The turbidity appeared in the amyl acetate layer is eliminated by leaning the flask carefully from vertical position to horizontal several times. Continue carefully, incline the flask if turbidity persist. The flask is kept in a dark place for 1 h.

3.11.4.4 BLANK TEST

Run the blank by taking 20 ml water instead of test milk sample under similar conditions using the same amount of reagents.

3.11.4.5 PHOTOMETRIC MEASUREMENTS

The amyl acetate layer is pipetted out into 10 mm spectrophotometer cell. The absorbance of the amyl acetate layer of the test solution and reagent blank is measured against H_2O at 436 nm. The reagent blank value is subtracted from the value of the test solution.

Note:

a. The absorbance value (AV) of reagent blank should correspond to < 0.4 µg Cu/50 ml. Check the reagents, if the concentration exceeds 0.4 µg Cu/50 ml; and

b. If the methyl acetate layer changes to colorless after mixing with KCN (5% m/m) solution, then it indicates that tellurium and bismuth are supposed to be absent. Otherwise, these can also be removed from amyl acetate layer by washing with 1 M NaOH solution.

3.11.4.6 CALIBRATION CURVE

The 0 (blank), 0.5, 1, 2, 3, 4, and 6 ml of $CuSO_4$ working solution is pipetted out in seven different digestion flasks; and each quantity is diluted to 25 ml with H_2O. Add H_2SO_4 (2.5 ml) to the flask and the contents are thoroughly mixed. Then, transfer Citrate-EDTA solution (5 ml) and phenolphthalein (0.1 ml) solution into the flask. Add NH_4OH solution with occasional swirl, till the mixture remains light red. Proceed with the same procedure as mentioned in the "Color Development" section, and absorbance of amyl acetate layer is recorded at 436 nm. A calibration curve is plotted between absorbance and amount of copper added. The calibration curve is checked monthly.

3.11.5 CALCULATIONS

The Cu concentration (mg/kg) is calculated as follows:

$$\text{Cu (mg/kg)} = \frac{m_1}{m_0} \quad (15)$$

where; m_1 is the weight of test sample (g); m_0 is the weight of Cu as calculated from the calibration curve (mg).

3.11.6 DATA COLLECTION SHEET

Title: Determination of copper content in milk by photometric method.

Location and Organization: ______________________________

Participants: ______________________________

Date (Month/Day/Year): ______________________________

Observations:

Parameter	Repeat 1	Repeat 2	Repeat 3	Average
Weight of the test sample (g) = m_0				
Weight in micrograms of copper, which is obtained from the standard curve corresponding with the absorbance of the test solution = m_1				
Copper content (mg/kg) = m_0/m_1				

Comments:____

3.12 CALCIUM CONTENT IN MILK

3.12.1 INTRODUCTION

Calcium (Ca) is a significant component of our health and plays a vital role in building and maintaining strong teeth and bones. It is also vital for other physiological functions, such as: muscle control and blood circulation. The adult human body has about 1,200 g of Ca, which accounts for about 1.5 to 2% of our body weight. There is a regional variation in the contribution of dairy products to overall Ca intake and it ranges from 52 to 75% [35, 36, 39]. The cow milk contains about 1,120–1,200 mg/liter of Ca [37]; and hence, milk serves as an important source of calcium. Methods available for calcium estimation in milk are based on atomic absorption spectroscopy (AAS), calcium specific electrodes, sequential injection analysis, energy dispersive X-ray fluorescence; as well as using complexometric titration using chelating agents (such as: EDTA, trichloroacetic acid (TCA), etc.). Method based on TCA involves the precipitation of Ca as oxalates, and the results obtained are within the acceptable limits, if proper procedural care is followed.

3.12.2 PRINCIPLE

The protein substances in a test portion are precipitated by TCA and are then filtered. The Ca in the filtrate is precipitated as Ca-oxalate and is separated by centrifugation. The washed and dissolved precipitate is titrated with potassium permanganate ($KMnO_4$) [38].

3.12.3 MATERIALS

3.12.3.1 REAGENTS

➢ **Note:** Use only analytical grade reagents.

- Acetic acid (CH_3COOH) solution (20%).
- Ammonia (NH_4OH) solution I: Equal volumes of NH_4OH solution (25% w/v) and H_2O are mixed.
- Ammonia (NH_4OH) solution II: 2 ml NH_4OH (25% w/v) is diluted to 100 ml with water.
- Ammonium oxalate [$(NH_4)_2C_2O_4$]: Cold saturated solution.
- Methyl red solution: 0.05% solution in ethyl alcohol (96% v/v).
- Standard $KMnO_4$ solution (0.004±0.0001 M): Check the titer value by normal laboratory procedure using oxalic acid or sodium oxalate.
- Sulfuric acid (H_2SO_4): 20 ml of H_2SO_4 (98% w/v) is added to 80 ml water.
- TCA solution I: 200 g of TCA is dissolved in 1,000 ml of H_2O.
- TCA solution II: 120 g of TCA is dissolved in 1,000 ml of H_2O.

3.12.3.2 APPARATUS

- Boiling water bath;
- Burette;
- Centrifuge (radial acceleration 1,400x g);
- Centrifuge tubes (capacity 30 ml);
- Filter paper, ashless, for slow filtration;
- Pipette (capacity 20 ml);
- Pipettes (capacity 2 and 5 ml);
- Suction device, with capillary tube;
- Volumetric flask (capacity 50 ml);
- Weighing balance.

3.12.4 PROCEDURE

3.12.4.1 PREPARATION OF MILK SAMPLE

Bring the milk sample or reconstituted milk to a temperature of 20°C and mix it properly. If fat is not homogeneously dispersed, then milk sample is heated gradually to 40°C in the water bath, then is gently mixed by repeated inversions; and then it is cooled to 20°C.

3.12.4.2 PRECIPITATION OF PROTEIN SUBSTANCES

Pipette out 20 g of test portion to the volumetric flask. Weigh the sample to the nearest of 0.01 g. Add gradually the TCA solution I to the test portion, while shaking, till volume is made to 50 ml. Shake it vigorously for few seconds and keep it undisturbed for half an hour. Using filter paper filter the solution and clear filtrate are obtained.

3.12.4.3 PRECIPITATION AND SEPARATION OF THE OXALATE

The 5 ml clear filtrate, 5 ml TCA solution II, 2 ml $(NH_4)_2C_2O_4$, 2 drops methyl red solution and 2 ml CH_3COOH solution are taken in a centrifuge tube; and mixed by swirling. Ammonia solution I is added drop by drop to the mixed solution in the centrifuge tube until the color changes to light yellow. Transfer few drops of CH_3COOH until color of the solution changes to pink. Keep it undisturbed for 4 h at ambient temperature. The content of centrifuge tube is diluted to 20 ml with H_2O and is centrifuged at 1,400 x g for 10 minutes. The clear supernatant is removed with a suction device. The wall of the centrifuged tube is rinsed with 5 ml ammonia solution II. The deposited calcium oxalate should not be disturbed. The tube is again centrifuged at 1,400 x g for five minutes. Again, discard the supernatant and repeat the washing twice.

3.12.4.4 TITRATION

The 2 ml H_2SO_4 and 5 ml water are added to the tube containing calcium oxalate deposit. The tube is placed in the water bath (boiling) to completely dissolve the deposit. The dissolved calcium oxalate is titrated with $KMnO_4$ solution, until a pink color appears. The solution temperature should be more than 60°C during titration. Record the volume of $KMnO_4$ (nearest to 0.01 ml) that is used during titration.

3.12.4.5 BLANK TEST

For the blank test, take 20 ml of water instead of milk sample under the same identical conditions; and record the volume of $KMnO_4$ used during the titration.

3.12.5 CALCULATIONS

$$\text{Ca content (W)} = 0.0004\,(V - V_0) \times \frac{(1000 \times f)}{m}$$

$$= 0.4\,(V - V_0) \times \frac{f}{m} \qquad (16)$$

where; W is the content of calcium (%); V is the amount of $KMnO_4$ solution used for the sample (ml); V_0 is the amount of $KMnO_4$ solution used for the blank test (ml); m is the mass of test sample (g); and f is the correction factor (CF) for the precipitate volume that is obtained from precipitation of TCA (Table 3.1).

TABLE 3.1 Correction Factor for Different Values of Fat Percentage in the Milk Sample

Fat Content of the Milk Sample (%)	Correction Factor (f)
3.5 to 4.5	0.972
3	0.976
2	0.980
1	0.985
< 0.1	0.989

3.12.6 DATA COLLECTION SHEET

Title: Determination of Ca content in milk sample by titrimetric method.

Location and Organization:______________________________

Participants: ____

Date (Month/Day/Year):________________________________

Observations:

Parameter	Repeat 1	Repeat 2	Repeat 3	Average
Amount of $KMnO_4$ solution used for the sample (ml) = V				
Amount of $KMnO_4$ solution used for the blank test (ml) = V_0				
Weight of sample (g) = m				
Correction factor = f				
Ca content (%)				

Comments:____.

3.13 ZINC CONTENT IN MILK BY AAS METHOD

3.13.1 INTRODUCTION

Zinc is an important mineral for the development and growth, wound healing and sexual maturation. It is also responsible for better functioning and different physiological functions of the immune system. Zn is a constituent of insulin and regulates various hormones taking part in reproduction, also involved in nucleic acid and protein synthesis and acts as a cofactor for most of the enzymes involved in metabolisms of our body [40]. Dairy products are a good source of Zn and contribute about 19 to 31% of overall Zn intake in Western countries [33, 36]. The cow milk contains about 3.9 mg/liter with a variable range of 2.0–6.0 mg/liter [37]. Selected methods for estimation of Zn in milk involve the use of X-ray fluorescence spectrometry, AAS, and inductive coupled plasma spectrometry. However, AAS gives the advantage of higher accuracy along with ease of operation compared to other methods.

3.13.2 PRINCIPLE

The test portion is ashed in a furnace and obtained ash is dissolved in conc. HCl followed by addition of $SrCl_2$ and dilution with water. The Zn content is estimated by flame AAS at 213.9 nm wavelength with Zeeman or deuterium background correction [42].

3.13.3 MATERIALS

3.13.3.1 REAGENTS

- Conc. HCl (sp. gr. 1.17–1.18 g/ml at 20°C).
- Conc. HNO_3 (sp. gr 1.42 g/ml at 20°C).
- $SrCl_2$ solution: 38 g of Strontium chloride hexahydrate ($SrCl_2 \cdot 6H_2O$) is dissolved in H_2O and is diluted to 250 ml with H_2O.
- Standard Zn solutions:
- Stock solution: 1,000 mg of Zn is dissolved in the 0.3 M HNO_3 solution (equivalent to 18.9 g/liter HNO_3).

- Working solution (100 mg/liter): One ml of HNO_3 is added to the stock solution (10 ml) and is diluted with H_2O to 100 ml.
- Zero-standard solution: Take 2.5 ml of HCl and 12.5 ml of $SrCl_2$ solution and this mixture is diluted with H_2O to 500 ml and is mixed thoroughly.

3.13.3.2 APPARATUS

- Boiling water bath;
- Drying oven (temperature set at 102±5°C);
- Hot plate (100°C);
- Pipettes (capacities 0.2, 1.0, and 5.0 ml);
- Programmable ashing furnace or equivalent (min. temperature 550°C, with heating rate of 50°C/h);
- Quartz crucibles, with lid (capacity 50 ml);
- Volumetric flask (capacity 100, 250, and 500 ml).

3.13.4 PROCEDURE

3.13.4.1 PREPARATION OF THE MILK SAMPLE

Bring the milk sample or reconstituted milk to a temperature of 20°C and mix it properly. If fat is not homogeneously dispersed, then the sample is heated gradually to 40°C in the water bath; and then is gently mixed by repeated inversions and it is cooled to 20°C.

3.13.4.2 ASHING

In a quartz crucible, weigh 5±1 g (nearest to 1 mg) of prepared milk sample (or an amount of the prepared milk product sample equal to 0.5±0.1 g of dry weight). The content is dried carefully in the oven at 102°C. The dried samples are transferred into the furnace and the contents in the crucible are heated to 500°C for 3 h at the rate of 50°C/h. The contents are kept in a Zn-free cabinet for cooling at room temperature. If the color of the ash is not white, then wet it with about 0.5 ml of water and 3 drops of HNO_3. Carefully dry it on a water bath (set at 100°C) or hot plate. Heat it again in the furnace at 500°C for 30 min.

3.13.4.3 PREPARATION OF THE TEST SOLUTION

The 0.5 ml of H_2O and 0.5 ml of HCl are added into the ash. The ash is dissolved with H_2O and is transferred into a volumetric flask (100 ml). The 2.5 ml of $SrCl_2$ is added by using a pipette (5 ml), and is diluted up to the mark with H_2O and mix properly.

3.13.4.4 AAS MEASUREMENT

Adjust the flame conditions and wavelength (213.9 nm) of AAS to get optimum sensitivity and precision.

1. **Calibration:** The 0.2, 0.4, 0.6, and 0.8 ml of Zn working solutions are transferred to four different volumetric flasks (100 ml). And each is diluted with zero standard solution up to the mark and is mixed properly. These solutions have 0.2, 0.4, 0.6, and 0.8 mg of Zn/liter, respectively. Further, aspirate the zero standard and 4 different calibration solutions four times each; and average AV is calculated. Deduct the average AV of zero standard solution from the average AV of each calibration solution. Draw the calibration curve between resulting net AVs and corresponding calibration concentrations.

- ➢ **Note:** By auto-zeroing, we can subtract the values, which depend on the instrument facility.

2. **Measurement of the Test Sample:** After the calibration, the test solution is instantly measured under identical conditions. If the signal of the test sample solution is higher than the highest signal of standard Zn solution, then dilute the sample solution with zero standard solution and measure it again. The instrument and calibration stability should be verified frequently with 0.4 mg of Zn/liter solution. Measurement is repeated four times, and the mean value is calculated. Subtract the zero-standard value from the mean value; and estimate the concentration of Zn from the standard value from the curve.

3.13.5 CALCULATIONS

$$\text{Zn content}, W_s = \frac{c_s \times f \times v}{m} \tag{17}$$

where; W_s is the Zn concentration in test sample (mg/kg); C_s is the Zn concentration of the test solution as calculated from the calibration curve (mg/liter); f is the dilution factor (DF); V is the volume of the test solution (ml); and m is the mass of the test portion (g).

3.13.6 DATA COLLECTION SHEET

Title: Determination of zinc content in milk by flame AAS method.

Location and Organization:______________________________

Participants: ______________________________

Date (Month/Day/Year):______________________________

Observations:

Parameter	Repeat 1	Repeat 2	Repeat 3	Average
Weight of sample (g) = m				
Zn concentration of test solution that is calculated from the calibration curve (mg/liter) = Cs				
Dilution factor = f				
Volume of test solution (ml) = V				
Zn concentration (mg/kg)				

Comments:______________________________.

3.14 MAGNESIUM CONTENT IN MILK BY AAS METHOD

3.14.1 INTRODUCTION

The role of magnesium (Mg) has been validated in a wide range of physiological activities, such as: nucleic acid and protein metabolism, muscle contraction and neuromuscular transmission, and acting as a cofactor for various enzymes. The average Mg content in bovine milk is about 110 mg/liter [37] and is not affected by the removal of fat [43], and dietary Mg intake in cow [46]. In milk, about 98–100% of the Mg is associated with skim milk and trace amount is observed in the cream portion [44]. Mg in milk and most milk products is estimated by using AAS (Atomic Absorption Spectrometric) method [48]; however, other titrimetric methods are also available.

3.14.2 PRINCIPLE

Wet digestion or dry ashing involves the use of HNO_3 for decomposition of organic matter in the test sample (either in a pressurized microwave-assisted (MWA)-wet digestion or an open MWA-wet digestion or in polytetrafluorethylene (PTFE) decomposition vessel or other suitable instrument for wet digestion). Mg containing ash is allowed to be soluble in HNO_3 in dry ashing method, whereas dilution of the digest is allowed in wet digestion. The calibration and test solutions are atomized into air-acetylene flame of AAS, and absorbance is measured at suitable wavelengths using AAS method [45].

3.14.3 MATERIALS

3.14.3.1 REAGENTS

- Concentrated HNO_3 (65% w/v).
- H_2O_2 (30% v/v).
- HNO_3 (25% v/v).
- Lanthanum trichloride ($LaCl_3$) solution (27 g/liter): 27 g of $LaCl_3$ heptahydrate is dissolved in H_2O and it is diluted to 1,000 ml.
- Mg ion stock solution: Mg ion of concentration one g/liter (Commercially available or equivalent).
- Petroleum ether (Boiling point 40–60°C): if required, it is distilled in a contaminant free distillation unit.
- Standard working solution (10 mg of Mg ion/liter): 1 ml of stock solution of Mg ion is pipetted into the volumetric flask (100 ml) and 5 ml HNO_3 solution is added into the flask. Volume is made up to the mark with H_2O and the mixture is mixed thoroughly. The working solution is stored in a HDPE bottle to avoid contamination.

3.14.3.2 APPARATUS

Before use, clean all the glass or plastic wares with HNO_3 (10%); Immerse for a minimum of 6 h in this solution followed by rinsing with H_2O (double distilled) thrice and then it is dried:

- Centrifuge (radial acceleration 2,500 x g).
- Centrifuge tubes (capacity 150 ml).
- Flame AAS: Contains air-acetylene burner and hollow cathode lamps.
- Furnace (min. temperature of 550±25°C; heating rate of 50°C/h).
- Grinding unit.
- High density polyethylene (HDPE) bottles: For storing samples and standard solutions.
- Measuring cylinder (Capacity 10 ml).
- Micropipette (capacity 1 and 5 ml) with plastic tips.
- Open-focused MWA wet digestion system: Provided with vessels of suitable capacity (50 ml) and MW power of 200 W.
- Oven (temperature set at 150°C).
- Pipettes (capacity of 1, 2, 5, and 10 ml).
- Pressurized wet decomposition PTFE vessel (23 ml) with screw cap: The vessel is made of stainless steel with inner PTFE materials.
- Sieve (aperture size of 0.5 mm).
- Silica crucibles (25 and 50 ml).
- Volumetric flask (capacity of 20, 50, 100, 250, and 1,000 ml).
- Water baths (each at temperature of 20±2°C, 40±1°C, 45±1°C and 65±1°C).
- Weighing balance.

3.14.4 PROCEDURE

3.14.4.1 PREPARATION OF SAMPLE

Bring the milk sample or reconstituted milk to a temperature of 20°C and mix it properly. If fat is not homogeneously dispersed, then the sample is heated gradually to 40°C in the water bath, and is gently mixed by repeated inversions and is cooled to 20°C.

3.14.4.2 DECOMPOSITION OF ORGANIC MATTER

1. **Dry Ashing:** The organic matter is decomposed either by using wet digestion or dry ashing. The 10 g (nearest to 1 mg) of prepared sample is weighed in a silica crucible. Transfer it into the furnace set at ambient temperature and then rise the temperature @ 50°C/h up to 550°C. Heat the crucible for six hours at 550°C. If the ash is still

gray in color after cooling, then dissolve it in 1 ml of HNO_3 solution and continue the dry ashing method.

2. **Wet Digestion:** Weigh 0.5–1.0 g (nearest to 1 mg) of prepared sample in PTFE or in a microwave vessel. For wet digestion process, use any of the following methods:

   i. **Open MWA Wet Digestion System:** Decomposition guide for MWA wet digestion system is given in Table 3.2.

TABLE 3.2 Decomposition Guide for MWA Digestion

Step	Chemical Added	Amount (ml)	Power (W)	Duration (min)
1	H_2O	2	–	–
2	Nitric acid	7	30	5
3	–	–	80	15
4	H_2O_2	1	60	5

   ii. **Pressurized MWA Wet Digestion System:** Transfer 3 ml of HNO_3 solution in the wet MW vessel and close it. The vessel is placed in the MW oven. Use the decomposition program given in Table 3.3; the decomposition and pressurized time can be changed according to the size and type of test sample.

TABLE 3.3 Decomposition Program of Pressurized MWA Digestion

Step	Outset Power (W)	Duration (min)	Final Power (W)	Cooling System
1	500	10	800	Low
2	800	20	1,000	Low
3	0	20	0	High

   iii. **Decomposition Vessel:** Transfer 3 ml of HNO_3 solution into the vessel and close it. Put the vessel in an MW oven set at ambient temperature. The temperature of the oven is increased to 150°C and it is kept at this temperature for a minimum of 3 h.

3.14.4.3 PREPARATION OF TEST SOLUTION

1. **For Dry Ashing:** Dissolve the ash in HNO_3 solution (1 ml), and this mixture is quantitatively transferred to volumetric flask (250 ml), and volume is made up to the mark with H_2O. Mix properly and the dilution is continued.

2. **For Wet Digestion:** Initially, the digested mixture is cooled to ambient temperature. Transfer the mixture quantitatively to volumetric flask (50 ml) while reducing to atmospheric pressure before. Volume is made up to the mark with H_2O. Mix properly and the dilution is continued.

3.14.4.4 DILUTION

Dilute the test solution depending on the type of test sample and the ions are estimated using micropipette in the required volumetric flask. Transfer 10% of $LaCl_3$ by using a measuring cylinder. Dilute with H_2O up to the mark of the flask.

Simultaneously, run the blank test following the same method; and the same quantity of reagents are used in the decomposition and estimation steps of the sample.

3.14.4.5 AAS MEASUREMENT

Adjust the AAS and their flame conditions based on the guidelines of the manufacturer so as to obtain optimal accuracy and sensitivity. Set the AAS at the desired wavelength for determination of Mg ions.

1. **Calibration:** Choose the volumes and corresponding concentrations that are within the linear range of the specific equipment in use. Transfer 0-, 1-, 2-, 3-, 4-, and 5-ml standard working solution into six different volumetric flasks (each 100 ml); and transfer $LaCl_3$ solution (10 ml) into each flask. Dilute with H_2O and the mixture in each flask is mixed thoroughly (Table 3.4).

TABLE 3.4 Calibration Solution of Mg

Flask No.	Mg Ion Solution
1	0
2	0.1
3	0.2
4	0.3
5	0.4
6	0.5

2. **Calibration Graph:** For measurement of Mg ion, aspirate the zero standard solution and 5 calibrating solutions thrice each. Average of the AV is calculated. Deduct the average AV of zero standard solution from the averages of AVs of three calibration solutions. Draw the calibration (standard) curve resulting from averages AVs against the corresponding concentration of Mg ion.

3. **Measurement:** Determine the AVs of the test solution and blank test directly after the calibration measurements under similar conditions. Test solution is diluted with zero standard solution, if its signal is greater than that of the higher concentration of standard. Transfer 10% $LaCl_3$ to each solution. Any deviation during measurement can be checked by repeating the measurements; at the end, do measurements for at least one standard solution.

   Repeat the measurement thrice for each test sample. Average of the AVs is calculated. Deduct the average AV of blank solution from the averages of AV of test solution. Corrected average AV is considered to get the corresponding concentration of Mg from the standard curve.

3.14.5 CALCULATIONS

Mg content (mg/g) of milk sample is calculated as follows:

$$W = \frac{(c \times v)}{m \times 1000} \times f_1 \times f_2 \qquad (18)$$

where; W is the concentration of Mg (mg/g); c is the Mg ion concentration (mg/liter) in the test solution as calculated from the calibration graph; v is the volume of the flask in which the digested solution or dry ash is quantitatively transferred (ml); m is the mass of the test portion (g); f_1 is the DF (dilution factor) of the test solution applied in preparation step; and f_2 is the DF of the test solution applied in the measurement step.

3.14.6 DATA COLLECTION SHEET

Title: Determination of Mg content in milk sample by AAS method.

Location and Organization:____________________________

Participants: __________________________________

Date (Month/Day/Year):______________________________

Observations:

Parameter	Repeat 1	Repeat 2	Repeat 3	Average
Mg ion concentration (mg/liter) in test solution as calculated from the graph = c				
Volume of the flask (ml) in which the digested solution or dry ash is quantitatively transferred = v				
Mass of the sample (g) = m				
DF of the test solution applied in preparation step = f_1				
DF of the test solution applied during measurement step = f_2				
Mg content (mg/g)				

Comments: ________________________________.

3.15 PHOSPHORUS CONTENT IN MILK BY SPECTROSCOPIC METHOD

3.15.1 INTRODUCTION

Phosphorus serves various essential biological functions in our body. Several biological molecules (e.g., proteins, fats/oils, nucleic acids and carbohydrates) contain phosphorus as vital component and these play essential role in metabolic processes. The cow milk contains about 890 mg/liter of phosphorus, which is considerably greater than that observed in human milk (140 mg/liter) [37]. In cow milk, 44% of phosphate (PO_4 inorganic) is linked with casein micelles in the form of $Ca_3(PO_4)_2$ while 56% is soluble as free PO_4 ions [48]. The colorimetric and titrimetric methods are available for the measurement of phosphorus in milk, and spectroscopic methods are most reliable and are widely used.

3.15.2 PRINCIPLE

It is based on wet digestion of milk using H_2SO_4 and H_2O_2, or dry ashing of milk. Addition of molybdate-ascorbic acid (AA) solution to the obtained ash forms a molybdenum blue and it is followed by spectrophotometric measurement of absorbance at 820 nm wavelength [49].

3.15.3 MATERIALS

3.15.3.1 REAGENTS

Analytical grade reagents and deionized or distilled water (free of phosphorus compounds) shall be used:

- Ascorbic acid ($C_6H_8O_6$, 0.25 M): Weigh 5 g of ascorbic acid in 100 ml volumetric flask. Transfer enough volume of H_2O to dissolve the ascorbic acid. Volume is made up to the mark with H_2O and it is mixed thoroughly. Prepare this solution fresh before use.
- Concentrated H_2SO_4 (density 1.84 g/ml at 20°C): 18 M.
- Diluted H_2SO_4 (about 5 M): Carefully add 278 ml of conc. H_2SO_4 to 722 ml H_2O.
- H_2O_2 (9 M): Free of phosphorus containing substances.
- HCl (1 M, for dry ashing): Dilute 83 ml of conc. HCl to 1,000 ml with H_2O.
- Molybdate-ascorbic acid solution: Transfer 25 ml of Na_2MoO_4 solution to 10 ml ascorbic acid solution in a volumetric flask (100 ml). Volume is made with H_2O, and it is mixed thoroughly. Prepare this reagent fresh before use.
- Sodium molybdate (Na_2MoO_4, 0.1 M): Weigh 2.5 g of sodium molybdate dihydrate and transfer it to a volumetric flask (100 ml). Transfer sufficient amount of dilute H_2SO_4 solution to dissolve the sodium molybdate. Volume is made with the same acid and it is mixed thoroughly.
- Standard solution A: Dry about 1 g of potassium dihydrogen orthophosphate (KH_2PO_4) for at least 48 h in a desiccator. Take 0.4394 g of dried phosphate in 1,000 ml volumetric flask and dilute it with H_2O and it is mixed thoroughly. The phosphorus level of this standard solution is 100 mg/liter.
- Standard solution B: 10 ml of standard solution A is pipetted into the flask (100 ml) and the volume is made with H_2O and mix well.

3.15.3.2 APPARATUS

All the glassware should be properly cleaned by using phosphorus-free detergent and then rinsed with water:

- Desiccator with efficient drying agent.
- Digestion flasks (Kjeldahl) or test tubes (50 ml capacity).
- Dish, made of platinum or silica, of approximately 55 mm diameter, and a suitable watch-glass.

- Electric furnace: operating temperature between 500°C and 550°C.
- Electric heater or micro-gas burner.
- Filter paper (medium grade).
- Glass beads (approximately 5 mm in diameter).
- Graduated cylinders (5 and 25 ml).
- Molecular absorption spectrometer, suitable for measurement at 820 nm and containing cells of 10 mm path length.
- Drying oven: capable of operating at 100°C.
- Pipettes (one mark: 1, 2, 3, 5, and 10 ml).
- Volumetric flasks (capacity 50, 100, and 1,000 ml).
- Water-bath, capable of operating at 100°C.
- Weighing balance.

3.15.4 PROCEDURE

3.15.4.1 PREPARATION OF MILK SAMPLE

Bring the milk sample or reconstituted milk to a temperature of 20°C and mix these properly. If the fat is not homogeneously dispersed, then the sample is heated gradually to 40°C in the water bath, then it is gently mixed by repeated inversions and it is cooled to 20°C.

3.15.4.2 DECOMPOSITION

1. **Wet-Digestion Method:** Weigh 1.5 g (nearest to 1 mg) of test sample in a digestion flask and add 3 glass beads and 4 ml of concentrated H_2SO_4. Operating under a well-ventilated fume hood provided with a water scrubbing system, incline the flask and heat it on a micro-gas burner or electric heater. Control the heating to limit the foam production and place the mixture on gentle boiling. Local overheating is avoided and the flask is heated above the surface of the liquid mixture. When the foaming stops, the mixture is cooled in air at ambient temperature. Add 2 ml of H_2O_2 solution carefully and it is heated again. The procedure is repeated until the content becomes clear and colorless. Occasionally, mix the contents while heating by swirling; and local overheating is avoided. Cool the mixture in air to room temperature and rinse the flask neck with 2 ml H_2O. Again, heat the contents till the water gets evaporated. Boil the liquid for 30 min. to remove the traces of H_2O_2, and localized overheating should be avoided. The mixture is cooled in air at ambient temperature. The

liquid content is transferred into a 100 ml volumetric flask. Volume is made with H_2O and it is thoroughly mixed. Transfer 2 ml of this solution into a volumetric flask (50 ml) followed by addition of about 25 ml of H_2O and 2 ml of molybdate-ascorbic acid solution into it. Make the volume up to the mark and it is mixed properly. Boil the flask contents for 15 minutes in a water bath. Cool the mixture to ambient temperature in cold water. Proceed for spectroscopic measurement.

> ***Note:*** The spectrometric determination should be carried out within 1 h.

2. **Dry-Ashing Method:** Take 10 g (nearest to 1 mg) of test sample in a platinum or silica dish. The sample is evaporated to dryness by placing it in an oven (temperature 100°C or lower) or in a water bath. After drying is complete, heat the test portion in the furnace at 500–550°C temperature till white (or nearly white) ash is obtained.

> ***Note:*** The dish should preferably be heated on a hot plate to burn-off ignitable contents before being placed in the furnace.

Allow the contents of the dish to cool, and then cover with a watch-glass. The ash is dissolved in 2 or 3 ml of HCl solution and is diluted with 3 ml of H_2O. The ash solution is transferred to volumetric flask (100 ml); rinse the dish and watch-glass with H_2O and the washings are transferred to the flask. The volume is made with H_2O, and it is mixed thoroughly. Filter it by using a medium grade filter paper. Pipette out 10 ml filtrate into volumetric flask (100 ml). The volume is made with H_2O and it is mixed properly. Take 2 ml of this solution into a volumetric flask (50 ml); add 25 ml of H_2O and 2 ml of molybdate-ascorbic acid solution into it. The volume is made with H_2O and thoroughly mixed. Boil the flask contents in the water bath for 15 minutes. Cool the mixture to room temperature in cold water. Proceed for spectroscopic measurement.

> ***Note:*** The spectrometric determination should be carried out within 1 h.

3.15.4.3 BLANK TEST

Simultaneously, prepare the blank test by taking 1.5 ml or 10 ml of phosphorus-free H_2O instead of test sample; and follow the similar procedure as for the test sample.

3.15.4.4 CALIBRATION GRAPH

Transfer 0, 1, 2, 3, and 5 ml of standard B solution into five separate volumetric flasks (50 ml capacity). Dilute contents in each flask to about 25 ml H_2O and then transfer 2 ml molybdate-ascorbic acid solution into each flask. Volumes of each flask is made up to the mark with H_2O and it is properly mixed. The resulting solutions will contain 0, 10, 20, 30, and 50 µg per 50 ml phosphorus, respectively. Boil the flask contents for 15 min. in the water bath. Solutions are cooled to ambient temperature in cold water. Within 1 h, the absorbance of each calibration solution is measured against the solution containing zero µg of phosphorus using a spectrometer at 820 nm wavelength with 10 mm cell.

- ***Note:*** If the AV of the solution containing 0 µg of phosphorus per 50 ml is high, then check the reagents. Plot the mean AVs thus obtained against the mass (micrograms (µg)) of phosphorus contained in the calibration solutions.

3.15.4.5 SPECTROMETRIC MEASUREMENT

Within 1 h, carry out the spectrometric measurements on the test solution against the blank using the spectrometer at 820 nm wavelength.

3.15.5 CALCULATIONS

Using the calibration graph, determine the mass of phosphorus corresponding to the average AV of the test solution. Total phosphorus content (W_P, expressed as % m/m) of test sample is calculated as follows:

$$\text{Wet-digestion method:} W_p = \frac{m_1}{200 \times m_0} \tag{19}$$

$$\text{Dry-ashing method}: W_p = \frac{m_1}{20 \times m_0} \tag{20}$$

where; m_0 is the weight of the test sample (g); and m_1 is the weight of phosphorus (µg), calculated from the calibration graph.

3.15.6 DATA COLLECTION SHEET

Title: Determination of phosphorus content in milk by spectroscopic method.

Location and Organization:____________________________________

Participants: ____

Date (Month/Day/Year):____

Observations:

Parameter	**Repeat 1**	**Repeat 2**	**Repeat 3**	**Average**
Weight of the test sample (g) = m_0				
Weight of phosphorus as calculated from the calibration graph (μg) = m_1				
Phosphorus content (% m/m)				

Comments:____.

3.16 CITRIC ACID (CA) CONTENT IN MILK BY GRAVIMETRIC METHOD

3.16.1 INTRODUCTION

CA (or citrate) is a vital organic salt constituent that is present naturally in milk. Cow milk contains about 130–160 mg citrate per 100 ml [51]. Citrate is also considered as part of the salt components of milk. It has a significant impact on milk processing and on some of the characteristics of certain dairy products specifically cheese and fermented products. The *de novo* synthesis of FAs by the mammary gland is related to the citrate concentration, which is affected by the diet [48].

3.16.2 PRINCIPLE

CA is oxidized by means of potassium permanganate ($KMnO_4$) to acetone dicarboxylic acid, which is then made to react with bromine to give penta-bromoacetone. Penta-bromoacetone is then weighed and correlated with CA content [52].

3.16.3 MATERIALS

3.16.3.1 REAGENTS

- Concentrated H_2SO_4 (sp. gr. 1.84);
- Dilute H_2SO_4 (normal);
- Ethanol (C_2H_5OH, 95% v/v);
- Ethyl ether (AR grade);
- Ferrous sulfate solution: Dissolve 40 g of $FeSO_4.7H_2O$ in 100 ml water containing 1 ml of concentrated H_2SO_4;
- $KMnO_4$ solution (5% w/v) ;
- Na_2SO_4 anhydrous;
- Phospho-tungstic acid (PTA, 20% w/v);
- Potassium bromide (KBr);
- Tartaric acid.

3.16.3.2 APPARATUS

- Beakers (150 ml);
- Desiccator;
- Erlenmeyer flask (250 ml);
- Gooch crucible (size G4);
- Ice bath for cooling;
- Steam bath;
- Volumetric flask (250 ml).

3.16.4 PROCEDURE

Take 50 g of prepared sample in a beaker (150 ml). Transfer 100 mg of tartaric acid and 6 ml of diluted sulfuric acid. Heat the beaker for 15 min. on a steam bath. Instantly, transfer 3 ml PTA solution, properly mix it and again heat for 5 min. on the steam bath. The contents are transferred to a 250 ml flask containing C_2H_5OH and are allowed to cool. Dilute with C_2H_5OH up to the mark, mix these properly and filter by using filter paper. The obtained filtrate (200 ml) is evaporated to around 20 ml and rinsed into 250 ml of tared Erlenmeyer flask (with glass stopper) and the total weight is adjusted to around 40 g with water. Transfer 2 g of KBr and 5 ml of conc. H_2SO_4 and heat to around 50°C if required and keep it undisturbed for 5 minutes.

Transfer 20 ml of $KMnO_4$ solution slowly through burette or pipette; after each addition, the flask is mixed for a few seconds. Let it stand for 5 minutes and cool to 15°C. $FeSO_4$ solution is added slowly with continuous mixing until the mixture starts to clear. Agitate for one minute and continuously add $FeSO_4$ solution till MnO_2 is completely dissolved (add a few ml in excess). While mixing, add 20 g of anhydrous Na_2SO_4 to assure solution (If Na_2SO_4 is not completely dissolved, estimation is repeated). The solution is cooled to 15°C and vigorously mixed for 5 minutes. While it is cold, the penta-bromoacetone is immediately collected on asbestos in a crucible (Gooch) and rinse the residual precipitate in the flask with filtrate. Lastly, the crucible is washed with cold H_2O (50 ml); the crucible is remained under suction for a few more minutes. Crucible is dried in a desiccator overnight containing H_2SO_4 and weighed, or dry the crucible in a drying tray, aerate it until the weight loss does not differ by 0.1 mg (maximum), take the first reading of the weight after 20 minutes. The penta-bromoacetone is eliminated from the crucible by ethanol followed by diethyl ether, and the crucible is filled three times with both solvents. Dry the crucible at 98–100°C for 10 minutes and take the weight after cooling. The difference between two weighing provides the penta-bromoacetone mass [52].

3.16.5 CALCULATIONS

$$\text{Mass of CA in aliquot (X)} = 0.424 \times \text{P}$$

$$\text{Anhydrous CA } (\%\, w/w) = \frac{125 \times X}{W} \quad (21)$$

where; P is the penta-bromoacetone weight (g); X is the weight of CA in aliquot (g); and W is the weight of the test sample (g).

3.16.6 DATA COLLECTION SHEET

Title: Determination of CA content in milk.

Location and Organization: ______________________________

Participants: ______________________________

Date (Month/Day/Year): ______________________________

Observations:

Parameter	Repeat 1	Repeat 2	Repeat 3	Average
Weight of sample (g) = W				
CA weight in aliquot (g) = X				
Penta bromoacetone weight (g) = P				
CA content (% w/w)				

Comments:____.

3.17 VITAMIN A CONTENT IN MILK BY HPLC METHOD

3.17.1 INTRODUCTION

Vitamin A (or retinol) belongs to the range of compounds named as retinoids. Generally, animal origin foods offer preformed vitamin A as retinyl esters, which get hydrolyzed in the gastrointestinal tract. Milk of some species also contains β-carotene, which acts as a precursor for the synthesis of vitamin A. Only carotenoids having β-ionone ring (e.g., β-carotene) are the precursors of retinol. Bovine milk contains about 20 μg of carotene and 40 μg of retinol per 100 g [47]. HPLC (high performance liquid chromatography) techniques are widely applied for estimation of vitamin A. However, sample preparation and extraction may differ depending on the column and other system requirements.

3.17.2 PRINCIPLE

Vitamin A (retinyl ester) is extracted from 2 ml of milk sample into hexane containing retinyl acetate (RA) as internal standard (IS). This extract is evaluated by liquid chromatography (LC) with UV detector (at 325 nm) using normal-phase silica column under controlled temperature. The content of retinyl palmitate (RP) is estimated by using relative response factor (RRF) that is measured with calibration standards [50].

3.17.3 MATERIALS

3.17.3.1 REAGENTS

➢ **Note:** All prepared standards must attain ambient temperature before use.

- 2,2-Dimethoxypropane.

- Absolute ethanol (100%).
- BHT (medium size) crystals.
- Deionized water (15 Mohm).
- Glacial CH_3COOH.
- Hexane, wet: Hexane (500 ml) and deionized H_2O (25 ml) are mixed and shaken in a separatory funnel for two minutes and kept undisturbed for a minimum of 24 h. The aqueous layer is drained from the bottom and wet hexane is transferred cautiously from the top of funnel into the flask (the water droplet should not come with the wet hexane, as its entry would enhance the mobile phase polarity and cause retention loss). This wet hexane can be stored in a stoppered flask for two weeks.
- Hexane.
- Isopropanol (LC grade).
- Methylene chloride (CH_2Cl_2, LC grade).
- Mixed calibration standards [RP (150–2,500 μg/liter) and RA (1,250 μg/liter)]: Mixed standards with five different RP concentrations are prepared for the calibration of the instrument. Prepare it during each analysis. Transfer 5 ml of RA I standard (2,500 μg/liter) solution into five separate volumetric flasks (10 ml). Dilute one of the flasks with standard RP I solution (5,000 μg/liter) to provide calibration standard C_1. Likewise, dilute with other standard RP solutions II to V in order to make calibration standard solutions C_2 to C_5.
- Mobile phase: Mixture of conditioned hexane and isopropanol [99.85: 0.15 (v/v)]. Conditioned hexane is prepared by mixing equal volumes of wet hexane and regular hexane.
- Regenerant solution: Mix CH_3COOH (10 ml) and 2,2-dimethoxypropane (8 ml) in one liter of hexane (LC grade).
- Standard RP (all trans).
- Standard RP stock solution (2,50,000 μg/liter): Weigh 0.02–0.03 g of RP standard into N_2-purged volumetric flask (100 ml) and volume is made with isopropanol and thoroughly mixed by inverting the solution several times. Transfer 5–6 crystals of butylated hydroxytoluene (BHT) into the flask to decrease the vitamin degradation. Addition of BHT will not interfere with vitamin A measurement. Store this solution in the dark place at 4°C and discard after two weeks. The concentration of stock solution is determined by adding 1 ml of stock solution into a volumetric flask (25 ml) and it is diluted with isopropanol. Prepare it in duplicate and note the

absorbance against isopropanol as a reference. Reading is recorded in triplicate for each dilution. The RP concentration is calculated as follows:

$$RP\ (\mu g\ /\ ml) = \frac{R_{abs} \times 25ml \times 10^{6} \times 1g}{E_{s} \times 1 \times 1g \times 100ml} \tag{22}$$

where; R_{abs} is the average absorbance of 6 readings; E_s is the absorptivity, 960 (L × m^{-1} × cm^{-1}) for RP at 325 nm; I is the cuvette path length (cm); 25 ml is the dilution for stock solution; 10^6 µg/liter is the conversion factor (CF) (g to µg); and 1 g/100 ml is the conversion of percent to concentration in g/ml.

- Standard RP working solutions (300–5,000 µg/liter):
- RP solution I (5,000 µg/liter): 2 ml of RP stock solution is added into a volumetric flask (100 ml), and it is diluted with hexane.
- RP solution II (2,500 µg/liter): 25 ml of RP solution I is added into a volumetric flask (50 ml), and it is diluted with hexane.
- RP solution III (1,250 µg/liter), IV (625 µg/liter), and V (313 µg/liter) are prepared by diluting RP II solution. These solutions are stable for one week when stored at 0–4°C in the dark place.
- Standard RA.
- Standard RA stock solution (ca 2,50,000 µg/liter): Weigh 0.02–0.03 g of RA into N_2-purged low-actinic volumetric flask (100 ml) and volume is made with isopropanol up to the mark. Calibration instructions should be followed, but use RA absorptivity (E_S) 1,530 (L x m^{-1} x cm^{-1}) to estimate ug/ml. This solution is stable for two weeks when it is stored at 0–4°C in the dark place.
- Standard RA working solution in hexane (2,500 µg/liter): 1 ml of RA stock solution is added into actinic volumetric flask (100 ml) and is diluted with hexane up to the mark. It is marked as an RA I solution and is used for the preparation of mixed calibration standards. It is stable for 7 days, when it is stored in the dark place at 0–4°C.
- Standard RA in ethanol (1,250 µg/liter; as IS): One ml of stock solution of RA is added into actinic volumetric flask (100 ml) and it is diluted with ethanol. From this solution, transfer 50 ml into volumetric flask (100 ml) and it is diluted with ethanol up to the mark. Mark it as RA II; and use it for IS addition to milk sample extracts. Prepare it during each analysis.
- Ultra-pure N_2 gas.

3.17.3.2 APPARATUS

- Amber vials (2 ml capacity) with caps.
- Centrifuge (capable of operating at 633 x g).
- Centrifuge tubes (50 ml) with screw caps.
- LC system that is equipped with UV detector (325 nm wavelength); 100 μl injection loop; unbounded silica column (250 x 4.6 mm, 5 μm, 300 Å); and column heater (maintained at 22°C). Column temperature > 22±2°C can cause changes in retention of peak and thus can lead to lower recovery.
- Measuring cylinders (25, 250, and 500 ml capacity).
- Pasteur pipettes (146 and 229 mm).
- Quartz cuvette (1 cm).
- Separating funnel (500 ml).
- UV-Visible spectrophotometer.
- Volumetric flask (low actinic with capacity 10, 25, 50, and 100 ml).
- Volumetric pipettes (capacity of 1, 2, 5, 25, and 50 ml).
- Vortex mixer.
- Weighing balance.

➢ ***Note:*** Use N_2 gas for purging centrifuge tubes and volumetric flasks before use.

3.17.4 PROCEDURE

3.17.4.1 PREPARATION OF LC SYSTEM

Before analysis, the column and HPLC system are equilibrated by passing the mobile phase through the instrument for 1 h. Inject one of the calibrating solutions before sample injection to ensure proper peak separation in the chromatogram. Use new or regenerated column, when the estimation is carried out for the first time.

3.17.4.2 REGENERATION PROCEDURE

Procedure for regeneration is required when peaks begin to broaden or when 300 to 400 injections are made into the column. Firstly, the direction of the column is reversed and then detaches the line connecting to the detector. The following reagents are used to submerge intake of solvent delivery system (SDS) and pump them over LC at 1 ml/min flow rate (FR) for a specified time: 33 min. CH_2Cl_2; 15 min. regenerant solution; 33 min. CH_2Cl_2; 15 min.

mobile phase. After conditioning, fix the column in the right direction. Do not reattach the end of the line connecting to the detector. Use mobile phase to flush the SDS and column for at least 2 h @ 1 ml/min. Now detector is attached to the end of the column; and mixed calibration standard solution (contains 1,163 μg/liter RP) is injected. There is a clear resolution of RP and RA peaks and there are no extra peaks beyond 10% of peak height of the analyte of interest in the chromatogram. If extra peaks are observed, then use mobile phase to flush the column for extra 1–2 hours, followed by injection of the same mixed standard solution.

3.17.4.3 EXTRACTION

All centrifuge tubes and volumetric flasks are purged with N_2 before use. Take 2 ml of milk sample into the centrifuge tube (50 ml) and transfer 5 ml of IS (i.e., RA in ethyl alcohol) into it. For blank, take 5 ml ethyl alcohol (100%) in the place of IS solution and follow the same procedure. Vortex it for 30 s and keep it undisturbed for 5 min. The 5 ml hexane is transferred into it, vortex it for half a minute and allow to stand for two minutes (minimum). Mix and let it stand for two more times. Add 3 ml deionized H_2O into the tube. Vortex it for 5 s (maximum) and centrifuge it at 633 × g/10 min. Transfer top hexane phase into an amber colored vial for HPLC analysis. If the hexane phase is not properly separated (emulsion formation), centrifuge it again.

3.17.4.4 CHROMATOGRAPHY

Set the FR of mobile phase @ 1 ml/min and detector at 325 nm. The column temperature is set at 22±2°C. The 100 μl of each calibration solution is injected and standard/calibration curve is drawn and it must be linear. The peaks of RA and RP should be separated by more than 0.5 min. The 100 μl of all test extract solutions are injected twice. Average of two injections is calculated.

3.17.5 CALCULATIONS

Peak areas (PAs) are used for calculation as follows:

(a)

$$\text{Relative response factor (RRF)} = \frac{PA_{RP} \times conc.RA(\mu g \,/\, liter)}{conc.RP(\mu g \,/\, liter) \times PA_{RA}} \qquad (23)$$

where; PA_{RP} is the peak area of standard RP, and PA_{RA} is the peak area of IS (or RA).

RRF is calculated for all the calibrating standard solutions and an average of all RRF (ARRF) values is calculated.

- Note: ARRF = 1.0, when standard concentration is determined in IU/qt. ARRF = 0.60, when standard concentration is calculated in µg/liter.

(b)

$$RP = \frac{PA_{RP} \times conc.RA(\frac{\mu g}{liter}) \times 5\ ml}{ARRF \times PA_{RA} \times 2\ ml} \tag{24}$$

where; 2 ml is the volume of the test sample in use; 5 ml is the hexane is used in extraction.

(c)

Conversion of results from µg/liter to IU/qt

$$RP(IU/qt) = \frac{\frac{\mu g}{liter} RP \times 946(ml/qt)}{0.55\frac{\mu g}{IU} \times 1000(ml/liter)} \tag{25}$$

$$RP(IU/qt) = \frac{\frac{\mu g}{liter} RP \times 946(ml/qt)}{0.34\frac{\mu g}{IU} \times 1000(ml/liter)} \tag{26}$$

3.17.6 DATA COLLECTION SHEET

Title: Estimation of Vitamin A (retinol) in milk by HPLC Method.

Location and Organization:______________________________

Participants: ______________________________________

Date (Month/Day/Year):________________________________

Observations:

Parameter	Repeat 1	Repeat 2	Repeat 3	Average
Volume of test portion used (ml)				
PA_{RA}				
PA_{RP}				
Concentration of RP (μg/liter)				
Concentration of RA (μg/liter)				
RRF				
ARRF				
RP (μg/liter) in test sample				
RP (IU/qt) in test sample				

Comments:____.

3.18 VITAMIN D CONTENT IN MILK BY HPLC METHOD

3.18.1 INTRODUCTION

Vitamin D is a member of sterols family, formed from precursors (such as: ergosterol) by the action of UV radiation. In humans, vitamin D_3 is synthesized in the skin from 7-dehydrocholesterol by the exposure to sun light. Variable stability is observed by pure vitamins D_2 and D_3 in powders, oily solutions or as emulsions. Where heat, O_2, light, or acidic conditions enhance the vitamin D loss. However, this vitamin is quite stable in the absence of O_2 and light [41]. Natural milk is a poor source of vitamin D, which ranges from 10–120 ng/100 g, with an average value of 60 ng/100 g of cow milk. Consumption of one liter of milk provides only 10 to 20% of the recommended daily intake requirement of Vitamin D. Therefore, the milk is commonly fortified with Vitamin D to include about 0.3–0.4 μg/100 g. Vitamin D content in most of the dairy products is exceptionally low. Since long time, biological methods were the only methods, which could assess the lower level of vitamin estimation. The use of new methods employing separation on HPLC and various detectors has been emerged recently, which rapidly estimates vitamin D in food sample with significant accuracy [13, 20]. In this section, AOAC method of determination of cholecalciferol in fortified milk is described.

3.18.2 PRINCIPLE

Vitamin D_3 is extracted with n-heptane after the addition of IS (ergocalciferol) and alkali hydrolysis. Normal phase-LC is used to separate the fraction that contains vitamin D_2 and D_3. Content of vitamin D_3 is measured by reverse phase (RP)-LC with UV detector after evaporation and dilution in methanol-CH_3CN. Absence of endogenous vitamin D_2 is confirmed by analyzing the separated test potion [53].

3.18.3 MATERIALS

3.18.3.1 REAGENTS

- Ascorbic acid (AA).
- BHT.
- C_2H_5OH (40% v/v).
- Cholecalciferol (Vitamin D_3): >98% purity; crystalline form.
- Cyclohexane-n-heptane mixture (1:1 ratio): Mix equal volume of cyclohexane and n-heptane.
- Ergocalciferol (Vitamin D_2): >98% purity; crystalline form.
- Ethanol (C_2H_5OH, 99.5%).
- Ether (peroxide free).
- KOH solutions: (1) KOH solution (1 M): 56 g KOH is dissolved in H_2O and volume made to 1,000 ml; (2) KOH solution (50% m/m): 50 g KOH is dissolved in 50 g H_2O.
- Mobile phase-I: Mix MTBE (20 ml) and isopropanol (5 ml) and dilute it to 1,000 ml with cyclohexane-n-heptane mixture (1:1).
- Mobile phase-II: Transfer 200 ml of isopropanol to one liter flask and volume is made with n-heptane.
- Mobile phase-III: Transfer 200 ml of CH_3OH to one liter flask and volume is made with CH_3CN.
- Phenolphthalein solution: 1% (w/v) solution in C_2H_5OH (99.5%).
- Solvents: n-heptane, cyclohexane, isopropanol, acetonitrile (CH_3CN), methyl-tert-butyl ether (MTBE), methanol (CH_3OH).
- Standard Vitamin D_2 solution:
- Stock standard solution (SSS, 1 mg/ml): 0.1 g of vitamin D_2 is dissolved in C_2H_5OH (99.5%), and it is diluted to 100 ml.
- Working standard solution (WSS, 20 μg/ml): 2 ml of SSS is diluted to 100 ml in C_2H_5OH (99.5%).

- Internal standard (IS) solution (0.8 μg/ml): 4 ml of WSS is diluted to 100 ml in C_2H_5OH (99.5%).
- IS solution (0.2 μg/ml): 1 ml of WSS is diluted to 100 ml in C_2H_5OH (99.5%).
- Standard Vitamin D_3 solution:
- Stock standard solution (SSS, 1 mg/ml): 0.1 g of vitamin D_3 is dissolved in C_2H_5OH (99.5%), and it is diluted to 100 ml.
- WSS (20 μg/ml): 2 ml of SSS is diluted to 100 ml in C_2H_5OH (99.5%).
- WSS (0.6 μg/ml): 3 ml of WSS (20 μg/ml) is diluted to 100 ml in C_2H_5OH (99.5%).
- Vitamins D_2 and D_3, WSS:
- Vitamins D_2 and D_3 (5 μg/ml each) WSS: Transfer 0.5 ml each of SSS to flask (round bottom) and evaporate it. The residue is dissolved in Cyclohexane-n-heptane mixture (1:1) and volume is made to 100 ml.
- Vitamins D_2 and D_3 (0.8 μg/ml each) WSS: Each WSS of vitamins D_2 (20 μg/ml) and D_3 (20 μg/ml) is diluted to 50 ml using CH_3CN.

➢ **Note:** SSS are stable in the dark place at –18°C. Discard after one year or if purity declines to less than 90% that is estimated by LC. Before use, SSS are brought to ambient temperature. Prepare WSS and IS solution monthly; and store these in dark place at 4–8°C.

3.18.3.2 APPARATUS

- Erlenmeyer flask (300 ml) with ground glass neck.
- LC for quantitative analysis: Equipped with injector (loop or auto), UV detector and data processor. Conditions: volume used for injection 100 μl; FR 1.3 ml/min; detector wavelength 265 nm; LC column – main column (C18, 250 × 4.6 mm, 5 μm) and pre-column (C18, 4 × 4 mm, 5 μm).
- LC for semipreparative cleanup: Equipped with injector (loop or auto), UV detector, an integrator, column thermostat and a fraction collector (manual collection can also be used). Conditions: volume used for injection 500 μL; detector wavelength 265 nm; temperature of column 35°C; FR 1.5 ml/min; LC column – main column (silica, 250 × 4.6 mm, 5 μm) and pre-column (silica, 30 × 4 mm, 5 μm).
- Quartz cuvettes (1 cm path length).
- Reflux condenser (50 cm long) with water bath: it can be fitted with an Erlenmeyer flask (300 ml).

- Rotary evaporator: it can be fitted with 50- and 500-ml flask with heated water bath.
- Round bottom flask (500 ml).
- Separating funnel (250-, and 500-ml capacity).
- UV spectrophotometer (capable of measuring absorbance at 265 nm).
- Volumetric flask (2, 100, and 1,000 ml).

3.18.4 PROCEDURE

3.18.4.1 ESTIMATION OF STANDARD CONCENTRATION

1. **LC Method:** Evaporate IS (vitamin D_2, 0.8 μg/ml) solution or WSS of vitamin D_3 (0.6 μg/ml) in rotary evaporator closely to dryness and it is dissolved in mobile phase-III (3 ml). Extract is analyzed by RP-LC (reverse phase-liquid chromatography). Purity value of standard is calculated as follows:

$$f_{LC} = \frac{PA_D}{A_{tot}} \tag{27}$$

   where; PA_D is the peak area for vitamins D_2 and D_3, respectively; A_{tot} is the overall area of all vitamin D peaks that are observed in the chromatogram (except void and solvent peaks). Purity value should range from 0.96 to 1.0.
2. **Spectrophotometric Method:** Using a spectrophotometer, scan the WSS of vitamins D_2 (20 μg/ml) and D_3 (20 μg/ml). Absorbance is measured at 265 nm against C_2H_5OH. Concentration of vitamin D is calculated as follows:

$$C(\mu g / ml) = \frac{A}{A_{tot}} \times 10000 \times f_{LC} \tag{28}$$

   where; A is the absorbance; E (1%, 1 cm) = 475 and 480 for vitamins D_2 and D_3, respectively; 10,000 = conversion factor.

3.18.4.2 SAPONIFICATION

Accurately, weigh 50 g (nearest to 0.01 g) of test milk sample in an Erlenmayer flask (300 ml). Transfer 0.5 of AA, 100 ml of C_2H_5OH (99.5%), 2 ml of IS (vitamin D_2) solution (0.2 μg/ml) and 25 ml of KOH (50%) solution. Connect the condenser to the flask and reflux it for half an hour on a boiling water bath. Remove the flask after refluxing and transfer 25 ml of H_2O through the condenser. Contents of the flask are cooled to room temperature.

3.18.4.3 EXTRACTION

Transfer the hydrolysate mixture quantitatively into a separating funnel (500 ml) along with 50 ml of C_2H_5OH (40%). Transfer n-heptane (50 ml) and vigorously shake the contents for one minute. It is allowed to separate the phases and transfer the solvent layer to another separating funnel (250 ml). Extraction procedure is repeated twice with n-heptane. Combine the extracts and wash with 50 ml of KOH (1 M) solution and two times with 50 ml of C_2H_5OH (40%) and lastly with 50 ml of H_2O till the solvent phase is free of alkali (check with phenolphthalein indicator; wash water should be colorless with indicator). Vigorously shake the funnel for half a minute in each washing step. The solvent phase is then transferred to a round bottom flask (500 ml) and add few crystals of BHT and 15 ml of C_2H_5OH (99.5%). Evaporate the content at 45°C in the rotary evaporator. Remove the flask immediately after the mixture is completely dried. Residue is dissolved in the solvent mixture of n-heptane-cyclohexane (1:1) and transfer the contents to a volumetric flask (2 ml), rinse 2 times and dilute up to the mark. Use this extract for semi-preparative cleanup of vitamin (Section 4.4). Store this extract at 4–8°C for overnight.

3.18.4.4 SEMI-PREPARATIVE CLEANUP

1. **System Setup:** Mobile phase-I is used for cleanup and the system is equilibrated. Inject the WSS (5 μg/ml each of D_2 and D_3) twice or thrice for checking the stability of retention time (RT). Usually, fluctuation of RT is $< \pm 1\%$. If the variation is larger, then check the mobile phase homogeneity, injection solution (water free), pumping unit and temperature stability. Under these conditions, the RT of D_2/D_3 is around 17 minutes. From the RT and peak volume (PV), the collection window is measured. The PV is the base width (3 ml) under the above conditions. Fraction is collected so as to confirm that the whole PV is collected without including various interfering substances.
2. **Purity of Mobile Phase:** Mobile phase (5 ml) is evaporated similarly as carried out for test extract and run it on RP-LC. If any interfering substances elute at the same time as vitamin D, then completely

evaporate mobile phase-I, determined by LC, and contaminated batch has to be changed.

3. **Chromatography:** Test extract is injected followed by collection of vitamin D fraction in a round bottom flask (50 ml). Evaporate at 45°C in a rotary evaporator until it becomes dry. Dried residue is dissolved in 0.5 ml mobile phase-III. Store this solution at 4–8°C until analysis.

➢ ***Note:*** Silica column is cleaned after each run for 15 min. using mobile phase-II to avoid polar lipid accumulation.

3.18.4.5 CHROMATOGRAPHIC DETERMINATION

Mobil phase-III is used for analysis and the system is equilibrated. Inject the WSS (0.8 µg/ml of each of vitamins D_2 and D_3) twice or thrice for checking the stability of RT. Usually, fluctuation of RT is $< \pm 1\%$. Text extracts are injected and peak areas (PA) are measured with the integrator. Using such conditions, RT for vitamin D_2 and D_3 are 10 and 11 min, respectively.

If required, the composition of the mobile phase may be changed to suit the column's quality. Peaks of vitamin D_2 and D_3 should be free from any interfering substances as measured either by spectral monitoring or wavelength ratio technique (260: 280 nm). Resolution value (R_s) should be >1.5.

3.18.5 CALCULATIONS

Vitamin D_3 content of milk sample is calculated as follows:

$$Vitamin\ D_3,\ \mu g/100\ g = \frac{PAD_3 \times mD_2 \times 100}{PAD_2 \times ms \times F} \tag{29}$$

where; PAD_3 is the peak area of vitamin D_3; PAD_2 is the peak area of vitamin D_2; mD_2 is the mass of IS (vitamin D_2) added to the test sample (µg); ms is the mass of test sample (g); and F is the response factor ($F = D_3/D_2$) at 265 nm.

Response factor (F) is determined by using WSS in the RP-LC system as follows:

$$F = \frac{PAD_3 \times CD_2}{PAD_2 \times CD_3} \tag{30}$$

where; CD_2 is the conc. of vitamin D_2 in the standard solution; and CD_3 is the conc. of vitamin D_3 in standard solution.

3.18.6 DATA COLLECTION SHEET

Title: Estimation of vitamin D_3 in milk by LC.

Location and Organization:______________________________

Participants: ____________________________________

Date (Month/Day/Year):_______________________________

Observations:

Parameter	Repeat 1	Repeat 2	Repeat 3	Average
Peak area (PA) of vitamin D_3 = PAD_3				
PA of vitamin D_2 = PAD_2				
Conc. of vitamin D_2 in standard solution = CD_2				
Conc. of vitamin D_3 in standard solution = CD_3				
Response factor = F				
Mass of IS (vitamin D_2) added to the test portion (µg) = mD_2				
Mass of test portion (g) = ms				
Vitamin D_3 content (µg/100 g)				

Comments:______________________________________.

KEYWORDS

- ***citric acid***
- ***lactose***
- ***minerals***
- ***moisture***
- ***protein***
- ***total solids***
- ***vitamins***

REFERENCES

1. AOAC Official Method 925.21: Preparation of milk test sample, Chapter 33. In: *Official Methods of Analysis of AOAC International* (21st edn., p. 5). Gaithersburg, MD – USA: AOAC International; (2019).
2. AOAC Official Method 990.19: Solid (Total) in milk, Chapter 33. In: *Official Methods of Analysis of AOAC International* (21st edn., pp. 38, 39). Gaithersburg, MD – USA: AOAC International; (2019).
3. Walstra, P., Wouters, J. T. M., & Geurts, T. J., (2006). *Dairy Science and Technology* (2nd edn., p. 808). Boca Raton – FL: CRC Press/ Taylor & Francis Group.
4. Medhammar, E., Wijesinha-Bettoni, R., Stadlmayr, B., Nilsson, E., Charrondiere, R. U., & Burlingame, B., (2011). Composition of milk from minor dairy animals and buffalo breeds from a biodiversity perspective. *Journal of the Science of Food and Agriculture, 92*(3), 445–474.
5. AOAC, (2019). AOAC official method, 2000.18: Fat content of raw and pasteurized whole milk: Gerber method by weight. Chapter 33. In: *Official Methods of Analysis of AOAC International* (21st edn., pp. 21–23). Gaithersburg, MD – USA: AOAC International.
6. BIS ISO 2446-2008. *Milk – Determination of Fat Content* (p. 22). London, UK: British Standards Institution (BIS); (2008).
7. IDF Standard 152A (provisional): Milk and milk products – determination of fat content. In: *General Guidance on the Use of Butyrometric Methods* (p. 4). Brussels, Belgium: International Dairy Federation (IDF); (1997).
8. *IS 1224-1: Determination of Fat by the Gerber Method, Part I – Milk* (pp. 1–10). New Delhi: Indian Standards Institution (IS); (1977).
9. *ISO-2446 (IDF-226). Milk – Determination of Fat Content* (2nd edn., pp. 1–10). Geneva, Switzerland: International Organization for Standardization (ISO); (2008).
10. *ISO-2446, Milk: Determination of Fat Content (Routine Method)* (1st edn., pp. 1, 2). Geneva, Switzerland: International Organization for Standardization (ISO); (1976).
11. Kleyn, D. H., Lynch, J. M., Barbano, D. M., Bloom, M. J., & Mitchell, M. W., (2001). Determination of fat in raw and processed milks by Gerber method: Collaborative study. *Journal of AOAC International, 84*(5), 1499–1508.
12. Sindhu, J. S., & Arora, S., (2011). Buffalo milk. In: Fuquay, J. W., Fox, P. F., & McSweeney, P. L., (eds.), *Encyclopedia of Dairy Sciences* (2nd edn., Vol. 3, pp. 503–511). London – UK: Academic Press.
13. Ye, W., Liu, L., & Li, H., (1992). The determination of the fat-soluble vitamins and beta-carotene in milk by high-performance liquid chromatography (HPLC). *Chinese Journal of Chromatography, 10*, 240, 241.
14. AOAC Official Method 989.05: Fat in milk: Modified Mojonnier. Chapter 33. In: Latimer, Jr. G. W., (ed.), *Official Methods of Analysis of AOAC International* (21st edn., pp. 18, 19). Gaithersburg, MD – USA: AOAC International; (2019).
15. AOAC official method 990.21: Solid-not-fat in milk. Chapter 33. In: Latimer, Jr. G. W. (ed.), *Official Methods of Analysis of AOAC International* (21st edn., Vol. II; pp. 39, 40). Rockville, Maryland – USA: AOAC International.(2019).
16. AOAC Official method, 896.01: Lactose in milk – Polarimetric method. In: Latimer, Jr. G. W., (ed.), *Official Methods of Analysis of AOAC International* (21st edn., Vol. II, pp. 16, 17). Rockville, Maryland – USA: AOAC International; (2019).

17. Fox, P. F., Uniacke-Lowe, T., McSweeney, P. L. H., & O'Mahony, J. A., (2015). *Dairy Chemistry and Biochemistry* (2nd edn., p. 584). Switzerland: Springer International Publishing.
18. Abd El-Salam, M. H., & El-Shibiny, S., (2011). Comprehensive review on the composition and properties of buffalo milk. *Dairy Science & Technology, 91*, 663–672.
19. AOAC Official Method 991.20 – nitrogen (Protein) in milk: Kjeldahl method-block digestion/ steam distillation method. In: Latimer, Jr. G. W., (ed.), *Official Methods of Analysis of AOAC International* (21st edn., Vol. II, pp. 10–12). Rockville, Maryland – USA: AOAC International; (2019).
20. Delgado-Zamarreno, M. M., Sanchez-Perez, A., Gomez-Perez, C., & Hernandez-Mendez, J., (1992). High-performance liquid chromatography with electrochemical detection for the simultaneous determination of vitamin A, D_3 and E in milk. *Journal of Chromatography, 623*, 69–74.
21. AOAC official method 998.05 – non-casein nitrogen content of milk: Kjeldahl method. In: Latimer, Jr. G. W., (ed.), *Official Methods of Analysis of AOAC International* (21st edn., Vol. II, pp. 56, 57). Rockville, Maryland – USA: AOAC International; (2019).
22. Misra, S. S., Sharma, A., Bhattacharya, T. K., Kumar, P., & Saha, R. S., (2008). Association of breed and polymorphism of α-s1 and α-s2 casein genes with milk quality and daily milk and constituent yield traits of buffaloes (*Bubalus bubalis*). *Buffalo Bulletin, 27*(4), 294–301.
23. AOAC official method 991.21: Nonprotein nitrogen in whole milk – Kjeldahl method. Chapter 33. In: Latimer, Jr. G. W., (ed.), *Official Methods of Analysis of AOAC International* (21st edn., Vol. II, pp. 12, 13). Rockville, Maryland – USA: AOAC International; (2019).
24. DePeters, E. J., & Ferguson, J. D., (1992). Nonprotein nitrogen and protein distribution in the cow milk. *Journal of Dairy Science, 75*(11), 3192–3209.
25. Fox, P. F., & McSweeney, P. L. H., (1998). *Dairy Chemistry and Biochemistry* (p. 478). New York: Kluwer Academic Publishers.
26. AOAC Official Method 991.23 – Protein nitrogen content of milk: Indirect Kjeldahl method. Chapter 33. In: Latimer, Jr. G. W., (ed.), *Official Methods of Analysis of AOAC International* (21st edn., Vol. II, p. 14). Rockville, Maryland – USA: AOAC International, 2019.
27. AOAC Official Method 945.46 – Ash of milk: Gravimetric method. Chapter 33. In: Cunniff, P., (ed.), *Official Methods of Analysis of AOAC International* (16th edn., Vol. II, p. 10). Washington, DC – USA: AOAC International; (1995).
28. Bhave, S. A., Pandit, A. N., & Tanner, M. S., (1987). Comparison of feeding history of children with Indian childhood cirrhosis and paired controls. *Journal of Pediatric Gastroenterology and Nutrition, 6*(4), 562–567.
29. Danks, D. M., (1988). Copper deficiency in humans. *Annual Review of Nutrition, 8*, 235–257.
30. *ISO 5738/IDF 76 – Milk and Milk Products: Determination of Copper Content by Photometric Reference Method* (2nd edn., p. 11). Switzerland: International Organization for Standardization (ISO); (2004).
31. Murthy, G. K., (1974). Trace elements in milk. *CRC Critical Reviews in Environmental Control, 4*(1–4), 1–37.

32. Pennington, J. A. T., Wilson, D. B., Young, B. E., Johnson, R. D., & Vanderveen, J. E., (1987). Mineral content of market samples of fluid whole milk. *Journal of the American Dietetic Association, 87*(8), 1036–1042.
33. Renner, E., Schaafsma, G., & Scott, K. J., (1989). Micronutrients in milk. In: Renner, E., (ed.), *Micronutrients in Milk and Milk-Based Food Products* (pp. 1–70). London: Elsevier Applied Science.
34. Rob, J. K., Bradley, R. L., Richardson, T., & Weckel, K. G., (1976). Removal of copper from milk. *Journal of Dairy Science, 59*, 382–385.
35. Block, G., Dresser, C. M., Hartman, A. M., & Carroll, M. D., (1985). Nutrient sources in the American diet: Quantitative data from the NHANES-II survey, Part I: Vitamins and minerals. *American Journal of Epidemiology, 122*(1), 13–26.
36. Hazell, T., (1985). Minerals in foods: Dietary sources, chemical forms, interactions, bioavailability. In: Bourne, G. H., (ed.), *World Review of Nutrition and Dietetics* (Vol. 46, pp. 1–123). Basel: Karger Publisher.
37. McCance, R. A., Holland, B., & Widdowson, E. M., (1991). *McCance and Widdowson's the Composition of Foods* (p. 462). London: Royal Society of Chemistry.
38. *ISO 12081/ IDF 36:2010. Milk: Determination of Calcium Content (Titrimetric Method)* (p. 6). Switzerland: International Organization for Standardization (ISO); (2010).
39. Schaafsma, G., (1984). The significance of milk as a source of dietary calcium. In: *Nutrition and Metabolism* (pp. 19–32). Brussels: International Dairy Federation (IDF); Document 166.
40. Hambidge, K. M., Casey, C. E., & Krebs, N. F., (1986). Zinc. In: Mertz, W., (ed.), *Trace Elements in Human and Animal Nutrition* (5th edn., Vol. 2, pp. 1–137). Orlando – FL: Academic Press.
41. Ball, G. F. M., (1988). *Fat – Soluble Vitamin Assays in Food Analysis: Comprehensive Review.* London: Elsevier Applied Science.
42. *ISO 11813/ IDF 156-2010: Milk and Milk Products (Determination of Zinc Content) by Flame Atomic Absorption Spectrometric Method* (p. 6). Switzerland: International Organization for Standardization (ISO); (2004).
43. Bruhn, J. C., & Franke, A. A., (1988). Protein and major cations in California market milks. *Journal of Dairy Science, 71*(4), 917–924.
44. Fransson, G. B., & Lönnerdal, B., (1983). Distribution of trace elements and minerals in human and cow's milk. *Pediatric Research, 17*(11), 912–915.
45. ISO-8070/ IDF-119:2007. *Milk and Milk Products – Determination of Calcium, Sodium, Potassium and Magnesium Contents by Atomic Absorption Spectrometric (AAS) Method* (p. 14). Switzerland: International Organization for Standardization (ISO); (2007).
46. Kirchgessner, M., Friesecke, H., & Koch, G., (1967). *Nutrition and the Composition of Milk* (p. 273). London: Crosby Lockwood & Son Ltd.
47. Morrissey, P. A., & Hill, T. R., (2009). Fat-soluble vitamins and vitamin C in milk and dairy products. In: McSweeney, P. L. H., & Fox, P. F., (eds.), *Advanced Dairy Chemistry: Lactose, Water, Salts and Minor Constituents* (3rd edn., Vol. 3, pp. 527–589). New York: Springer.
48. Holt, C., (1985). The milk salts: Their secretion, concentrations and physical chemistry. In: Fox, P. F., (ed.), *Developments in Dairy Chemistry – 3: Lactose and Minor Constituents* (pp. 143–181). London: Elsevier Applied Science Publishers.

49. ISO-9874/ IDF-42:2006. *Milk – Determination of Total Phosphorus Content – Method Using Molecular Absorption Spectrometry* (p. 7). Switzerland: International Organization for Standardization (ISO); (2006).
50. AOAC official method 2002.06: Retinyl palmitate (vitamin A) in fortified fluid milk. Chapter 45. In: Latimer, Jr. G. W., (ed.), *Official Methods of Analysis of AOAC International* (21st edn., Vol. III, pp. 3–6). Gaithersburg, MD – USA: AOAC International; (2019).
51. Peaker, M., & Linzell, J. L., (1975). Citrate in milk: A harbinger of lactogenesis. *Nature, 253*(5491), 464–466.
52. IS 1479 (Part II)-1961. *Methods of Test for Dairy Industry, Part II: Chemical Analysis of Milk* (p. 52). New Delhi: Indian Standards Institution (IS); (1961).
53. AOAC official method, 2002.05: Cholecalciferol (vitamin D3) in selected foods. Chapter 45. In: Latimer, Jr. G. W., (ed.), *Official Methods of Analysis of AOAC International* (21st edn., Vol. III, pp. 34–37). Gaithersburg, MD – USA: AOAC International, 2019.

CHAPTER 4

ANALYSIS OF PHYSICOCHEMICAL ASPECTS OF MILK

CHITRA GUPTA, PRASAD PATIL, ANURADHA KUMARI, and HEMANT GAWANDE

ABSTRACT

Milk constituents and their concentration in milk affect various physical properties like density, specific gravity, viscosity, surface tension (ST), freezing point (FP), boiling point, pH, and acidity of milk. Physical property determination is important to collect the data for designing of dairy equipment, for estimating the amounts of various components and assessing the extent of physical and chemical changes. Some of these properties (e.g., specific gravity, density, pH, titratable acidity, etc.) are being affected by the nature of the molecule rather than the number of molecules. However, FP and boiling point are considered as colligative properties which are being affected by the number of molecules rather than the properties of molecules. In this chapter, various methods used for the analysis of physicochemical properties of milk have been discussed.

4.1 DETERMINATION OF SURFACE TENSION (ST) OF MILK

4.1.1 INTRODUCTION

Surface tension (ST, γ) is the attraction force exerted on the surface molecules of fluid by the underneath molecules which resulted in the shrinkage of fluid surface to minimum surface area. ST is expressed as force per unit length and its unit is N/m (or dyne/cm). Molecule of liquid which is present in the interior portion of liquid is completely surrounded by the other similar molecules and intermolecular attraction forces are applied from all the sides. Hence,

Analytical Methods for Milk and Milk Products, Volume 1: Sampling Methods and Chemical and Compositional Analysis. Megh R. Goyal, N. Veena, and Santosh Kumar Mishra (Eds.)

molecule remains stable in the interior portion of liquid. However, molecule at interface does not have the similar molecules from all the sides and attraction force is implemented by the underneath molecules only which resulted in the film formation on the surface of liquid. Due to this film, piercing of liquid surface is more difficult and this cohesive force is called ST [1].

ST of pure water is higher than the milk, as milk contains various surface-active compounds, i.e., fat, protein, free fatty acids (FFAs) and phospholipids. It is also being affected by various parameters, i.e., temperature, fat content, homogenization, oxidation of fat, aging, and heat treatment, including sterilization. ST of milk and water is 40–60 dynes/cm and 72.75 dynes/cm at 20°C [2]. It can be measured by capillary rise method, stalagmometer, du Nouy ring, pendant drop, Wilhelmy plate, and bubble pressure method.

Milk ST can be measured by a Stalagmometer method. Stalagmometer determines the number of drop or weight of the liquid formed from the given volume of liquid. It has two graduation marks: one is above the bulb (graduation mark A), and the other one is below the bulb (graduation mark B), which is further connected to a capillary [3] (Figure 4.1).

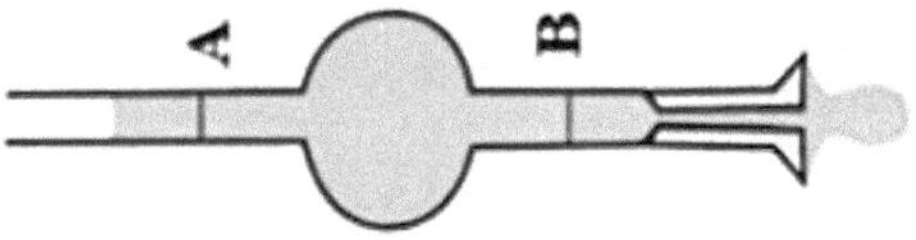

FIGURE 4.1 Stalagmometer.

4.1.2 *PRINCIPLE OF OPERATION*

If a liquid flow through the small orifice which is followed by a flat surface, then it will lead to the formation of a drop. Liquid will spread on the clean and polished flat surface area. Water is collected at the particular point, lead to the formation of a drop. Drop is held at that particular point due to ST. Drop will grow until the weight of the drop exceeds the force due to ST. ST can be calculated using the weight of the drop as follows:

$$\text{Surface tension} = \frac{\textit{Weight of the drop in g} \times 980.365}{2\pi r} \quad (1)$$

ST acts at the interface of two phases; therefore, it is also known as interfacial tension. Due to ST liquid tends to have minimum surface area for a given volume of a liquid, droplets of liquid are spherical. ST of milk can be determined by comparing numbers of drops of definite volume and densities of milk with a standard liquid (water), whose ST value at the given temperature is known [3].

4.1.3 MATERIALS

4.1.3.1 APPARATUS

- Stalagmometer;
- Small bottle;
- Beaker;
- Thermometer;
- Rubber tubing with a screw pinch cock.

4.1.4 PROCEDURE

4.1.4.1 PREPARATION OF SAMPLE

Properly mix the milk sample to ensure the proper mixing of the fat layer if formed. Heat the milk to 40°C to ensure the removal of dissolved gases. Cool the milk sample to the temperature of analysis.

4.1.4.2 EXPERIMENTATION

Clean the stalagmometer with chromic acid followed by distilled water and further dry it in a hot air oven. After proper cooling, connect the rubber tubing with pinch cock at the stalagmometer upper end. Dip the stalagmometer lower flat end in the beaker filled with distilled water and suck the distilled water just above the upper graduation mark (A) on the stalagmometer. Adjust the meniscus of water on the upper graduation mark. Release the screw pinch cock and count the number of drops formed from the water present in the stalagmometer between the graduation mark A to B. Note down the reading as n_1.

Clean and dry the same stalagmometer and repeat the same process with milk. Note down the number of drops formed by the milk as n_2. Density of water and milk at 20°C can be estimated using a pycnometer or specific gravity bottle. ST can be calculated using the following Eqn. (2) [3].

4.1.5 CALCULATION

$$\gamma_2 = \frac{n_1}{n_2} \times \frac{d_1}{d_2} \times \gamma_1 \tag{2}$$

where; γ_1 is the surface tension of water; γ_2 is the surface tension of milk; n_1 is the number of drops of water; n_2 is the number of drops of milk; d_1 is the density of water; and d_2 is the density of milk.

4.1.6 DATA COLLECTION SHEET

Title: Determination of Surface tension of milk

Location and Organization:______________________________

Participants: ______________________________________

Date (Month/Day/Year):________________________________

Observations:

Parameters	**Repeat 1**	**Repeat 2**	**Repeat 3**	**Average**
Number of drops of water at 20°C (n_1)				
Number of drops of milk at 20°C (n_2)				
Density of water at 20°C (d_1)				
Density of milk at 20°C (d_2)				
Surface tension of water at 20°C (γ_1)				
Surface tension of milk (γ_2)				

Comments:____.

4.2 DETERMINATION OF FREEZING POINT (FP) OF MILK

4.2.1 INTRODUCTION

Colligative properties are dependent upon a number of molecules rather than the nature of the molecule and freezing point (FP) is one among these properties. FP is the temperature at which both water and ice exist in equilibrium with each other. It is the temperature at which both water and ice have the same vapor pressure. Addition of non-volatile solute to the solvent decreases the FP of the solution and it is known as FP depression. During freezing, molecules slow down and rearrange themselves and intermolecular bonds are formed. The presence of any other solute can interfere with the intermolecular bond formation and thus resulted in a decrease in the FP. According to Roult's law, the addition of solute to the solvent decreases the vapor pressure. Lowering of vapor pressure leads to the lowering of the temperature at which the vapor pressure of liquid and frozen form of the solution will be equal, hence FP depression [1].

Among all physical properties, FP is the most constant property. It is mainly affected by the soluble constituents, i.e., lactose and soluble salts in milk. Milk FP is not affected by casein micelles, whey protein and fat globules due to their high molecular weight. FP for bovine milk is –0.512 to –0.550°C, with a mean value of –0.540°C [4]. FP of milk varies with the increase and decrease in the solute concentration. Water addition has a substantial influence on the milk FP; hence FP can be used as a one of the parameters to evaluate the adulteration of milk with water. As milk is isotonic with blood, FP of milk remains constant until there is a change in the mammary gland physiology. Milk FP can also vary with seasonal change, animal breeds, stage of lactation, feed, water consumption and heat stress. However, all these parameters are interrelated and had little influence on the FP of milk. Cooling and moderate heating show a very small effect on milk FP; however, during direct UHT heating, water is mixed with milk sample and affects the milk FP [1].

In 1921, Hortvet described the technique to measure the FP of milk. It was the first apparatus used for analyzing milk FP. International Organization for Standardization (ISO) and International Dairy Federation (IDF) have considered the thermistor cryoscope (Figure 4.2) method as the reference method for analyzing the FP of milk [5, 6].

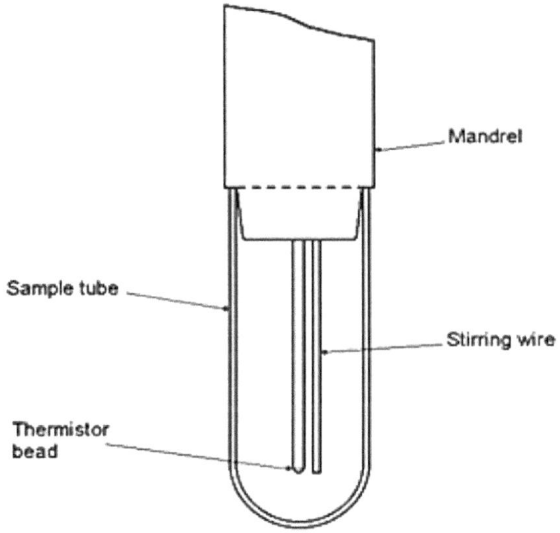

FIGURE 4.2 Thermistor cryoscope.

4.2.2 PRINCIPLE OF OPERATION

FP is the temperature at which water and ice are in equilibrium with each other and due to that there will not be any change in the temperature. Firstly, the test sample is supercooled below the temperature of crystallization. Initiation of crystallization instantly releases the heat which is sufficient to raise the temperature of the sample to the temperature plateau. Plateau is the point at which temperature rise is not more than 0.5 milli°C for the last 20 sec. This temperature is generally represented as the FP of the sample. Thermistor cryoscope is calibrated using two different concentrations of either sucrose or salt solution using the same procedure as described for the test portion of milk [5, 6].

4.2.3 MATERIALS

4.2.3.1 REAGENTS

- Distilled water.
- Sodium chloride: Dry the finely grounded sodium chloride in electric furnace at 300±25°C for 5 h or alternatively dry it in hot air oven (130±2°C for at least 24 h). Cool it to ambient temperature in the desiccator.
- Standard sodium chloride (NaCl) solution: Different quantity of dry NaCl will be needed for preparing the solution on a weight basis or volume basis. However, a solution made as gm per one kilogram is preferred. Quantity given in the column 1 (Table 4.1) needed to be dissolved in 1,000 gm of water and quantity given in column 2 (Table 4.1) needed to be dissolved in 1,000 ml of water. Dry NaCl needs to be weighed nearest to the 0.1 mg. Depending upon the quantity of dry NaCl weighed, dissolved it in 1,000 g or 1,000 ml. Solution can be stored in polyethylene bottles of capacity not more than 250 ml at 5°C. Freshly prepared sample is preferred; however, sample without any mold growth can be used for the period of 2 months. Prior to use, gently invert the bottle several times without incorporation of air. Pour the standard solution from the bottle in the beaker and never use the pipette for the said purpose. Use fresh NaCl standard solution for reference method; however, solution preserved with fungicidal and fungistatic agents can be used for routine analysis [5, 6].

TABLE 4.1 FP of Standard Sodium Chloride Solution

Sodium Chloride (g/kg)	Sodium Chloride (g/liter) at 20°C	Freezing Point (milli°C)
6.763	6.731	–400.0
6.901	6.868	–408.0
7.625	7.587	–450.0
8.489	8.444	–500.0
8.662	8.615	–510.0
8.697	8.650	–512.0
8.835	8.787	–520.0
9.008	8.959	–530.0
9.181	9.130	–540.0
9.354	9.302	–550.0
9.475	9.422	–557.0
10.22	10.161	–600.0

4.2.3.2 APPARATUS

1. **Cryoscope:** Thermistor cryoscope consist of thermostatically controlled cooling device, sample agitator, read out device, crystallization device and thermistor probe. Cooling device can be of immersion type, circulation type or cooling block type. Cooling devices maintain the temperature of cooling liquid at –7.0±0.5°C around the sample tube. Propylene glycol (33% v/v aqueous solution) is used as a suitable cooling liquid. Glass probe type thermistor with diameter 1.60±0.4 mm and electrical resistance 3–30 kΩ at 0°C is used. In any case, it should not transfer the heat more than 2.5 x 10^{-3} J/s to the sample.

In operational condition, thermistor bead should lie in the axis of sample tube, and it should also maintain the equal distance from the internal wall and sample tube inner bottom too. Thermistor and associated circuitry should be able to discriminate 1 milli°C in the range of –400.0 to –600.0 milli°C. If the instrument is working properly, then the error in the circuit shall not be more than 1 milli°C. Read out device should be able to discriminate 1 milli°C in the range of 0 milli°C to 1,000 milli°C.

Stirring wire is used to stir the test sample during cooling. It is mounted vertically and vibrational amplitude should be adjustable. It should be able to vibrate with the amplitude of 2–3 mm for uniform cooling of the test portion and in any case, it should not strike the glass probe type thermistor and sample tube wall.

Crystallization device initiate the freezing of test portion, as soon as temperature of the sample reaches to –3°C. Sample tube should be made up of borosilicate glass with a length of 50.5±0.2 mm, internal and external diameter 13.7±0.3 mm and 16.0±0.2 mm, respectively [5, 6].

- Analytical balance (accuracy of 0.1 mg);
- Weighing scoop;
- Volumetric flasks (Capacity 1,000 ml);
- Electric furnace or drying oven;
- Desiccator;
- Polyethylene bottles.

4.2.4 PROCEDURE

4.2.4.1 PREPARATION OF SAMPLE

Test sample should be the true representative of the whole sample. Preferably, the sample should be analyzed immediately after sampling. In case of raw milk, it should be stored at 0–6°C for 48 h and processed milk can be analyzed during its shelf life. If necessary, remove the solid butter fat or fat from the sample using filter paper or by centrifugation.

4.2.4.2 EXPERIMENTATION

4.2.4.2.1 Calibration of Thermistor

Ensure cryoscope should be in the working condition and check for the probe position, stirring wire vibrational amplitude and cooling device temperature. Two standard NaCl solutions should be selected depending upon the expected FP of the test sample; however, the difference between the FP of the two standard NaCl solutions should not be less than 100 milli°C. During the analysis temperature of standard NaCl solution and test sample should be same during the analysis. About 2.5±0.1 ml of NaCl standard solution is taken in the sample tube. Calibrate the instrument as per the instruction of the manufacturer. Measure the FP of standard NaCl solution until the difference between the two consecutive readings is less than 1 milli°C. If the actual reading of FP of standard NaCl solution differs more than 2 milli°C than the expected reading, then the instrument should be recalibrated following the same procedure as described above. During continuous use of cryoscope, routine calibration should be carried out after every one hour.

4.2.4.2.2 Analysis of Sample

Sample container must be inverted several times without incorporation of any air to the sample. Transfer 2.5±0.1 ml of test sample into the sample tube (as used for calibration). Properly clean the stirring wire and probe with soft tissue paper. The sample tube is inserted in the calibrated thermistor cryoscope. Cool the test sample and start the freezing at –30±0.1°C. Wait until the sample reaches the temperature plateau. Temperature plateau is the temperature in milli°C where temperature will not raise more than 1 milli°C in last 20 second of analysis. This temperature plateau point is considered as the FP of the test sample. Repeat the analysis for analyzing the repeatability of the results.

4.2.5 CALCULATION

Reading on the display of instrument = Freezing point of the sample.

4.2.6 DATA COLLECTION SHEET

Title: Determination of freezing point of milk.

Location and Organization:________________________________

Participants: ____

Date (Month/Day/Year):__________________________________

Observations:

Parameters	**Repeat 1**	**Repeat 2**	**Repeat 3**	**Average**
Freezing point of test sample				

Comments:____.

4.3 DETERMINATION OF REFRACTIVE INDEX (RI) OF MILK

4.3.1 INTRODUCTION

Velocity of the light varies with the density of the medium. As light is passed through the medium of different densities, it is refracted. Air is the less dense medium as compared to the water. Refractive index (RI) is expressed as the ratio between the velocity of light under vacuum and the velocity of light under sample medium. It is commonly defined as the sine angle of incidence light to the sine angle of the refraction light. RI is denoted with the symbol neta (η) [1, 4, 8] as follows:

$$\eta = \frac{\sin i}{\sin r} \tag{3}$$

where; i is denoted as the angle of incidence; and r is denoted as the angle of refraction.

Dissolution of substance in water also increases the RI. Due to opacity, it is difficult to determine the RI of milk. Whole milk is centrifuged, and fat is separated prior to analyzing the RI of milk, as fat interfere in the clear separation of two layers. RI of milk changes with the temperature. Rise in temperature increases the RI of the sample. RI for water is 1.33299 and for milk fat is 1.4527 to 1.4800 at 20°C. RI of cow milk and buffalo milk is 1.3450 to 1.3472 (mean value of 1.3460) and 1.3460 to 1.3490 (mean value of 1.3477) at 40°C. The milk RI is not affected by milk fat as the refraction arises at the air and continuous phase (milk serum/plasma) interface. Casein, whey protein, lactose, and miscellaneous constituents contribute to the RI of milk [1, 8].

4.3.2 PRINCIPLE

Abbe refractometer was discovered by Ernst Abbe (1840–1905) and is the high precision device to measure the liquid RI. RI differs according to the light wavelength, hence monochromatic light or D line of sodium spectrum (wavelength of light 589 nm) is being used as incidence light (Figure 4.3). Light from the radiation source is first reflected by the mirror and then passed through the double prism. Two prisms are adjusted over each other and the sample is applied in between these two prisms. Incidence light passed through the double prism or sample only if the incident angle at the interface is less than the critical angle of total reflection. A mirror and microscope are adjusted to light/dark boundaries using a suitable mechanism. Rotary knob is rotated until the light/dark boundary matches the intersection of the microscope crosshair and refractive indices can be read through the vernier scale.

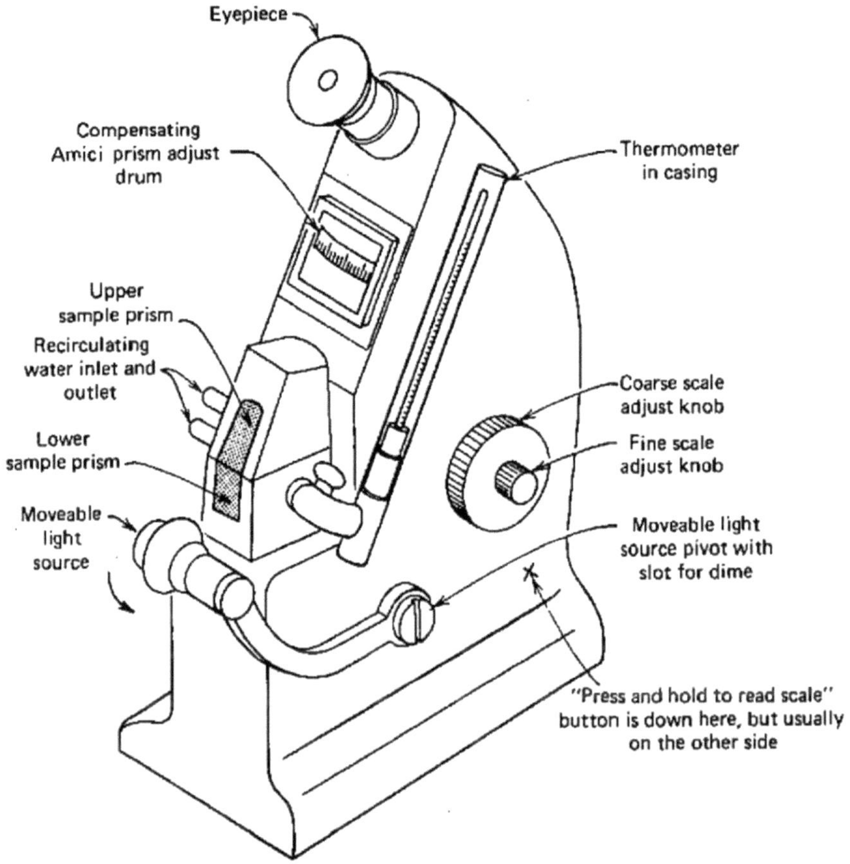

FIGURE 4.3 Abbe refractometer.

Manually, it is difficult to differentiate between the light/dark boundaries due to the low contrast. Various readings need to be noted down, and an average of these values is considered as the RI of the sample [7].

4.3.3 MATERIALS

4.3.3.1 REAGENTS

Any standard solution, i.e., ethanol, acetone, dichloro-methane, etc., may be used and RI of these solutions must be known.

4.3.3.2 APPARATUS

- Abbe refractometer;
- Sodium lamp;
- Water bath maintained at 20°C;
- Centrifuge.

4.3.4 PROCEDURE

4.3.4.1 PREPARATION OF SAMPLE

Fat present in milk interfere in the determination of RI of milk. Milk sample is centrifuged at 3,000 rpm for clear separation of the fat layer. Cream plug should be removed, and a lower layer of milk (or skim milk) is used for the determination of RI of milk [9].

4.3.4.2 EXPERIMENTATION

Check the precision of the equipment using the standard solutions of known RI. Apply the drop of the standard solution in between the double prism. Drop of solution needs to be applied in such a way that there should not be formation of any air bubble. Wait for the observation until the temperature of the sample reaches to 20°C. The circulatory water bath is used for maintenance of temperature. Monochromatic light or D line of sodium spectrum (wavelength of light 589 nm) is used as a light source. If natural light is used as a light source, then equipment must be fitted with the selective filters for selection of one wavelength light.

Light is first reflected through the mirror and then passed through the prism where it comes in contact with standard solution. Position of mirror and prism is adjusted in a way that light/dark boundary matches the intersection of microscope crosshair. At that point, the RI of the standard solution can be read through the vernier scale. Repeat the same process as above for the prepared milk sample. Note down the reading as RI at 20°C.

Due to the low contrast of light/dark boundaries, it is important to take a large number of readings for the accuracy of results at the fourth decimal. Average of the observed reading gives the highly accurate results.

4.3.5 CALCULATION

RI increases with a fall in temperature and decreases with rise in temperature. If RI has been observed at a lower temperature as compared to desired temperature then X factor needs to be subtracted from the observed RI reading.

$$\text{Reading of Abbe refractometer} = \text{RI of milk sample}$$

$$\text{Accurate RI value} = \text{Observed RI} - X, \text{ or}$$

$$\text{Accurate RI value} = \text{Observed RI} + X \quad (4)$$

where; X = 0.000365 for each degree fall; If RI has been observed at higher temperature as compared to desired temperature then X factor needs to be added in the observed RI reading.

4.3.6 DATA COLLECTION SHEET

Title: Determination of refractive index of milk.

Location and Organization:____________________________

Participants: ____________________________________

Date (Month/Day/Year):______________________________

Observations:

Parameter	**Repeat 1**	**Repeat 2**	**Repeat 3**	**Average**
Refractive index of milk at 20°C				

Comments:____.

4.4 DETERMINATION OF PH OF MILK BY COLORIMETRIC METHOD

4.4.1 INTRODUCTION

Negative logarithm of hydrogen ion concentration is defined as pH. Firstly, it is logarithm; unit change in the pH depicts the 10-fold change in hydrogen ion concentration. Secondly, it is negative, lower pH depicts the higher acidity. pH value lesser than 7.0 is considered as acidic environment and pH value greater than 7.0 is considered as basic or alkaline environment.

pH of milk ranges from 6.6 to 6.8, with an average value of 6.7 at 20°C. As milk is isotonic with blood, pH of milk is near to the pH of blood, i.e., 7.4. Mastitis infection and late lactation leads to the damage to the mammary cell membrane which leads to the leakage of blood constitutes in milk and resulted in rise in pH of milk. pH of mastitis milk is around 6.9 to 7.0 and late lactation milk approaches to pH 7.2. The pH of the colostrum is 6.0. Bacterial growth or fermentation of milk also leads to the reduction in pH of milk due to conversion of milk into lactic acid [1].

The titratable acidity is a simple method to determine the amount of acid formed, however the pH is an important parameter for acidity characterization. pH of milk can be determined by colorimetric and potentiometric method. Colorimetric estimation can be done using indicator strip or comparator method. Indicator strips consist of indicator dye coated paper that changes the color at specific pH. These indicator papers give an approximate value of pH. A specific disk with the different shades of indicator at different pH is used in the comparator method. Bromocresol purple (pH 5.2–6.8) and bromothymol blue (BTB) (pH 6.0–7.6) disks are used for estimation of pH of milk sample. This method is only considered as an aid for detecting the abnormal milk; however, results are not conclusive for such abnormalities. Accuracy of more than 0.2 pH is not expected using this method [10].

4.4.2 PRINCIPLE OF OPERATION

In the comparator method (Figure 4.4), the indicator is added to milk. Indicator addition leads to the development of color. Color development in the milk sample is compared with the color shown in the disk. Milk sample without addition of indicator is used as a blank for the reading. Comparing the color of the disk with milk sample gives the value for the pH of the milk sample.

4.4.3 MATERIALS

4.4.3.1 REAGENT

- BTB solution (0.2%) (pH range 6.2 to 7.6, color yellow to blue): Dissolve 0.2 g of BTB in 100 ml of 47.5% alcohol.

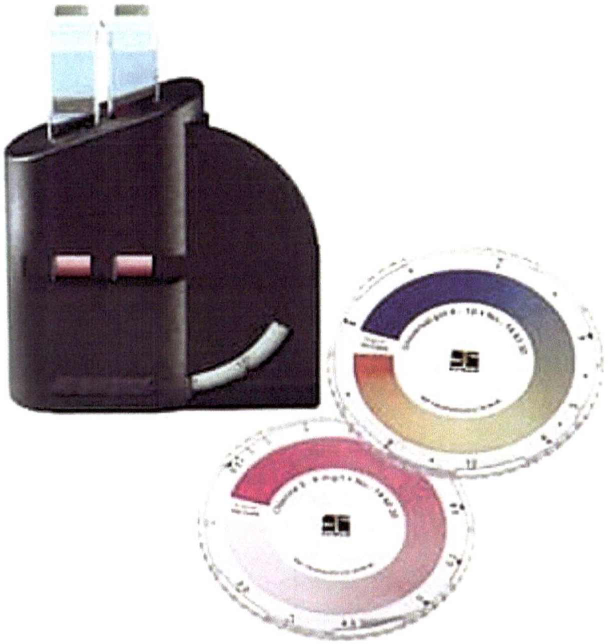

FIGURE 4.4 pH comparator.

4.4.3.2 APPARATUS

- Lovibond comparator (with bromothymol blue disc);
- Erlenmeyer flask (250 ml);
- Pipette (10 ml);
- Pipette (graduated 1 ml);
- Test tubes.

4.4.4 PROCEDURE

4.4.4.1 PREPARATION OF SAMPLE

Bring the temperature of the milk to about 26 to 28°C and mix it properly to get a homogeneous mixture. If it contains lumps of fat then warm the sample to 40°C, cool it to 26 to 28°C after properly mixing. If measured volume is required, then bring the sample temperature to 27°C before pipetting.

4.4.4.2 EXPERIMENTATION

Take one ml milk sample in Erlenmeyer flask and 10 ml of distilled water is added in it. Transfer 10 ml of this diluted sample in the test tube and mix it after adding 0.5 ml indicator. Keep this test tube in the sample side of the comparator. Transfer the diluted sample in another test tube and keep in the blank side. Rotate the disc in a way that color of the disc should match with the color of the sample. Observe the reading of pH on the disc label. Note down the reading as a pH of the milk sample [10].

4.4.5 CALCULATION

Reading on the Lovibond comparator disc = pH of milk.

4.4.6 DATA COLLECTION SHEET

Title: Determination of pH of milk by Lovibond comparator.

Location and Organization:____________________________

Participants: ____________________________________

Date (Month/Day/Year):______________________________

Observations:

Parameters	**Repeat 1**	**Repeat 2**	**Repeat 3**	**Average**
pH of milk				

Comments:____.

4.5 DETERMINATION OF PH OF MILK BY POTENTIOMETRIC METHOD

4.5.1 INTRODUCTION

The potentiometric method is a more reliable and accurate method for determination of pH of milk. It measures the difference in potential between two electrodes, i.e., reference electrode (generally calomel electrode) and glass electrode [11]. Both electrodes are hollow bulbs which contain KCl solution with AgCl wire immersed in it. The glass sensing electrode has a glass membrane which is sensitive to specific H^+ and this electrode measures the pH as the concentration of H^+ ion. In this chapter, pH is determined by the official method described by Bureau of Indian Standards (BIS) [10].

4.5.2 PRINCIPLE

The pH meter (Figure 4.5) works under the principle of difference in potential between the two electrodes, i.e., glass electrode (affected by test solution's pH) and the reference electrode (whose potential is independent of test solution pH). This difference in potential is measured and pH value is read directly.

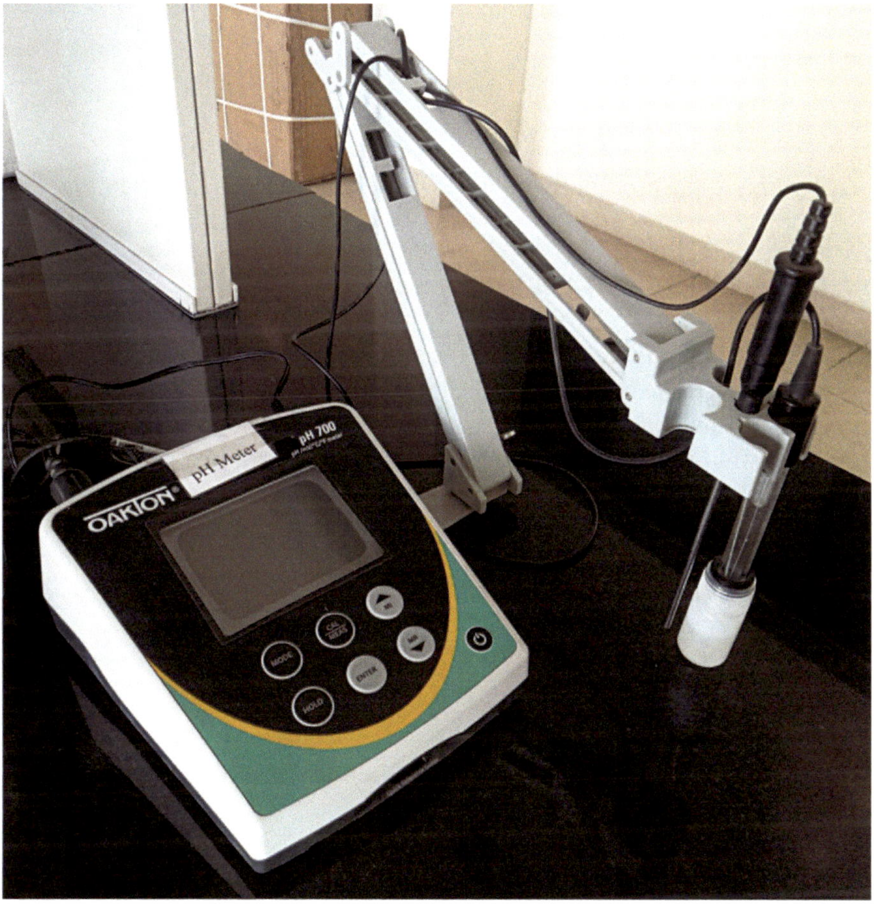

FIGURE 4.5 pH meter for determination of pH.

4.5.3 MATERIALS

4.5.3.1 REAGENTS

- Standard buffer solutions of known pH (4.0, 7.0, 9.2, or 10.0).

4.5.3.2 APPARATUS

- Beaker;
- Wash bottle filled with distilled water.

4.5.4 PROCEDURE

4.5.4.1 PREPARATION OF SAMPLE

According to the procedure described in Section 4.4.1 in this chapter.

4.5.4.2 EXPERIMENTATION

The pH meter is calibrated against the standard buffer solution of known pH. Wash the electrode end with distilled water. Take milk sample of appropriate quantity in the beaker enough to immerse the electrode probe end. Put the electrode into it and wait for the stabilization of the reading on the pH meter display. The instrument will indicate the pH of the milk sample. Note down this reading as pH of milk sample [10].

4.5.5 CALCULATION

pH of milk sample = Value on display of pH meter

4.5.6 DATA COLLECTION SHEET

Title: Determination of pH of milk by potentiometric method.

Location and Organization: ______________________________

Participants: ____

Date (Month/Day/Year): ________________________________

Observations:

Parameters	Repeat 1	Repeat 2	Repeat 3	Average
pH of milk				

Comments: ____

4.6 DETERMINATION OF ACIDITY OF MILK

4.6.1 INTRODUCTION

Acidity is mainly determined to check milk freshness or to control the manufacturing of fermented dairy products. Titration method is generally used as it is very fast and simple. The titratable acidity indicates the buffering capacity of milk between the initial pH of milk and point where phenolphthalein color changes to pink, i.e., pH 8.3 [13]. The ionizable group of protein, phosphates, and citrates present in milk contribute to natural milk acidity. Some of the ionizable groups of protein are buried inside the protein and not available during titration. All ionizable groups of casein is available however all groups of whey proteins are not available [16]. Developed acidity of milk is due to the conversion of lactose into lactic acid by microorganisms.

The initial acidity of individual cow milk varies from 0.08 to 0.25% lactic acid. Generally, it ranges from 0.14% to 0.16% lactic acid [13]. The higher initial acidity (with no developed acidity) of milk is due to a higher amount of protein and other buffering components in milk. Milk with acidity more than 0.18% is not suitable for the production of heat treated products. Milk acidity is affected by the amount of sample taken, dilution of sample, concentration and amount of indicators used and speed of titration [12]. On adding a similar volume of water into milk decrease the titratable acidity by 0.02% and adding nine volumes of water, decreases the acidity by 0.06% [20]. The titration results are greatly affected by the titration speed. Slow titration leads to the formation of insoluble calcium phosphate which releases hydrogen ions and decreases pH, which is called fading of phenolphthalein end point [13]. In this chapter, titratable acidity of milk measured as per the method described by BIS [14].

4.6.2 PRINCIPLE OF OPERATION

The titratable acidity is estimated by the titration method (Figure 4.6), where the sample is titrated with the standard NaOH solution by using an indicator (phenolphthalein). During titration, the pH of the sample reached to 8.3 or 8.4, the color of phenolphthalein changed from colorless at the acidic condition to pink in alkaline condition. The acidity is expressed in the form of percent lactic acid.

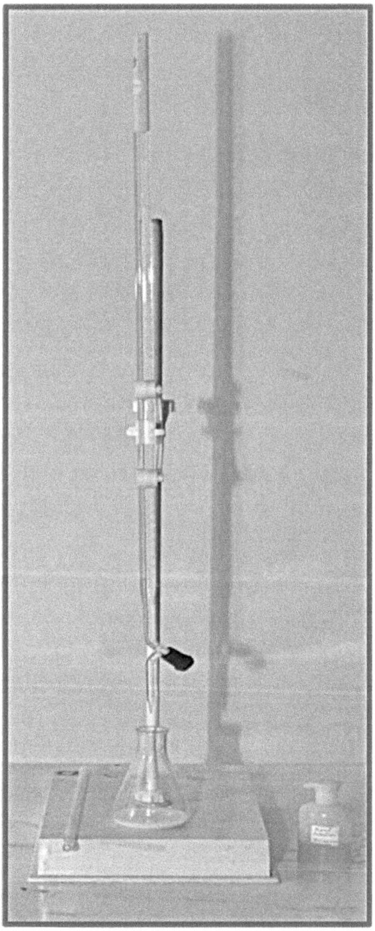

FIGURE 4.6 Assembly for determination of acidity.

4.6.3 MATERIALS

4.6.3.1 REAGENTS

- NaOH (0.1 N).
- Phenolphthalein solution (0.5%): 1 g phenolphthalein is dissolved in 100 ml ethyl alcohol solution (95%). Add 0.1 N NaOH until faint pink color observed. Dilute the solution up to 200 ml by adding water.
- Rosaniline acetate solution (stock solution): 0.12 g Rosaline acetate is dissolved in approx. 50 ml of rectified spirit containing 0.5 ml of glacial CH_3COOH. Volume is made to 100 ml with spirit.
- Bench solution: 1 ml of stock solution is diluted to 500 ml with the spirit and water mixture (1:1 ratio).
- Note: Store the stock and bench solution in dark brown color bottle.

4.6.3.2 APPARATUS

- White porcelain dish: 60 ml capacity;
- Pipette: 10 ml;
- Burette;
- Measuring cylinder (25 ml);
- Glass rod.

4.6.4 PREPARATION OF SAMPLE

According to the procedure described in Section 4.4.1 in this chapter.

4.6.4.1 EXPERIMENTATION

Take 10 ml sample in two separate porcelain dish and dilute it by adding 10 ml water (previously boiled and cooled). Add 1 ml indicator to one dish and add 1 ml Rosaline acetate bench solution to another dish. Titrate the sample containing phenolphthalein against sodium hydroxide (NaOH) until the color matches the pink color of the solution in the dish having Rosaline acetate. Titration should be completed within 20 seconds.

4.6.5 CALCULATION

$$Acidity\ \% = \frac{9NV_1}{V_2} \tag{5}$$

where; N is the normality of NaOH; V_1 is the titer value (ml) of 0.1 N NaOH used for sample; and V_2 is the volume of milk taken in ml.

4.6.6 DATA COLLECTION SHEET

Title: Determination of acidity of milk by the titrimetric method.

Location and Organization:______________________________

Participants: ____

Date (Month/Day/Year):_________________________________

Observations:

Parameter	Repeat 1	Repeat 2	Repeat 3	Average
Volume of the sample (ml)				
Volume of 0.1 N NaOH (ml)				
Acidity (% lactic acid)				

Comments:____.

4.7 DETERMINATION OF SPECIFIC GRAVITY OF MILK USING SPECIFIC GRAVITY BOTTLE AND PYCNOMETER

4.7.1 INTRODUCTION

Mass per unit volume is defined as the density of a substance at a specified temperature. It is expressed in g/ml at a given temperature. The density, ρ, is known as the elementary physical property of matter. Density of a substance gives idea about how closely its particles are packed. If density is higher, it means particles are more closely packed. Ratio of mass by volume of homogenous liquid is defined as density [1, 4] as follows:

$$\rho = \frac{m}{v} \tag{6}$$

SI unit of density is kg/m^3 or g/cm^3 ($1\ g/cm^3 = 1{,}000\ kg/m^3$).

During heating, the volume of an object increases due to the expansion of the object. As per the Eqn. (6), object's density is affected by its temperature. A rise in temperature leads to an increase in volume and decrease in the density. It is also dependent upon the temperature history of material, temperature of material at the time of measurement, inclusion of air in the viscous fluids and material and composition of milk (mainly affected by % fat content). Density of milk ranges from 1,027 to 1,033 kg/m^3 at 20°C [4].

Specific gravity (SG) is the ratio of the density or weight of a substance to the density or weight of distilled water at a similar temperature. Thus, it has no units and distilled water is normally used as a reference substance (specific gravity of water is one at 4°C). Therefore, the addition of water to milk decreases the milk SG. Fat has lower density as compared to milk or water and rise in the fat content leads to a decrease in milk SG. Temperature also has a major influence on the milk SG. SG decreases with rise in temperature, which might be due to the hydration of protein at higher temperature.

SG of milk can be measured by pycnometer, SG bottle or various types of hydrometers specifically lactometer. Pycnometer and SG bottle have fixed volume. Weight of fixed volume of liquid (milk) and distilled water is measured. Density of liquid (milk) is compared with the density of distilled water and specific gravity is calculated. During analysis temperature of both the fluid should be at 20°C [4].

4.7.2 PRINCIPLE

Water at 4°C is taken as a standard substance for the solids and liquids because that has a unique density at that temperature. The density of milk is calculated by directly measuring a particular volume of milk taken in pycnometer/SG bottle and by determining the volume taken. Similarly, SG of milk is estimated by taking the ratio of weights of a particular volume of milk to the same volume of distilled water at the same temperature taken in a pycnometer/SG bottle [4].

4.7.3 MATERIAL

- Specific gravity bottle or any other type of pycnometer;
- Analytical balance;
- Thermometer.

4.7.4 PROCEDURE

4.7.4.1 PREPARATION OF SAMPLE

According to the procedure described in Section 4.4.1 in this chapter.

4.7.4.2 EXPERIMENTATION

Clean the SG bottle/pycnometer and dry it cautiously with a current of cold dry air. Take the weight of the SG bottle/pycnometer with the stopper. Fill SG bottle/pycnometer with milk and put the stopper, allow overflowing some milk; taking care that no air bubbles should be present in the milk. Clean the milk from outside the bottle with the help of cotton and take the weight of SG bottle/pycnometer with milk. Empty the SG bottle/pycnometer, wash it thoroughly with detergent, diluted acid and distilled water. Later fill it with distilled water without the formation of air bubbles. Wipe off water from outside the bottle and take a weight accurately. Take a note of the milk and water temperature [4].

Note: SG bottle/pycnometer should not be heated to dryness.

4.7.5 CALCULATION

$$\text{Relative specific gravity of milk at t°C} = [(W_1 - W)/(W_2 - W)]$$

$$\text{Absolute specific gravity of milk at t°C} = [(W_1 - W)/(W_2 - W)] \times d_w \quad (7)$$

where; W is the empty weight of the pycnometer/SG bottle (g); W_1 is the weight of milk at t°C occupying the volume of the pycnometer/SG bottle plus empty weight (g); and W_2 is the weight of water at t°C occupying the volume of pycnometer/SG bottle plus empty weight (g).

4.7.6 DATA COLLECTION SHEET

Title: Determination of specific gravity milk using specific gravity bottle or pycnometer.

Location and Organization:________________________________

Participants: ______________________________________

Date (Month/Day/Year):_________________________________

Observations:

Parameters	Repeat 1	Repeat 2	Repeat 3	Average
Weight of empty SG bottle/pycnometer with stopper (W), g				
Weight of SG bottle with stopper and milk at t°C (W_1), g				
Weight of SG bottle with stopper and distilled water at t°C (W_2), g				
Density of water at t°C (d_w)				
Relative specific gravity of milk at t°C				
Absolute specific gravity of milk at t°C				

Comments:____.

4.8 DETERMINATION OF SPECIFIC GRAVITY OF MILK USING LACTOMETER

4.8.1 INTRODUCTION

Lactometer and hydrometers are based upon the Archimedes principle, which states that if an object is partially or fully immersed in a liquid then force due to buoyancy is equivalent to the weight of the liquid displaced by that object. Volume displaced by the standard weight object can be used for calculating the weight per unit volume [18, 20].

Hydrometer is an instrument used to measure the relative density of a liquid consisting of a sealed graduated tube with standard weight in the bulb portion. As it immerses in a liquid, it displaces the weight of liquid equivalent to its own weight. Hydrometer sinks to a more depth in the low-density liquid as compared to high density liquid [20]. Results obtained by hydrometer are not as accurate as pycnometer or SG bottle; however, the time required for the analysis is very less which is a crucial factor. The SG of milk can be conveniently measured by means of the lactometer, which is a special type of hydrometer calibrated over the range 1.020–1.035 (25.0–35.0 as lactometer degrees in case of ISI type) or 0.000–0.040 (0–40.0 as lactometer degrees in case of zeal type). Lactometers are used for rapid determination of SG [4].

Quevenne, zeal, and ISI lactometers are used to determine milk SG at 15.5°C (60°F), 29°C (84°F) and 27°C (80.6°F), respectively [2]. The sensitivity of the lactometer relies on the selective size of the stem to that of the body. A lactometer with a small stem and large body shall give more accurate results than one with a large stem and small body. Measurement of the lactometer reading at a temperature below milk fat melting point leads to inaccurate results [19].

4.8.2 PRINCIPLE

Lactometers are used for rapid determination of SG and commonly used at milk processing units and dairies. Law of floatation principle is used for this method, according to which the free-floating body immersed in a liquid displaces the same quantity of liquid as the weight of the floating body. Lactometers are variable immersion type hydrometers. They are calibrated before handle with liquids of known specific gravity so that they can be used for directly reading the specific gravity of a liquid. In liquid of low specific gravity, the lactometer sinks more and vice-versa [18, 20].

4.8.3 MATERIALS

4.8.3.1 APPARATUS

- Lactometer;
- Thermometer;
- Lactometer jar or cylinder;
- Beaker.

4.8.4 PROCEDURE

4.8.4.1 PREPARATION OF SAMPLE

Bring the temperature of the milk to about 26 to 28°C and mix it properly to get a homogeneous mixture. If it contains lumps of fat then warm the sample to 40°C, cool it to 26 to 28°C after properly mixing. Then bring the sample temperature to 27°C before pipetting [10].

4.8.4.2 EXPERIMENTATION

Heat the milk sample to 40°C for 5 min. to liquefy the milk fat. Cool the milk and adjust the temperature as nearly as possible to that at which the lactometer is calibrated (84°F for Zeal lactometer). The milk should be thoroughly and gently mixed until a homogeneous mixture is obtained, but at the same time avoid air bubble formation in milk. Pour enough milk into the lactometer jar or cylinder. Some milk should overflow when lactometer is inserted. Insert the lactometer gently to wet the stem not more than a short length, about 3 mm beyond the position of equilibrium. Let the lactometer float freely, but not touching the cylinder side. Add the milk to the brim of the jar/cylinder. Allow the lactometer to come to rest in the milk and note the reading at the top of the meniscus within about 30 seconds. Record the temperature of the milk [4, 19].

4.8.4.3 PRECAUTIONS

- Lactometer should be clean and dry;
- No air bubble should be sticking to the lactometer;
- The diameter of the cylinder should be at least 1 inch greater than the largest diameter of the lactometer;
- Lactometer should float in the liquid without touching the sides of the cylinder in which they float.

4.8.5 CALCULATION

$$\text{Specific gravity of milk (at 84°F)} = 1 + \frac{CLR}{1000} \tag{8}$$

where; LR is the observed lactometer reading (at a given temperature); CLR is the corrected lactometer reading (at 84°F) = observed LR + correction factor.

4.8.6 DATA COLLECTION SHEET

Title: Determination of specific gravity of milk using lactometer.

Location and Organization:________________________________

Participants: ____

Date (Month/Day/Year):__________________________________

Observations:

Parameters	Repeat 1	Repeat 2	Repeat 3	Average
Lactometer reading				
Temperature of milk				
Corrected lactometer reading				
Specific gravity of milk				

Comments:____.

4.9 DETERMINATION OF VISCOSITY OF MILK

4.9.1 INTRODUCTION

Viscosity plays a vital function in food systems and is an extremely important physical property of fluids. It is associated to the fluid flow property that governs many dairy industry production operations. Viscosity is defined as flow strength when either shear or tensile stress deforms the flow. It is a measure of the friction between molecules as they slide past one another. Internal frictional forces decrease the rate of flow. It is also defined as a "thickness" or "internal friction." Thus, thickness is directly proportional to viscosity and vice-versa. The viscosity of milk depends upon its nature, composition, and the physical state of its colloidally dispersed substances, including milk fat [4].

The viscosity of milk and milk products regulates the flow and heat and mass transfer in various dairy processes. Major constituent contributing to the viscosity of milk are milk protein and fat, however, lactose and salts contributes to a minor extent. In addition to their concentration, proteins often influence milk viscosity through their number of particles, size, and degree of hydration. Size of fat globule and degree of their clustering also influences the viscosity of milk. Viscosity is commonly expressed as poise or centipoise (cP) [8]. Poise is the unit of dynamic viscosity, which is equivalent to dyne.sec/cm^2. It is defined as the force needed between two parallel planes a centimeter apart to maintain a relative speed of 1 cm/s. Dynamic viscosity is denoted with a symbol neta (η) and kinematic viscosity with symbol nu (υ). The ratio between dynamic viscosity and its density (ρ) is defined as kinematic viscosity as follows:

$$\upsilon = \eta / \rho \tag{9}$$

SI unit of dynamic viscosity is Pascal.sec (or Pa.s), which is equivalent to kg per m-s (Kg/m-s). In CGS unit, dynamic viscosity is expressed as poise or centipoise (cP), which is widely used. The viscosity of water is almost one centipoise (1.0020 cP) at 20°C. Conversion factors are as follows:

1 poise = 100 centipoise = 1 $gcm^{-1}s^{-1}$ = 0.1 Pa.s

1 centipoise = 1 mPa.s (milli Pascal sec)

Kinematic viscosity is expressed in m^2s^{-1}. In honor of George stoke, widely used unit for kinematic viscosity is stoke (St). Conversion factors are:

1 stokes = 100 centistokes = 1 cm^2s^{-1} = 0.0001 m^2s^{-1}

4.9.2 PRINCIPLE

Ostwald viscometer (Figure 4.7) is used for determining the viscosity of Newtonian fluids. In 1844, Hagen–Poiseuille has worked concerning to the flow of liquid through a tube and based upon his work, Poiseuille's equation was derived which express the viscosity of Newtonian liquid in terms of time of flow, volume of liquid, hydrostatic pressure and distance traveled by the liquid.

$$\eta = \frac{\pi r P t}{8VL} \tag{10}$$

where; η is the viscosity; t is the time in second for flow of liquid; P is the hydrostatic pressure; V is the volume of liquid; and L is the distance moved by the liquid in time t.

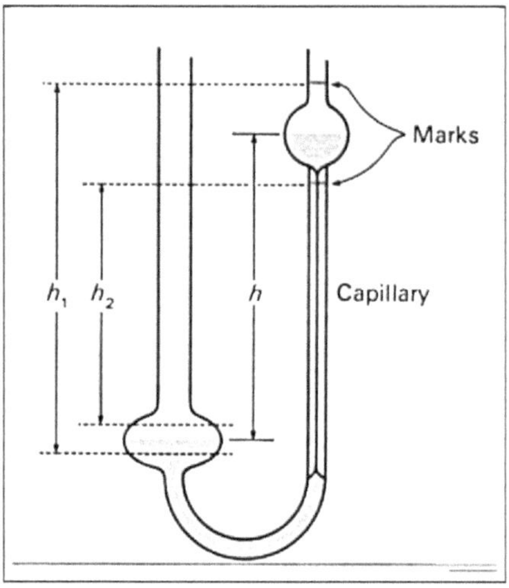

FIGURE 4.7 Ostwald viscometer.

Ostwald viscometer is a type of capillary viscometer, which has a U-shape glass tube. Viscometer is held vertically in the controlled temperature water bath for controlling the temperature of fluid. Two marks are etched on the Ostwald viscometer and a fixed volume of liquid is allowed to flow from one point to another point. From Eqn. (10), viscosity can be expressed as:

$$\eta = KPt \tag{11}$$

where; $K = \pi r^4 / 8VL$

In Ostwald viscometer, internal volume and radius of a viscometer is always constant. As per the markings, distance traveled from one point to another point is also constant; therefore, K is a constant for a specific Ostwald viscometer. Hydrostatic pressure is directly proportional to the fluid density, and we can measure the mass of the fixed volume using either pycnometer or specific gravity bottle.

$$\eta\ a\ \mathrm{mt} \tag{12}$$

4.9.3 MATERIALS

4.9.3.1 APPARATUS

- Ostwald viscometer (1–3 cP);
- Stand with clamp to fix the viscometer;
- Thermometer with graduation of 1°C;
- Stopwatch;
- Water bath maintained at 20°C±0.1°C;
- Pipettes (10 or 25 ml);
- Hair dryer.

4.9.4 PROCEDURE

4.9.4.1 PREPARATION OF SAMPLE

Bring the temperature of the milk to about 26 to 28°C and mix it properly to get a homogeneous mixture. If it contains lumps of fat then warm the sample to 40°C, cool it to 26 to 28°C after properly mixing.

4.9.4.2 EXPERIMENTATION

Clean the viscometer with NaOH (1%) solution, tap water, chromic acid and tap water, and with distilled water four or five times, followed by drying with a dust-free air current. In the constant temperature bath of 20°C (±0.1°C), fix a clean, dry viscometer and keep the marking above the bulb beneath the surface of water. Fill in about two-third of viscometer reservoir bulb by transferring a fixed volume of distilled water of 10–25 ml using a pipette, as needed. Permit the viscometer to stand at 20°C (±0.1°C) for 15 min. Connect the soft rubber tubing to the viscometer arm holding the marked bulb and draw the distilled water over the upper mark. Cover the rubber tube with your fingers and release the water pressure slowly. Start the stopwatch when the water meniscus leaves the upper mark and calculates the time to hit the lower mark.

Repeat steps for a minimum of 5 times with water and note down the correct time (in seconds) in an observation table. Remove the viscometer, drain the water and dry it with a dust-free air current. Repeat the same procedure for milk samples by replacing the water with milk. Record the observations (time) for minimum 5 times in the table. As may be the case, calculate the density or mass for the fixed volume of the milk sample with respect to the distilled water used in the tests at 20°C using pycnometer or specific gravity bottle [3].

The scale and error sources in Ostwald viscometer are as follows:

- Loading errors arising because the driving fluid head depends on how much liquid the viscometer contains. If too much liquid is loaded into the device, the lower reservoir level is too high, which reduces the drive head by that amount.
- Kinetic energy corrections mainly due to losses of contraction and expansion at the capillary entrance and exit.
- An apparent increase in capillary length due to the propensity of the fluid to maintain the form of the capillary for a finite distance through the fluid medium, as it emerges from the capillary.
- Drainage errors due to the lack of easy drainage of all liquids from a surface.
- Errors arising from incorrect viscometer alignment.
- Errors attributable to temperature fluctuations and gravity constants.
- However, the most significant error occurs when the viscometer is applied to non-viscous liquids in the flow.

4.9.5 CALCULATION

If viscosity of a standard fluid is known then by finding out the time required for traveling a specific distance by the standard fluid and liquid under analysis, the viscosity of liquid under analysis can be calculated as follows [17]:

$$\eta/\eta_r = mt/m_r t_r$$

and

$$\text{Viscosity of milk } (\eta) = (mt\eta_r)/m_r t_r \qquad (13)$$

where; η is the viscosity of sample or milk; η_r is the viscosity of reference liquid or water; m is the mass of the sample or milk; m_r is the mass of reference liquid or water; t is the time of flow for sample, or milk; t_r is the time of flow for reference liquid or water.

4.9.6 DATA COLLECTION SHEET

Title: Determination of viscosity of milk using Ostwald viscometer.

Location and Organization:____________________________

Participants: ____________________________________

Date (Month/Day/Year):______________________________

Observations:

Parameters	**Repeat 1**	**Repeat 2**	**Repeat 3**	**Average**
Mass of fixed volume of distilled water (m_r), g				
Mass of fixed volume of milk (m), g				
Time of flow for distilled water (t_r), s				
Time of flow for milk (t), s				
Viscosity of distilled water at t°C (η_r), cP				
Viscosity of milk at t°C (η), cP				

Comments: ______________________________________.

KEYWORDS

- **density**
- **freezing point**
- **pH**
- **refractive index**
- **specific gravity**
- **surface tension**
- **titratable acidity**
- **viscosity**

REFERENCES

1. Fox, P. F., Uniacke-Lowe, T., McSweeney, P. L. H., & O'Mahony, J. A., (2015). *Dairy Chemistry and Biochemistry* (pp. 321–344). Switzerland: Springer International Publishing.
2. Mathur, M. P., Datta, R. D., & Dinakar, P., (2005). *Textbook of Dairy Chemistry* (pp. 19–21). New Delhi: Indian Council of Agricultural Research (IARI).
3. Thakur, M., & Wadhwa, H., (2017). Experimenting with stalagmometer and viscometer on day-to-day drinking liquids. *International Journal of Applied Research, 3*(7), 478–481.
4. Jenness, R., & Patton, S., (1959). *Principles of Dairy Chemistry* (p. 445). New York: John Wiley & Sons, Inc.
5. IDF. *IDF-108 (E). Milk — Determination of Freezing Point — Thermistor Cryoscope Method (Reference Method)* (3rd edn., pp. 1–17). International Dairy Federation (IDF) Bulletin; (2009).
6. ISO. *ISO-5764 (E). Milk — Determination of Freezing Point — Thermistor Cryoscope Method (Reference Method)* (3rd edn., pp. 1–17). Geneva – Switzerland: International Organization for Standardization (ISO); (2009).
7. Anonymus. *Abbe Refractometer*, 9. https://fac.ksu.edu.sa/sites/default/files/abbe_experiment.pdf (accessed on 18 January 2022).
8. Sherbon, J. W., (1988). Physical properties of milk. Chapter 8. In: Wong, N. P., Jenness, R., Keeney, M., & Marth, E. H., (eds.), *Fundamental of Dairy Chemistry* (pp. 409–460). New York –USA: Aspen Publishers, Inc.
9. Rangappa, K. S., (1947). Studies on refractive index of milk. Part I observation on genuine samples. *Proceedings of the Indian Academy of Science-Section B, 25*, 86–94.
10. IS, (1961). *IS 1479-2: Method of Test for Dairy Industry, Part II: Chemical Analysis of Milk* (pp. 12–14). New Delhi: Bureau of Indian Standards.
11. Atherton, H. V., & Newlander, J. A., (2003). *Chemistry and Testing of Dairy Products* (4th edn., pp. 246–267). New Delhi – India: CBS Publishers & Distributors.

12. Caulfield, W. J., & Riddell, W. H., (1936). Some factors influencing the acidity of freshly drawn cows' milk. *Journal of Dairy Science, 19*(4), 235–242.
13. Fox, P. F., & McSweeney, P. L. H., (1998). *Dairy Chemistry and Biochemistry* (1st edn., pp. 437–461). London–UK: Blackie Academic and Professional.
14. IS. SP 18. Rapid examination of milk. In: *ISI Handbook of Food Analysis (Part XI) Dairy Products* (pp. 5–20). New Delhi–India: Bureau of Indian Standard; (1981).
15. Sommer, H. H., & Menos, J., (1931). The effect of dilution on the titratable acidity of cows' milk. *Journal of Dairy Science, 14*, 2–136.
16. Walstra, P., & Jenness, R., (1984). *Dairy Chemistry and Physics* (pp. 186–210). New York: John Wiley & Sons.
17. *Determination of Viscosity of Organic Solvent* (p. 2). https://vlab.amrita.edu/?sub=2&brch=190&sim=339&cnt=1#:~:text=Ostwald%20viscometer%20is%20a%20commonly,shaped%20glass%20tube%20held%20vertically.&text=It%20is%20also-%20known%20as,is%20measured%20using%20a%20stopwatch (accessed on 18 January 2022).
18. Wright, J. D., Bean, V. E., & Aguilera, J., (2020). *NIST Special Publication 250-78 r1: NIST Calibration Services for Hydrometers* (pp. 1–10). Gaithersburg, MD – USA: National Institute of Standards and Technology.
19. IS. *IS 9585: Lactometers* (pp. 1–14). New Delhi: Bureau of Indian Standards; (1980).
20. Lorefice, S., & Malengo, A., (2006). Calibration of hydrometer. *Measurement Science and Technology, 17*, 2560–2566.

CHAPTER 5

DETERMINATION OF THE QUALITY OF MILK

SNEHA SHENDE, CHITRA GUPTA, N. VEENA, SUVARTAN RANVEER, and PRAVIN SAWALE

ABSTRACT

Quality control of milk by performing various tests to screen its quality is considered as an essential component of the milk-based processing industry. Throughout the world, milk is being adulterated by unscrupulous and unfaithful workers with various additives, preservatives, or subtraction of one of the important components in milk. It is also important to check the quality during the processing of milk and further convert it into milk products. Moreover, milk is highly nutritious as it comprises different components such as fat, protein, lactose, minerals, vitamins, etc. Therefore, there are more chances for the rapid multiplication of microorganisms, and majorly it happens under unhygienic production and storage practices at room temperature. Milk handlers will only be assured of the quality of raw milk when the quality of raw milk is screened with various basic tests during the stage of milk's transportation from farm to fork and fork to the consumer. This chapter mainly focuses on the quality tests for milk.

5.1 ORGANOLEPTIC TEST

5.1.1 INTRODUCTION

Basically, milk consumer judges the quality of the milk and milk product by the flavor, color, and appearance of milk. Thus, organoleptic test is used to determine poor or good quality of milk very quickly. This method is

Analytical Methods for Milk and Milk Products, Volume 1: Sampling Methods and Chemical and Compositional Analysis. Megh R. Goyal, N. Veena, and Santosh Kumar Mishra (Eds.)

quite simple, rapid, and cost-effective method. To perform the experiment, an experienced person must have required with good sense of sight, smell, and taste and reject the milk which may possess objectionable flavor with a high degree of accuracy. Visual appearance of milk is an important factor to determine the changes in color or containing any foreign material on the surface of milk [1].

5.1.2 MATERIALS

No special equipment and testing reagents are required for this experiment.

5.1.3 PROCEDURE

Open the cover or lid of the milk can and smell the milk immediately after opening of cover lid of milk can so as to perceive any off flavor, before it escapes out. Further, stir the received milk thoroughly while smelling to confirm the sensation of smell. Chances of acidity development may occur due to unhygienic practices during milking and production process. Also off-flavor may also be developed due to type of feed provided, exposing of milk with foreign material might develop off odors after milking, metallic contamination caused oxidized flavor due to exposure of milk to untinned container. If milk graders perceived such type of abnormality in flavor and color to milk, then milk should be rejected or held for further confirmatory tests.

Temperature measurement of milk is also another factor in organoleptic test which is used to determine the bacterial quality of milk especially during transit of milk from chilling to the main processing plant. Raising of milk temperature leads to the growth and multiplication of the microflora of milk. Therefore, milk should be immediately and rapidly cooled to a temperature of not greater than 7°C, before it reaches to main processing plant. This can improve the bacteriological quality of milk [1].

5.1.4 DATA COLLECTION SHEET

Title: Organoleptic test and temperature of milk.

Location and Organization: ____________________________________

Participants: ____

Date (Month/Day/Year):____________________________________

Observations:

Parameters	Remarks
Smell	
Appearance of color	
Extent of contamination of extraneous matter	
Abnormal flavor (such as: acidic, metallic, putrid, rancid, etc.)	
Temperature of milk (°C)	

Comments: ____.

5.2 ALCOHOL TEST

5.2.1 INTRODUCTION

Fresh milk gets coagulated due to acidity development, disturbed salt balance and the presence of albumin and globulin in higher amount. Alcohol test is applied for fresh milk to check the stability of milk to processing and converting into dairy products mainly evaporated milk (EM), UHT milk, and milk powder. Milk being poor quality, it has a tendency to curdle milk due to mastitis milk, high salt content, and colostrum containing high level of albumin usually gives positive test.

5.2.2 PRINCIPLE

It is used to determine the heat stability of milk at thermal processing treatment. Coagulation of milk is initiated due to instability of protein because of acidity development, disturbance of salt balance ratio or high level of albumin which results in positive test and milk then should be rejected [1].

5.2.3 MATERIALS

5.2.3.1 REAGENTS

- Ethanol (0.8675 g/ml density at 27°C): 75% by volume or 68% by weight basis.

5.2.3.2 APPARATUS

- Test tubes;
- Pipette: capacity of 5 ml.

5.2.4 PROCEDURE

Measure exactly five milliliter of test milk sample in a pipette and it is transferred into the test tube. Then, add an equal amount of ethyl alcohol. Shake the tube; and observe the flakes or curd formation on the side of the test tube.

5.2.5 DATA COLLECTION SHEET

Title: Alcohol test.

Location and Organization:___________________________________

Participants: ___________________________________

Date (Months/Day/Year): ___________________________________

Observations:

Results	**Remarks**
No coagulation or flake formation	Indicates low acidity of milk/milk with good heat stability.
Flakes formation or curd on the sides of the test tube	Indicates abnormal milk and not suitable for the manufacture of EM or UHT milk.

Comments:___________________________________.

5.3 ALIZARIN-ALCOHOL TEST

5.3.1 INTRODUCTION

This test is usually performed for the same purpose as of alcohol test; however, it resulted in more confirmatory test of whether milk is acidic or alkaline. The quality of milk is judged on the basis of curd flakes formation due to coagulation and color of the mixture after reacting to the alizarin alcohol. Alcohol alizarin acting as an indicator which owes to changes the color according to developed acidity and flakes formation into the milk.

5.3.2 PRINCIPLE

It is used to assess the heat stability of milk at thermal processing treatment. Coagulation of milk is initiated due to instability of protein because of acidity development; disturbance of salt balance ratio or high level of albumin which results in positive test and milk then should be rejected.

5.3.3 MATERIALS

5.3.3.1 REAGENT

- Alcohol-alizarin (0.2% w/v) solution: 0.2 g of alizarin is dissolved in 75% by volume or 68% by weight of ethanol having density 0.8675 g/ml at 27°C.

5.3.3.2 APPARATUS

- Test tube (18×150 mm);
- Pipette: capacity of 10 ml.

5.3.4 PROCEDURE

Measure five milliliters of test milk sample in pipette and transfer into test tube. Add same amount of prepared 0.2% alcohol-alizarin solution into test milk sample. Further, shake well by inverting the test tube and observe the test sample for its flake's formation as well as color development [1].

5.3.5 DATA COLLECTION SHEET

Title: Alcohol-alizarin test.

Location and Organization: ______________________________

Participants: ______________________________

Date (Month/Day/Year): ______________________________

Observations:

Appearance of Color	Flake Size	Range of Titratable Acidity (% Lactic Acid)	Remarks
Lilac	No flakes formation	Up to 0.14	Normal milk
Pale red	No flakes formation	0.14–0.17	Slightly acidic milk
Brown to reddish brown	Smaller flakes	0.17–0.20	Slightly high acidic milk
Brownish yellow to yellow	Larger flakes	Over 0.20	Highly acidic milk

Comments:____.

5.4 CLOT ON BOILING (COB) TEST

5.4.1 INTRODUCTION

High acidity in milk is due to heavy bacterial action and disturbance of salt content leads to coagulation when heat treatment is allowed at the boiling temperature. This test is quite simple and rapid to judge the quality of milk based on developed acidity of milk. If the milk sample contains less than pH 5.8 or poor milk stability or abnormal milk (such as mastitis milk), then it will not withstand high thermal heat treatment and must therefore be rejected.

5.4.2 MATERIALS

5.4.2.1 APPARATUS

- Test tube (18×150 mm);
- Test tube holder;
- Spirit lamp or gas flame;
- Boiling water bath.

5.4.3 PROCEDURE

Collect representative sample of milk after sampling. Pipette out five milliliters of milk into a test tube. To check the susceptibility of milk, keep the tube in a boiling water bath for five minutes or provide direct heat treatment to the milk by keeping over spirit lamp or on a gas till boiling. Further, take out the tube from the water bath and observe the milk sample by smelling if any acidic flavor found as well as clot or precipitated particles on the sides of the test tube after mixing. The presence of clots indicates COB (clot-on-boiling) positive test [1].

5.4.4 DATA COLLECTION SHEET

Title: Clot on boiling test.

Location and Organization: ______________________________

Participants: ______________________________

Date (Month/Day/Year): ______________________________

Observations:

COB Test	Interpretation
Positive test	Appearance of any clots formation or precipitated particles due to highly developed acidity (> 0.17% lactic acid) and thus milk is not fit for further processing.
Negative test	No clots or flakes formation

Comments: ______________________________

5.5 SEDIMENT TEST

5.5.1 INTRODUCTION

The sediment test is used to represents the cleanliness of milk w.r.t. visible dirt. Visible foreign materials into milk are an objectionable for several reasons. During receiving of milk at the chilling center or at the collection center of the processing plant, many visible insoluble materials have not been removed by using single straining of milk. Presence of any visible foreign matter in raw milk indicates the carelessness in handling and lack of sanitary dairy practices may be carried out during procurement of milk. This test is a simple, rapid, and quantitative measure for revealing the extent of visible dirt has entered into milk or the extent of visible insoluble materials that have not been separated from milk by strainers. Best ideal condition of cleanliness during milk production is expressed with the absence of visible dirt or sediments into milk.

5.5.2 PRINCIPLE

A measured amount of fresh milk is passed through the fixed area of a filter disc. The collected sediment over filter disc is compared with standard sediment discs.

5.5.3 MATERIALS

5.5.3.1 APPARATUS

- Long barrel suction type sediment tester (diameter of filtering surface is 2.5 cm);
- White lintine cotton disc (32 mm diameter; area for filtration is 28 mm diameter);
- Sampling dipper (500 ml capacity);
- Standard sediment disc ratings (representing 0.0, 0.2, 0.5, 1.0, 2.0, 3.0 mg or higher per 500 ml of milk);
- Plunger.

5.5.4 PROCEDURE

Weigh 500 ml of milk in a glass beaker accurately from well mixed sample of freshly dumped milk in the weighing vat and simultaneously raising the plunger in both upward and downward movement. Before filtration of milk, warm up the milk up to 30–40°C by keeping beaker over a water bath or keep on gas flame with gentle heating for uniformly distributing of fat throughout the mass. If the colder milk is used, it will tend to clog the sediment discs while filtration. Then, fix the suction type of sediment tester with white lintine cotton disc filter (rough side facing milk). Due to suctioning action of sediment tester, milk will be forcefully filtered through the exposed area (28 mm in diameter) of the disc and retain sediment on filtered disc. After filtration, remove the disc from the tester and place the disc over suitable moisture absorbing cardboard. Allow it to dry properly. Then, note down the retention amount of sediment over the disc and compared it with the standard disc [1].

5.5.5 DATA SHEET COLLECTION

Title: Sediment test.

Location and Organization: ______________________________

Participants: ______________________________

Date (Month/Day/Year): ______________________________

Observations:

Quality (Grading) of Milk	Sediment Content (mg) in Prepared Standard Discs for Specific Requirement of dairies
Excellent	0.0
Good	0.2
Fair	0.5
Bad	1.0
Very bad	2.0

Comments: ______________________________.

5.6 ADULTERANTS: DETECTION OF STARCH

5.6.1 INTRODUCTION

Adulterants are commonly added in milk with numerous purposes. Mostly, it is added in milk for gaining of profit. Basically, the addition of any exterior agent to milk is strictly prohibited according to food laws, otherwise it leads to punishable. Starch or cereal flours are mostly available in the market at cheap rate. Being cheaper, it is generally added into milk by the unscrupulous person to make up density to prevent detection of added water or to increase the solids content.

5.6.2 PRINCIPLE

Iodine solution gives intense blue color with starch to form an iodine-starch complex.

5.6.3 MATERIALS

5.6.3.1 REAGENTS

- Iodine reagent: 2.6 g of iodine is dissolved in a sufficient amount of distilled water and mix it well by adding 3 g of KI. Volume made to 200 ml with water.

5.6.3.2 APPARATUS

- Test tube;
- Gas burner.

5.6.4 PROCEDURE

Measure three ml of test milk sample in a pipette and transfer into a test tube. Keep the test tube over a flame to bring the temperature up to boiling and allowed to cool at ambient temperature. Then, add 2–3 drops of prepared iodine reagent into milk sample. The presence of starch in the milk sample forms a complex with iodine and blue color can be seen visually. This color will be disappeared on boiling of milk sample and redevelop the same color again after cooling of milk sample [1].

5.6.5 DATA COLLECTION SHEET

Title: Detection of starch.

Location and Organization:______________________________

Participants: ______________________________________

Date (Month/Day/Year):________________________________

Observations:

Observation	**Interpretation**
Distinct blue color	Presence of starch
Yellow color	Absence of starch

Comments:____.

5.7 ADULTERANTS: DETECTION OF CANE SUGAR

5.7.1 INTRODUCTION

Cane sugar is generally added into milk for rising lactometer reading to prevent added water detection. Addition of cane sugar can be easily detected by using Seliwanoff's reagent.

5.7.2 PRINCIPLE

Resorcinol in acidic environment reacted to the cane sugar and it produces red color solution after heating the mixture for 5 min.

5.7.3 MATERIALS

5.7.3.1 REAGENTS

- Resorcinol;
- Concentrated hydrochloric acid (HCl).

5.7.3.2 APPARATUS

- Test tubes;
- Water bath: to maintain the temperature up to boiling.

5.7.4 PROCEDURE

Pipette out 15 ml of milk in a clean and dried test tube. Transfer 1 ml of HCl and 0.1 g of resorcinol into the tube and mix the contents thoroughly. Then keep tube over boiling water bath for five minutes. In the presence of acidic media and heat treatment, red color may develop if cane sugar is present in test milk sample. No such red color observed in the pure milk sample [1].

5.7.5 DATA SHEET COLLECTION

Title: Detection of cane sugar.

Location and Organization:________________________________

Participants: ____

Date (Month/Day/Year):___________________________________

Observations:

Observation	Interpretation
Red or pink color	Presence of cane sugar
No change in color	Absence of cane sugar

Comments:______________________________________.

5.8 ADULTERANTS: DETECTION OF UREA

5.8.1 INTRODUCTION

Urea is present in milk as a natural constituent which varies from 20 to 70 mg/100 ml. However, excess urea is used during preparation of synthetic milk to increase the concentration of non-protein nitrogen (NPN) content of milk. Generally, unauthorized man or unfaithful people prefer to prepare synthetic type of milk for gaining profit. However, it is essential to detect the excess amount of urea (> 70 mg/100 ml indicates added urea in milk), which may be hazardous to human.

5.8.2 PRINCIPLE

The presence of urea in milk sample forms a complex with para-dimethyl amino benzaldehyde (DMAB) in the presence of an acidic medium and it results in formation of yellow color at room temperature [1].

5.8.3 MATERIALS

5.8.3.1 REAGENTS

- Para-dimethyl amino benzaldehyde reagent (1.6%, w/v): 1.6 g of para-DMAB is dissolved in 90 ml of ethanol in a 100 ml volumetric flask. Make up remaining volume with concentrated HCl.

5.8.3.2 APPARATUS

- Test tube.

5.8.4 PROCEDURE

Measure five ml of test sample into test tube. Add equal amount of DMAB reagent to the test tube. Shake the test tube well and observe the color. Distinct yellow color in a milk sample shows the milk is adulterated with urea. However, slight yellow color represents the control sample owing to the presence of natural urea.

5.8.5 DATA COLLECTION SHEET

Title: Detection of urea.

Location and Organization:________________________________

Participants: ____

Date (Month/Day/Year):__________________________________

Observations:

Observation	Interpretation
Dark yellow color	Milk is adulterated with urea
Slight yellow color	Pure milk

Comments:____.

5.9 PRESERVATIVES: HEHNER TEST FOR FORMALDEHYDE

5.9.1 INTRODUCTION

As per Public Health Laws, milk should be free from any of preservatives. The formaldehyde is very poisonous chemical; and therefore, it should never be added to milk except in case of samples which have to be preserved for chemical analysis purpose. At room temperature, the milk can be kept for a longer time using formaldehyde without diminishing its quality. Its presence in the test sample is carried out as per Hehner test.

5.9.2 PRINCIPLE

Milk is a good source of tryptophan which is an important amino acid. This amino acid contains an indole group, which reacts with formaldehyde and gives characteristic color reaction in the presence of ferric chloride and oxidizing agent (concentrated H_2SO_4).

5.9.3 MATERIALS

5.9.3.1 REAGENTS

- Concentrated sulfuric acid (90%): specific gravity 1.84.
- Ferric chloride.

5.9.3.2 APPARATUS

- Test tube.

5.9.4 PROCEDURE

Transfer 10 ml of test sample into a test tube. Add 5 ml of concentrated sulfuric acid containing a small amount of ferric chloride into the test tube. Acid should be added slowly and carefully from the side of the test tube in such a way that separate layer is formed at the bottom. Formation of blue or violet color ring at the middle of two layers shows the presence of formaldehyde in the milk sample [1].

5.9.5 DATA COLLECTION SHEET

Title: Detection of formaldehyde.

Location and Organization:______________________________

Participants: ______________________________

Date (Month/Day/Year):______________________________

Observations:

Observation	Interpretation
Blue or violet color at the junction of two liquids	Presence of formaldehyde
No color is found at the junction	Pure milk

Comments:______________________________.

5.10 PRESERVATIVES: DETECTION OF HYDROGEN PEROXIDE (H_2O_2)

5.10.1 INTRODUCTION

Hydrogen peroxide (H_2O_2) is not permitted to add into milk sample. It lost in milk after heat treatment or storage for longer interval.

5.10.2 PRINCIPLE

In the presence of H_2O_2 in raw milk, the molecule of H_2O_2 is broken down into oxygen and water. Liberated oxygen reacts to paraphenylenediamine and develops a blue color.

5.10.3 MATERIALS

5.10.3.1 REAGENTS

- Para-phenylenediamine: Dissolve 2 g of para-phenylenediamine in enough distilled water and volume is made up to 100 ml using water.

5.10.3.2 APPARATUS

- Test tube;
- Pipette: 10 ml capacity.

5.10.4 PROCEDURE

Measure five ml of test milk sample in a pipette and transfer into test tube. Add raw milk in equal quantity. Then, add five drops of 2% paraphenylenediamine into milk sample. Further, mix the content well and observe the blue color development in the test sample by comparing with control. Take boiled milk as a control sample [1].

5.10.5 DATA COLLECTION SHEET

Title: Detection of hydrogen peroxide.

Location and Organization: ____________________

Participants: ____________________

Date (Month/Day/Year): ____________________

Observations:

Observation	Interpretation
Blue color formation	Presence of H_2O_2
No change in color	Pure milk

Comments:____.

5.11 DETECTION OF NEUTRALIZERS

5.11.1 INTRODUCTION

Lime water, sodium carbonate and bicarbonate are used as neutralizers. This neutralizer is added to milk to neutralize the developed acidity before milk is going to be processed into further products. However, such a practice of adding neutralizer to milk is not permissible.

5.11.2 PRINCIPLE

Rosalic acid is an indicator for pH range of 7.0 to 8.0. It changes color on addition to alkaline milk. This test works only in alkaline milk and when an excess amount of neutralizers is added in milk. However, this test fails to detect when added neutralizers are neutralized with sour milk below the normal pH of milk.

5.11.3 MATERIALS

5.11.3.1 REAGENTS

- Rosalic acid (0.1% w/v): 0.1 g of rosalic acid is dissolved in 100 ml of ethyl alcohol (95% v/v);
- Ethyl alcohol (95% v/v).

5.11.4 PROCEDURE

Transfer 5 ml of test sample into a test tube. Add 5 ml of prepared 95% alcohol into a test tube. Mix the content well by shaking. Then add two to three drops of Rosalic acid solution. Shake the contents again and observe the color in the test tube. The appearance of rose red color into milk sample indicates the presence of carbonates [1].

5.11.5 DATA COLLECTION SHEET

Title: Detection of neutralizer.

Location and Organization:______________________________

Participants: ______________________________

Date (Month/Day/Year):______________________________

Observations:

Observation	Interpretation
Rose-red color	Presence of neutralizer (carbonates)
Brownish color	Pure milk

Comments:______________________________.

5.12 DETECTION OF MASTITIS INFECTION

5.12.1 INTRODUCTION

Inflammation of the udder is one of the most common diseases in dairy cattle. It occurs due to infection of one or more quarters of the udder by certain species of Streptococci (mainly *S. agalactiae*) and staphylococci. The causative organism decreases the milk yield and may cause permanent damage to the injured quarters [1]. There are chances in alteration of the chemical composition of milk, which makes it unsuitable for further utilization into dairy products. Also, it causes economic loss. It is, therefore, essential to regularly examine all milking animals and even dry cows in a herd for the evidence of mastitis infection [1]. Suitable medical treatments should be provided to the infected animals [2].

5.12.2 PRINCIPLE

The normal pH of milk ranges from 6.5 to 6.8. Due to inflammation of udder, there is an increase in size of blood capillaries, which permit alkaline blood elements, such as, sodium, chloride, and bicarbonate ions to enter the milk. Therefore, there is an increase in the alkalinity of milk or the pH value of milk up to 7.4. This test for mastitis infection is detected by observing the changes in color by adding suitable pH indicator to the milk.

5.12.3 MATERIALS

5.12.3.1 REAGENTS

- Bromothymol blue (BTB) indicator: 0.5 g of BTB powder is dissolved in 80 ml of 0.01 N NaOH solution. Further dilute this solution with 295 ml of water.

5.12.3.2 APPARATUS

- Test tube;
- Pipettes.

5.12.4 PROCEDURE

Measure five milliliter of test sample in a pipette and transfer it into a test tube. Add one ml of BTB indicator into a test tube. Mix well and observe the color.

5.12.5 DATA SHEET COLLECTION

Title: Detection of mastitis infection.

Location and Organization:____________________

Participants: ____________________

Date (Month/Day/Year):____________________

Observations:

pH Range	Color Indication	Interpretation
6.4–6.8	Yellow color	Normal milk
7.0–7.4	Blue or greenish blue color	Mastitis milk

Comments:____________________.

5.13 DETECTION OF MELAMINE IN MILK

5.13.1 INTRODUCTION

Adulteration of milk with melamine is a new threat to the milk quality. Melamine is a synthetic compound, containing a rich source of non-protein nitrogen (NPN). Melamine adulteration of dairy products is illegal, even if the amount is much lower than the MRL (maximum residue limit). It may cause kidney failure in infants. Thus, the presence of melamine in milk and milk products indicates either adulteration or contamination. As melamine represents NPN content of 66%, therefore, its addition into milk and dairy products can misleadingly increase the protein content, when determined by Kjeldahl method.

5.13.2 PRINCIPLE

The homogenous sample should be used and optionally reconstituted in case of the powdered sample. To precipitate protein in milk, a suitable solvent is added, and melamine is extracted from the matrix. Supernatant is collected after centrifugation and further analyzed using electrospray ionization (ESI) liquid chromatography (LC)-tandem mass spectrometry (LC-MS/MS). Based on chromatographic separation, hydrophilic interaction-LC ensures better separation of melamine followed by ionizing the substance is attained by ESI and operates in selected reaction monitoring (SRM) mode. Quantification is based on isotope dilution approach using labeled ($^{13}C_3$, $^{15}N_3$)-melamine as internal standard (IS) [3].

5.13.3 MATERIALS

5.13.3.1 REAGENTS

Analytical grade reagents should be used. Similarly, demineralized or distilled water should be used:

- Water: for chromatography use.
- Acetonitrile (CH_3CN): GR grade for liquid chromatography.
- CH_3CN and H_2O (ratio 70:30) solution.
- Ammonium acetate: GR for analysis.
- Melamine ($C_3H_6N_6$): >99% mass fraction.
- Melamine stock solution (MSS, 0.25 mg/ml): Weigh accurately 62.5 mg of melamine into a beaker and dissolve it in sufficient water (i.e., up to 100 ml distilled water). Finally made the volume to 250 ml using water. Mixing of content is carried out using a sonicator for 5 min. Avoid contact with direct sunlight during its storage at room temperature.
- Melamine working solution (MWS)-I (20 μg/ml): Pipette out 4 ml from MSS and transfer into a volumetric flask (50 ml) and add water up to the mark. Mix well and store at room temperature.
- MWS-II (2 μg/ml): Pipette out 1 ml from MWS-I in a volumetric flask (10 ml) and add water up to the mark. Mix well and store at room temperature.

- MWS-III (200 ng/ml): Pipette out 0.5 ml from MWS-II in a volumetric flask (50 ml) and add water up to the mark. The solution should be mixed properly and store at room temperature and avoid exposure to light.
- Labeled melamine [$^{13}C_3N_3(^{15}NH_2)_3$]: >98% mass fraction, 100 μg/ml water.
- Labeled melamine stock solution (LMSS, 100 μg/ml): It is available commercially. For preparing both calibrations solution and spiking the extracts use the same batch of IS.
- Labeled melamine working solution (LMWS-I, (20 μg/ml)): Pipette out one ml from LMSS and transfer into a volumetric flask (5 ml) and add water up to the mark. Mix well and store at room temperature.
- LMWS-II (2 μg/ml): Pipette out 0.5 ml from LMWS-I and transfer into a volumetric flask (5 ml) and add water up to the mark. Mix well and store at room temperature.
- LMWS-III (200 ng/ml): Pipette out 0.5 ml from LMWS-II and transfer into a volumetric flask (50 ml) and add water up to the mark. Mix well and store at room temperature.
- Mobile phase-A (aqueous ammonium acetate solution (10 mmol/liter)): Dissolve 0.77 g of ammonium acetate in sufficient water and then transferred to 1,000 ml of volumetric flask. Add water up to the mark and filter through 0.22 μm filter.
- Mobile phase-B (CH_3CN): HPLC grade.

5.13.3.2 APPARATUS

- Analytical balance.
- Automatic micropipette: accurate, precise, and calibrated by the user.
- Centrifuge: containing set of 50 ml tubes rotating at 4,000 g rpm.
- Degassing system.
- HPLC Column: TSKgel Amide-80 (2 mm × 250 mm × 5 μm).
- LC-MS/MS equipment (HPLC coupled with triple stage quadrupole MS equipped with ionization source).
- Nylon syringe filter: 0.22 μm pore size.
- Polypropylene Conical tubes: 50 ml.
- Sonification device.
- Vials.
- Volumetric flasks of different capacities from 5, 10, 50, 100, 250, and 1,000 ml.
- Volumetric pipettes: capacities 0.5, 1, and 4 ml.

5.13.4 PROCEDURE

5.13.4.1 PREPARATION OF CALIBRATION CURVES

Prepare separately 5 different (using 5 ml volumetric flask) melamine standard solutions (1 to 5) in a concentration varies from 0 to 2 mg/kg and corresponding ISs added at a level of 1 mg/kg is shown in Table 5.1.

TABLE 5.1 Standard Solution of Melamine (concentration of 0 to 2 mg/kg)

Item	**1**	**2**	**3**	**4**	**5**
MWS II, µl	0	25	50	75	100
LMWS III, µl	50	50	50	50	50
CH_3CN: H_2O solution, µl	4,950	4,925	4,900	4,875	4,850
Equivalent Concentration of:					
Melamine content, ng/ml	0	10	20	30	40
Labeled melamine content, ng/ml	20	20	20	20	20
Melamine content (corresponding in sample), mg/kg	0	0.5	1	1.5	2
Labeled melamine content (corresponding in sample), mg/kg	1	1	1	1	1

Similarly prepare separately 5 different melamine standard solutions (6 to 10) in a concentration varies from 0 to 0.2 mg/kg and corresponding ISs added at a level of 0.1 mg/kg as shown in Table 5.2.

TABLE 5.2 Standard Solution of Melamine (concentration of 0 to 0.2 mg/kg)

Item	**6**	**7**	**8**	**9**	**10**
MWS III, µl	0	25	50	75	100
LMWS III, µl	50	50	50	50	50
CH_3CN:H_2O solution, µl	4,950	4,925	4,900	4,875	4,850
Equivalent Concentration of:					
Melamine content, ng/ml	0	1	2	3	4
Labeled melamine content, ng/ml	2	2	2	2	2
Melamine content (corresponding in sample), mg/kg	0	0.05	0.1	0.15	0.2
Labeled melamine content (corresponding in sample), mg/kg	0.1	0.1	0.1	0.1	0.1

5.13.4.2 EXTRACTION

- Weigh one g of cow milk into a 50 ml conical tube.
- For fraction level of melamine content up to 2 mg/kg with 1 mg/kg level of added IS (Table 5.1), spike the milk sample with 50 μl of LMWS-I. Mix the content well until all spiked volume is completely absorbed into the matrix.
- For fraction level of melamine content up to 0.2 mg/kg with 0.1 mg/kg level of added IS (Table 5.2), spike the milk sample with 50 μl of LMWS-II. Mix the content well until all spiked volume is completely absorbed into the matrix.
- Further add 5 ml of water in either of the above samples prepared and ensure that no clumps to be found. Add 5 ml of acetonitrile into a test sample and shake vigorously for at least one minute.
- Transfer 30 ml of CH_3CN and 10 ml of H_2O and swirl the content vigorously by using automatic shaker for at least 5 min. This is required to ensure no clump is left in the sample.
- After following a vigorous shaking, centrifuge the sample at 4,000×g for 10 minutes at room temperature.
- If the supernatant layer is not clear then transfer this aliquot to another centrifuge tube and again centrifuge for 10 minutes at 8,500×g using benchtop centrifuge at room temperature. Then, collect one ml of clear supernatant into HPLC vials and analyze it under LC-MS/MS.
- Similarly prepare a reagent blank without adding the matrix.

5.13.4.3 CHROMATOGRAPHIC ANALYSIS

Following conditions should be maintained during analysis of test solution:

- Mobile phase A: Ammonium acetate solution (10 mmol/liter);
- Mobile phase B: Acetonitrile;
- Injection volume: 5 μl;
- TSKgel Amide-80 column;
- Temperature of column: ambient temperature;
- Flow rate: 0.25 ml per minute;
- Gradient: Mobile phase gradient (MPG) as mentioned in Table 5.3.
- HPLC flow is directed into detector (MS) through diverter valve for 3 to 13.5 min.

With these conditions retention time (RT) for melamine will be about 11.5 min.

TABLE 5.3 Mobile Phase Gradient (MPG)

Time (in minutes)	0	8	13	14	15	15.5	25
A%	10	10	65	90	90	10	10
B%	90	90	35	10	10	90	90

5.13.4.4 MS ANALYSIS

To quantify the melamine content in the test solution, it is carried out by syringe-infusing 20 μg/ml concentration of MWS-I at a flow rate (FR) of 5 μl per minute into the HPLC flow at 0.25 ml per minute using a T-connector. The flow of HPLC consists of 50% mobile phase A and 50% mobile phase B. Tables 5.4 and 5.5 represent the operating conditions of the instrument for the estimation of melamine.

TABLE 5.4 Mass Spectrometry Parameters for Melamine

Parameters	Triple Quadrupole Instrument
Type of ionization	ESI
Source block temperature	500°C
Polarity	Positive ionization
Spray voltage	3.5 kV
Gas	Curtain gas: 10 psi/69 kPa
	Ion source gas 1: 30 psi/207 kPa
	Ion source gas 2: 30 psi/207 kPa
Collision exit potential	15 V
Entrance potential	10 V
De-clustering potential	60 V
CID gas pressure (SRM)	0.8 Pa, nitrogen
Resolution	Unit (both quadrupoles)
Width of peak	0.5 Th (for every transition)
Negative/positive switching at T	8 minutes
Scan time (for each transition)	100 ms

TABLE 5.5 Transition Reactions (TRs) and Peak Area Ratio (PAR) for Melamine and Internal Standard

Sample	TRs (m/z)*		PAR±Limit (%)
	Quantification	Analyte Confirmation	
Melamine	127 → 85.1 (26)	127 → 68.0 (45)	0.28±25
Labeled melamine	133 → 89.1 (26)	133 → 71.1 (45)	0.19±25

**Collision energies in eV (given in bracket).*

5.13.4.5 CALIBRATION AND ANALYSIS OF SAMPLE

External calibrations are used for quantification of melamine. Inject the standard solutions along with a sequence of samples.

Relative retention time (RRT: it is the ratio of RT of analyte to that of corresponding IS) of the analyte parallels to that of mean RT of calibration solutions within ±2.5%. The PAR from the different TRs noted for melamine is within tolerance limit as fixed by EU (presented in Table 5.4).

5.13.5 CALCULATION

Using relevant software to construct the calibration curve [y = m x + c; where; y is the peak area ratio; c is the intercept; m is the slope; and x is the mass fraction]; and calculate the PARs obtained from different TRs (shown in Table 5.4) and results are expressed in mg/kg.

$$\text{Melamine content } (w_m, mg / kg) = \frac{(A_m / A_{IS}) - c}{m} \times \frac{m_{IS}}{m_a \times 1000} \quad (1)$$

where; A_m is the peak area of melamine in the test sample; A_{IS} is the peak area of labeled melamine (IS) in test sample; m_{IS} is the amount of labeled melamine (IS) added to test sample (ng); and m_a is the weight of test sample (g).

5.13.6 DATA COLLECTION SHEET

Title: Detection of melamine in milk.

Location and Organization:______________________________

Participants: ____________________________________

Date (Month/Day/Year):________________________________

Observations:

Parameter	**Repeat 1**	**Repeat 2**	**Repeat 3**	**Average**
Weight of test sample (g) = m_a				
Peak area of melamine in test sample = A_m				
Peak area of labeled melamine (IS) in test sample = A_{IS}				
Conc. of labeled melamine added to test sample = m_{IS}				
Melamine content (mg/kg)				

Comments:____.

5.14 DETECTION OF FOREIGN FAT IN MILK

5.14.1 INTRODUCTION

Milk constituted various important components. Among those, milk fat is one of most important constituents and based on fat content, pricing of milk is decided. To collect maximum profit, some of unscrupulous persons prefer to add vegetable or refined type of oil by replacing milk fat. Most of vegetable fat/oil possess high butyro-refractometer (BR) reading when compared to milk fat. Therefore, it is possible to detect milk fat adulterated with foreign lipids if any increase in BR reading observed as compared to the reference value. If the BR reading of milk fat exceeds 45 then it represents the presence of vegetable oils/foreign fat. Modified Gerber butyrometer can be used for detection of foreign fat by isolating fat from milk followed by measuring the BR reading.

5.14.2 PRINCIPLE

After separating fat from milk by Gerber test, fat is collected by opening of rubber stopper from stem side of butyrometer with the help of a syringe. Liquidity of fat is maintained by adjusting the temperature to 40°C and subjected the extracted fat on the prism of BR. During milk fat test, Gerber sulfuric acid is used, which causes hydrolysis of fat and BR is calculated by multiplying a factor to obtain the corrected BR reading [1].

5.14.3 MATERIALS

5.14.3.1 REAGENT

- Gerber sulfuric acid (85% v/v);
- Iso-amyl alcohol.

5.14.3.2 APPARATUS

- Milk butyrometer: open mouth at both sides;

- Pipette (10.75 ml capacity);
- Syringe;
- Rubber stopper (acid resistant);
- Lock stopper;
- Butyro-refractometer;
- Water bath (maintained at 40°C).

5.14.4 PROCEDURE

1. **Isolation of Fat Using Modified Gerber Test:** A specially designed butyrometer (open mouth at both sides), can be used for the detection of foreign/vegetable fat in the milk fat. One end of butyrometer from the stem side should be closed with acid resistant rubber stopper. On the other end of the butyrometer, add 10 ml of 85% sulfuric acid by using the automatic dispenser. Measure 10.75 ml of test milk sample in the graduated pipette and is slowly added into the butyrometer. Then, add 1 ml isoamyl alcohol. Close the end from the neck side of butyrometer with stopper and mix well by inverting. Centrifuge the butyrometer at 1,200 rpm for five minutes, which will separate the fat portion from the acid mixture. Note down reading of fat percentage on the butyrometer column.
2. **Determination of BR Reading:** After separating fat from milk by Gerber test, fat is collected by opening of rubber stopper from stem side of butyrometer with help of syringe. BR reading of the fat is determined by using BR maintained at 40°C by circulating water through the prisms using a water bath. Apply the extracted fat between the prisms after calibrating the BR with standard liquid. Note down the observed BR reading after attaining sample temperature to 40°C [1].

5.14.5 CALCULATION

$$\text{Corrected BR reading of milk fat} = \text{Observed BR reading} + 0.08 \times \text{Observed BR reading} \quad (2)$$

5.14.6 DATA SHEET COLLECTION

Title: Detection of foreign fat in milk.

Location and Organization:______________________________

Participants: ______________________________________

Date (Month/Day/Year):________________________________

Observation:

BR Reading at 40°C	Remarks
40–44	Pure milk fat
Above 45	Presence of foreign fat such as vegetable oil

Comments:_______________________________________.

5.15 DETECTION OF PESTICIDE RESIDUES IN MILK

5.15.1 INTRODUCTION

Crops like fruits, vegetables, and cereals are treated with different types of synthetic chemicals, which are known as pesticides used to control insects on plants. These pesticides are applied in pre-harvest, post-harvest, and during storage stages. Therefore, they have the ability to transfer from lower plants and animals to the higher plants and animals. Ultimately, both routes (plants and animals) lead to the accumulation of pesticides in the animal products like milk and further conversion into milk products. Milk being nutritional and supplementary value is mostly consumed by different age groups. Therefore, this issue conquers the immediate attention of the researcher. Since milk contains a considerable amount of fat, organochlorine (OC) type of pesticides is mainly associated with the fat phase. It depends on the percentage of fat content so that it will be retained in the fat. To measure these OC residues, it is necessary to extract the fat from milk sample.

5.15.2 PRINCIPLE

The pesticide residues in milk are extracted by solvent, so as to get maximum extraction efficiency of residues and minimum co-extraction of any materials which may interfere in the estimation. OC pesticide residues are extracted from milk sample using a mixture of solvents. Extracted fat is partitioned with DMF. Na_2SO_4 solution is added for further separating OC residues into hexane. The solvent phase is purified by column chromatography on Al_2O_3

(neutral and activated) using hexane as eluting solvent. Further concentrated eluate is examined by TLC (thin layer chromatography).

Apply the purified extract of the sample along with sequence of standards on thin layer aluminum oxide (Al_2O_3). Petroleum ether is used as mobile phase to develop a chromatogram in ascending direction and silver nitrate ($AgNO_3$) solution is sprayed on TLC plate to identify the separated compounds followed by exposure to UV lamp [4].

5.15.3 MATERIALS

5.15.3.1 REAGENTS

- n-Hexane (68°C–70°C boiling range).
- Acetone.
- Dimethylformamide (DMF).
- n-Hexane saturated with DMF.
- DMF saturated with n-hexane.
- Petroleum ether (40°C–60°C boiling range).
- Sodium sulfate (Na_2SO_4, anhydrous): heated at 500°C for 4 h and cooled in desiccator before use.
- Na_2SO_4 solution (2%).
- Al_2O_3 (neutral and activated): Before use heated to 500°C for 4 h and cooled in a desiccator. For clean-up process, mix Al_2O_3 and water at a proportion of 93:7 (w/w) in a closed vessel for at least 90 minutes. Stoppered the vessel tightly and use within 10 days.
- $AgNO_3$ solution: Transfer 0.5 g of $AgNO_3$ in 100 ml volumetric flask containing 1 ml water. Volume is made up to the mark with 95% ethanol.
- Pre-coated TLC plate containing Type E (neutral) Al_2O_3, F254.

5.15.3.2 APPARATUS

- Rotary evaporator with evaporation flask (500 ml).
- Centrifuge with glass tubes (capacity 250 ml).
- High speed blender.
- Separating funnel.
- Vacuum flasks.
- Graduated tube with a ground-glass neck (15 ml).

- Chromatographic column (12 mm inner diameter and 300 mm length).
- Cotton and glass wool (extract with hexane before use and store in a flask under hexane).
- Micro-Snyder columns.
- TLC plate.
- Ultraviolet (UV) lamp.

5.15.4 PROCEDURE

5.15.4.1 SAMPLING

Mix the milk sample uniformly and collected around one liter as a representative sample. It should be collected in polythene bags or wide mouth glass bottles. After collecting of representative sample, it should be analyzed immediately. If it is taking a long time for analysis, then store the sample either in a deep freezer at –15°C or in a refrigerator at 4°C so that the sample will not absorb any moisture during its storage.

5.15.4.2 PREPARATION OF SAMPLE

Bring the temperature of the milk sample between 35°C and 40°C by placing on a water bath. Mix the sample thoroughly by inverting the bottle repeatedly without causing any churning or frothing and immediately cooled to 20°C.

5.15.4.3 EXTRACTION OF FAT AND OC RESIDUES

Transfer 40 ml of milk, 80 ml each of acetone and hexane in a vortex beaker (250 ml) in the same order and homogenize the mixture for three minutes. Pour the content to centrifuge tube (250 ml) immediately after homogenizing. Use 10 ml of hexane to wash the mixer blade followed by 5 ml of water and then transfer the washings to the centrifuge tube. Centrifuge the tube at 2,500 rpm for 5 min. and then separate solvent layer. The separated solvent layer is passed through a short column Na_2SO_4 (anhydrous). Using 25 ml portions of hexane wash the content of centrifuge tube twice and pass the washing through the column. Concentrate the combined extract to around 15 ml using a rotary evaporator. Pour this extract in separating funnel (100 ml) graduated at 25 ml and make up the volume to 25 ml with hexane.

5.15.4.4 PARTITIONING WITH DMF

Transfer 10 ml of DMF saturated with hexane into a separating funnel containing 25 ml solvent mixture and shake the mixture to extract the fat portion. After two to three minutes, separate and transfer the lower DMF layer into another separating funnel (100 ml). Repeat the same extraction procedure twice by adding 10 ml of DMF into hexane solution. Pool the DMF extract followed by washing with 10 ml of hexane saturated with DMF.

Separate the hexane (10 ml) layer and further wash with 10 ml of DMF. Discard the hexane and transfer the washings to previously collected DMF extract (30 ml) in a separating funnel (500 ml). Add 6 ml of hexane and 200 ml of Na_2SO_4 followed by vigorous shaking for 2 minutes. Keep the separating funnel undisturbed for 20 min. By gentle swirling collect the hexane phase. Discard the aqueous phase and transfer the hexane into the graduated tube. Rinse the funnel with little amount of hexane and transfer the washing to graduated tube. Further concentrate the hexane extract to two ml by fitting micro-Snyder column with tube.

5.15.4.5 CLEAN-UP (AL_2O_3 WITH N-HEXANE)

In a column (having a hexane washed cotton wool plug), pour slurry of prepared Al_2O_3 in hexane and allow Al_2O_3 to settle in the column. Then, cover the column with 30 mm layer of Na_2SO_4 (anhydrous). Drain the hexane until the meniscus touches the top of the Na_2SO_4 layer. Hexane extract is added to the column followed by washing the column with 2 ml of hexane. Elute with hexane (50 ml) at a FR not more than 5 ml per minute and collect the eluate in evaporating flask. Concentrate the eluate in rotary evaporator to approx. 5 ml. Transfer the concentrate into the graduated tube and further fitted with micro-Snyder column to concentrate the eluate to 1 ml.

5.15.4.6 DETECTION BY TLC

Transfer the suitable volume of concentrated sample as well as blank extract into graduated tubes. Make a spot of sample extract by using micropipette on pre-coated Al_2O_3 E TLC plates, containing an adequate quantity of OC pesticide compounds ranging between 0.025 and 0.25 μg in each spot. Similar volume of blank extract is taken to make a spot. Also apply different concentrations (OC compounds corresponds to 0.025, 0.05, 0.1, 0.15, 0.2, and 0.25 μg, respectively) of standard solution to make spot on plate and make sure that diameter of spot is as small as possible for better results.

Using petroleum ether as mobile phase, develop a chromatogram by ascending direction over a distance of around 15 cm. Take out the plates from the tank when solvent approached the front line and evaporate the excess solvent adhered to the plate.

The plate is sprayed with sufficient amount of alcoholic $AgNO_3$ solution. Wait for 10 minutes, then carefully examine the spots' color. Black or brown color of spots at this time represents the absence of OC pesticides. Using pencil position of spots is marked and placed it under UV lamp for 10 minutes. Remove the plate and moistened the plate by spraying water lightly. Again, keep the plate under UV lamp. Appearance of violet-black spots within one to two min. indicates the presence of OC pesticide compounds. Repeat the above procedure (10 min. irradiation → moisten → 2 min. irradiation) if the visibility of the spot is not satisfactory and by doing so lowest conc. of standard can be clearly visible. Compare the spots of the sample with those given by standard compounds. Consider only those spots into account that are absent in the blank [4].

5.15.5 CALCULATION

Determine the response factor (f_r) for each of reference OC pesticides. Compare the spots for reference compounds as given by different concentrations with sample spots for determination of approximate amount of OC pesticides.

5.15.6 DATA SHEET COLLECTION

Title: Detection of OC pesticides in milk.

Location and Organization: ______________________________

Participants: ______________________________

Date (Month/Day/Year): ______________________________

Observation:

f_r **Value of Sample**	f_r **Value of Reference Standard**

Comments: ______________________________.

5.16 DETECTION OF ANTIBIOTICS IN MILK

5.16.1 INTRODUCTION

Several types of antibiotics are used to treat the milch animals. Excess antibiotics contamination in milk impacts our health due to allergic reaction or make resistant to strains of microorganism. It also causes significant economic losses to milk producers and manufacturers of milk and milk products because of excess dosages of antibiotics for a short period, which is given either by injections or orally. When an antimicrobial is administered for animal in veterinary medicine, the obtained milk is not safe for human consumption if the specified withdrawal period has expired, or prolonged drug clearance, unable to observe or lack of guidance on withdrawal time.

The Delvo test SP is a microbial inhibition assay used for the detection of antibiotic residues and sulfonamides in milk as per IDF reference method [5]. This reference method is used for the detection of veterinary drug residue in both raw milk and processed milk using *Bacillus sterothermophillus*. This test involves a change in color based on the growth of *B. Sterothermophillus*. If this organism is unable to grow due to the presence of antibiotics, then it results in a negative test.

5.16.2 PRINCIPLE

At a certain favorable temperature of 64°C, the growth of the *B. sterothermophillus* spores is followed by the acidification process, which leads to changes in pH indicator color from purple to yellow. The agar medium has a standardized number of spores of bacterial strains (*B. sterothermophillus*). This strain is very sensitive to the antibiotics particularly, penicillin and other β-lactam compounds. After adding selected nutrients and providing favorable temperature of 64°C, the spores will germinate, and germinated spores will be multiplied with an initiating acid production. In the sufficient acidic media, the color changes from purple to yellow. However, if the presence of any inhibitory substances like antibiotics in milk, causes to inhibit or delay in their growth and multiplication of spores. Therefore, no changes in purple color is observed [5].

5.16.3 MATERIALS

5.16.3.1 APPARATUS

- Ampoules: 25 ampoules with 4 blocks containing a purple solid agar medium;

- Standardized *B. sterothermophillus* spores;
- Bottle with 110 nutrients tablets;
- 100 disposable pipettes;
- Dosing syringe;
- Pairs of tweezers.

5.16.4 PROCEDURE

First switch on the water bath or block heater and set the temperature to 64°C. Take ampoules and cut off from the block side with scissor. Then, punch a hole with the syringe in an aluminum foil. For favorable growth of spores, add nutrient tablets to the ampoules by using a pair of tweezers. Care is to be taken so that nutrient tablet temperature is maintained at room temperature. Then, infuse the milk sample in a syringe and the milk sample is mixed by pushing down and upside. Collect exactly 0.1 ml of milk sample in a disposal pipette by keeping the tip of disposal pipette on the syringe and transfer the milk sample into ampoules. Then, keep the ampoules in the block heater or water bath, which is preheated at 64±0.5°C. After the incubation time, take away the ampoules from the block heater. Read the results immediately after its incubation period. If the purple color is observed in the test result, then the incubation time is increased till the proper yellow color is observed.

5.16.5 DATA SHEET COLLECTION

Title: Detection of antibiotics in milk.

Location and Organization:______________________________

Participants: ______________________________________

Date (Month/Day/Year):________________________________

Observation:

Color to Observe	Remarks
Purple color	Positive test shows the occurrence of antibiotics
Yellow color	Negative test shows the absence of antibiotics

Comments:____.

5.17 DETECTION OF ALKALINE PHOSPHATASE IN MILK

5.17.1 INTRODUCTION

This test is used for monitoring the pasteurization efficiency of milk. Alkaline phosphatase enzyme is a well-known indigenous enzyme that is mainly located on the fat globule membrane in raw milk. Pasteurization is a heat treatment, carried out by holding method (63°C for 30 minutes) or by the high temperature-short time (72.2°–72.8°C for 15 s) method. At such higher temperature, alkaline phosphatase may get destroyed. Therefore, it is well known index of the efficiency of pasteurization of milk. Pasteurization treatment indicates the destruction of various pathogenic bacteria, *Mycobacterium tuberculosis* or *Coxiella burnetti*, are the heat resistant pathogens normally present in milk. These are causative organism for tuberculosis and Q-fever, respectively. Some of the phosphatase (alkaline phosphatase, ALP) will remain active if pasteurization is not proper and will release phenol from disodium phenyl phosphate as a substrate prepared in a buffered solution at a pH of 9.5. The released phenol is measured by spectrometry method. This test should be carried out within 48 h after pasteurization and temperature of the sample during storage shall not be more than 5°C [2, 6].

5.17.2 PRINCIPLE

In this test, milk is diluted with a suitably buffered substrate of p-nitrophenyl phosphate having a pH of 10.6, is hydrolyzed by the enzyme ALP to form para-nitrophenol, which results in yellow color. Phenol liberation can also be determined at 405 nm wavelength of spectrophotometer. The resulted yellow color shows the presence of ALP, whereas the absence of yellow color represents completely pasteurized milk.

5.17.3 MATERIALS

5.17.3.1 REAGENTS

1. **Buffer Solution:** Dissolve sodium bicarbonate and anhydrous sodium carbonate with the concentration 1.5 g and 3.5 g into 100 ml of distilled water separately, and volume is made to 1,000 ml in a volumetric flask. Store the solution into refrigerator for 1 month.
2. **Substrate:** Anhydrous disodium p-nitrophenyl phosphate (95% purity).

3. **Buffer-Substrate:** 0.15 g of disodium p-nitrophenyl phosphate is dissolved in buffer solution and make up volume to 100 ml in a volumetric flask. Store the solution in a refrigerator for 1 week.

5.17.3.2 APPARATUS

- Weighing balance.
- Water bath: Operating temperature at 37±1°C.
- Test tubes: Use clean and dry test tube with the length 150 mm and diameter 16 mm.
- Graduated pipettes.
- Incubator: Operating temperature at 37±1°C.
- Lovibond comparator.
- Fused cells: Made of glass (25 mm).
- Standard discs: Gives 0, 6, 10, 14, 18, 25, and 42 readings.

5.17.4 PROCEDURE

Five milliliter of buffer substrate solution is pipetted into a test tube followed by transfer it in a water bath at 37°C for 5 minutes. After reaching desired temperature, transfer 1 ml of the test sample, close the tubes with a rubber stopper and thoroughly mix by inversion. Blank test shall be carried out by taking boiled milk of the same type as that under test. Then, place the tubes in a water bath. Measure the developed yellow color after 30 min. Keep the tubes again in a water bath for 90 minutes and note the second reading using a comparator fitted with calibrated disc in μg of para-nitrophenol. In Lovibond comparator, place blank sample on the left-hand side, and test sample on right hand side. By facing north day light source, readings can be taken by comparing the two apertures with the comparator and rotate the disc till the test sample is matched.

Note: Serious fault in pasteurization will be revealed by 30 min. test whereas minor errors during pasteurization can be detected while taking the readings after 2 h of incubation.

5.17.5 DATA COLLECTION SHEET

Title: Detection of alkaline phosphatase in milk.

Location and Organization: ____________________________________

Participants: ______________________________

Date (Month/Day/Year): ______________________________

Observations:

Disc Reading (After 30 min. Incubation)	Disc Reading (After 2 h Incubation)	Remarks
Zero or trace: ___	Zero or trace: ___	Properly pasteurized milk
6: ____	–	Doubtful
Above 10: ____	Above 10: ____	Under pasteurized milk

Comments: ______________________________

5.18 TURBIDITY TEST FOR STERILIZED MILK

5.18.1 INTRODUCTION

When milk is heated at high temperature of 104° to 110°C for 20 min. to 40 min., the albumin gets completely precipitated at such high temperature. This test is carried out based on albumin denaturation in the test sample.

5.18.2 PRINCIPLE

The test is carried out by adding a sufficient amount of ammonium sulfate to precipitate the casein in milk followed by filtration of the mixture and heating the filtrate. If milk is not properly sterilized, the albumins present in the filtrate will be observed by turbidity in the heated filtrate [2].

5.18.3 MATERIALS

5.18.3.1 REAGENTS

- Ammonium sulfate [$(NH_4)_2SO_4$]: analytical grade reagent should be used.

5.18.3.2 APPARATUS

- Beaker (capacity of 500 ml);

- Boiling water bath;
- Conical flask (capacity of 50 ml);
- Funnels;
- Graduated cylinder (capacity of 25 ml);
- Pipette (capacity of 10 ml);
- Test tubes (150×16 mm diameter);
- Whatman filter paper no. 12 (12.5 cm in diameter).

5.18.4 PROCEDURE

Measure 20±0.5 ml of milk in a measuring cylinder and transfer into conical flask (50 ml). Add 4 g of $(NH_4)_2SO_4$ into milk and swirl the content until its complete dissolution. Keep the mixture as such for < 5 min. to settle down the content. Then, filter the content through filter paper and 5 ml of filtrate is collected in a clean, dry test tube. Keep the tube for 5 minutes in a boiling water bath followed by cooling the filtrate in cold water. After cooling, observe the tube for formation of turbidity under daylight electric lamp or north light [2].

5.18.5 DATA SHEET COLLECTION

Title: Turbidity test for sterilized milk.

Location and Organization: ______________________________

Participants: ______________________________

Date (Month/Day/Year): ______________________________

Observations:

Observation	Remarks
Turbidity observed in a test tube	Not sterilized properly
No turbidity observed in a test tube	Sterilized properly

Comments: ______________________________.

KEYWORDS

- **adulteration**
- **alcohol**
- **alkaline phosphatase**
- **antibiotics**
- **foreign lipid**
- **neutralizers**
- **pesticides**
- **preservatives**

REFERENCES

1. IS 1479 (Part-I):1960, (2003). *Methods of Test for Dairy Industry: Part I – Rapid Examination of Milk* (pp. 1–40). New Delhi: Indian Standard Institution (ISI).
2. IS 1479 (Part-II), (1961). *Methods of Test for Dairy Industry: Part II – Chemical Analysis of Milk* (pp. 1–52). New Delhi: Indian Standard Institution.
3. ISO/TS-15495|IDF/RM-230:2010(E), (2010). *Milk, Milk Products, and Infant Formulae: Guidelines for the Quantitative Determination of Melamine and Cyanuric Acid by LC-MS/MS* (p. 36). Geneva: International Organization for Standardization (ISO).
4. ISO 3890-2|IDF 75-2:2009, (2009). *Milk and Milk Products: Determination of Residue of Organochlorine Compound (Pesticide), Part 2: Test Methods for Crude Extract Purification and Confirmation* (2nd edn., p. 30). Switzerland: International Organization for Standardization (ISO).
5. International Dairy Federation (IDF), (1991). *Detection and Confirmation of Inhibitors in Milk and Milk Products* (p. 13). IDF Bulletin 258; Brussels, Belgium: International Dairy Federation.
6. BIS, (1977). *IS 8479 (Part II): Methods for Determination of Phosphatase Activity in Milk and Milk Products, Part I – Routine Method* (p. 6). New Delhi: Bureau of Indian Standards; (1977).

PART II

COMPOSITIONAL ANALYSIS OF FROZEN AND FAT-RICH PRODUCTS

CHAPTER 6

COMPOSITIONAL ANALYSIS OF CREAM

PAYAL KARMAKAR and RONIT MANDAL

ABSTRACT

The cream is a type of liquid milk product that is substantially rich in milk fat and is found in the form of an oil-in-water emulsion of fat in milk serum. It is obtained by mechanical separation from the milk. Other types of cream that are prevalent in the market are whipped cream, table cream, coffee cream, and plastic cream. Usually, the cream contains all the components as in the milk, not with different compositions. Compositional analysis is important for getting information on the components in the cream samples. In this chapter, standards protocols are discussed for the determination of total solids, fat, solids-not-fat (SNF), lactose, ash content, total nitrogen (TN) content, water-insoluble fatty acids (FAs), titratable acidity of fat, and pH in cream.

6.1 DETERMINATION OF TOTAL SOLIDS IN CREAM

6.1.1 INTRODUCTION

After a specific substance is subjected to complete drying, the mass of the remaining components is defined as the 'total solids' portion of that particular substance. It includes all the dry matter of the cream sample: typically, fat, proteins, lactose, minerals, and vitamins. It reflects the moisture content of cream and determines the texture and consistency of cream products. The procedure described below is the reference method used to determine the

Analytical Methods for Milk and Milk Products, Volume 1: Sampling Methods and Chemical and Compositional Analysis. Megh R. Goyal, N. Veena, and Santosh Kumar Mishra (Eds.)

total solids content of the cream sample [2, 4]. The repeatability and reproducibility should be within the limits as mentioned in the standard protocols [3].

6.1.2 PRINCIPLE

The principle of determination of the total solids content of a product involves drying a test sample in a drying oven maintained at a temperature of 102°C after being preheated in a boiling water bath. Drying removes the free water of the cream, leaving behind the dry solid matter.

6.1.3 MATERIALS

6.1.3.1 APPARATUS

- Desiccator;
- Boiling water bath;
- Drying oven, maintained at 102°C;
- Flat-bottomed dishes – preferably of aluminum, steel, or nickel;
- Water bath, maintained at 35–40°C.

6.1.4 PROCEDURE

6.1.4.1 PREPARATION OF SAMPLE

- Preheat the cream sample briefly to 35–40°C in a water bath and mix thoroughly by continuous mild stirring [1];
- Do not overheat to prevent churning;
- Cool it down to 20°C quickly.

6.1.4.2 DETERMINATION

- Weigh the pre-dried dish (by keeping it inside desiccators for about an hour) and record the data.
- Take 1–5 g of the sample (depending upon probable solid content in it) in the dish and spread it thoroughly. Again, weigh the dish containing the sample and keep the data.

- Now heat the dish containing the sample in a vigorously boiling water bath for 30 minutes.
- Transfer the plate to the drying oven and continue heating for 2–3 h. Keep constant attention so that there is no chance of churning.
- Quickly transfer it to desiccator and keep it there for 20–30 min. Then, keep the record of the weight of the dish.
- Repeat the process until the difference between two consecutive readings is not more than one mg, and take the final reading.

6.1.5 CALCULATION

The total solids content of the sample (expressed as mass %) is calculated by:

$$\text{Total solids content (\%)} = \frac{M_2 - M_o}{M_1 - M_o} \times 100 \quad (1)$$

where; M_o is the mass of the empty dish (g); M_1 is the mass of the dish and test sample (g); and M_2 is the mass of the dish and test sample after drying (g).

6.1.6 DATA COLLECTION SHEET

Title: Determination of total solids content in cream.

Location and Organization: ______________________________

Participants: ______________________________

Date (Date/Month/Year): ______________________________

Observations:

Parameter	**Repeat 1**	**Repeat 2**	**Repeat 3**	**Average**
Weight of dish and test sample (M_1)				
Weight of dish and dried sample (M_2)				
Weight of empty dish (M_o)				
Total solids content (%)				

Comments: ______________________________.

6.2 DETERMINATION OF FAT CONTENT

6.2.1 INTRODUCTION

Fat or butterfat content of cream determines its richness. The milk fat content of cream should have a minimum value of 10% (w/w) [5]. The Weibull-Berntrop method described in this section helps in the determination of fat content (in terms of mass fraction) of cream sample [2, 6, 7]. This is a gravimetric (reference) method for fat content determination. This method is rapid and reproducible for fat content determination [3].

6.2.2 PRINCIPLE

Diethyl ether and light petroleum is used to extract an ammoniacal-ethanolic solution of a test portion. The milk fat is soluble in solvents used like ether. The solvents are removed by evaporation or distillation procedure. Gravimetrically determine the total mass of the extracted fat.

6.2.3 MATERIALS

6.2.3.1 REAGENTS

- Ammonia solution (25% concentration);
- Ethanol (94% v/v);
- Congo red solution ($C_{32}H_{22}N_6Na_2O_6S_2$): Dissolve one g of Congo red in water and made the volume up to 100 ml;
- Diethyl-ether ($C_2H_5OC_2H_5$);
- Light petroleum ($CH_3(CH_2)_3CH_3$): having boiling point of 36°C;
- Mixed solvent: Prepare by mixing equal volumes of light petroleum and diethyl ether.

6.2.3.2 APPARATUS

- Analytical balance;
- Boiling aids;
- Centrifuge;
- Distillation or evaporation apparatus;
- Drying oven;
- Fat collecting vessels;
- Measuring cylinders;

- Mojonnier type fat extraction flask;
- Pipettes;
- Rack;
- Tongs;
- Volumetric flasks;
- Wash bottles;
- Water bath, maintained at 35–40°C.

6.2.4 PROCEDURE

6.2.4.1 PREPARATION OF SAMPLE

According to the procedure described in Section 1.4.1 in this chapter.

6.2.4.2 BLANK TEST

Perform the test on blank simultaneously by employing all other reagents except the test sample. Use 10 ml water instead of the test sample.

6.2.4.3 PREPARATION OF FAT COLLECTION VESSEL

Let the fat collection vessel dry at 102°C for 1 h using an oven drier, assisted by a few boiling aids which help in gentle boiling. Cool it to room temperature. Place the vessel to be used for fat-collecting on the balance with the help of a tong. Record the mass of the vessel.

6.2.4.4 DETERMINATION

- Add 10 ml of water at 50°C into the fat extraction flask and wash off the test sample into the small bulb present. Thoroughly mix it and then cool it to room temperature using running tap water.
- Add to it, 2 ml of ammonia solution and mix.
- Then add 10 ml of ethyl alcohol and mix properly. Two drops of Congo red solution can be added (optional).
- Mix vigorously for one minute after adding 25 ml diethyl ether with it, by closing the stopper. While shaking, keep the flask in a horizontal

position. Rinse the neck of the flask with mixed solvent such that the rinsing runs into the flask.

- Then, to it, add 25 ml petroleum ether in the same manner by putting off the cork and shaking vigorously for 30 s.
- Centrifuge it for 1–5 min. at 80–90 g till a clear supernatant layer is obtained which is distinct from the aqueous layer.
- Remove the stopper, followed by rinsing the neck of the flask with mixed solvent. The interface should not be below the bottom of the flask.
- Carefully decant the supernatant layer and collect it into the prepared fat collecting vessel.
- Rinse the neck of the flask with mixed solvent such that the rinsing runs into the flask.
- Take 5 ml ethanol and use it to rinse the neck and mix the rinsing.
- Now, perform another extraction just as before but use 15 ml each of diethyl ether and light petroleum in this case.
- Again, carry out the extraction procedure for the third time and here, do not use alcohol.
- Remove the used solvent almost totally from the fat collection vessel by evaporation or distillation procedure.
- Dry the vessel for 1 h in the oven at 102°C. Check the fat collection vessel to see if the fat is clear or not. Allow cooling till attainment of room temperature. Weigh the fat-collecting vessel using a balance.
- Again, heat the vessel for 30 min, which follows cooling and then finally weighing.
- Repeat the test until the difference between two consecutive readings is maximum 1 mg. Take the lowest weight.

6.2.5 CALCULATION

Calculate the fat content, W_{fat}, expressed as a percentage mass fraction of the sample with the help of the following equation:

$$W_{fat} = \frac{(m_1 - m_2) - (m_3 - m_4)}{m_o} \times 100 \quad (2)$$

where; m_o is the mass of the test portion (g); m_1 is the mass of fat-collecting vessel plus extracted matter (g); m_2 is the mass of the prepared fat-collecting vessel (g); m_3 is the mass of fat collection vessel used during blank test plus any extracted matter (g); and m_4 is the mass of fat collection vessel that is used during blank test (g).

6.2.6 *DATA COLLECTION SHEET*

Title: Determination of fat content in cream.

Location and Organization:____________________________________

Participants: ____

Date (Date/Month/Year): ______________________________________

Observations:

Parameter	**Repeat 1**	**Repeat 2**	**Repeat 3**	**Average**
Mass of the test portion (m_0)				
Mass of the fat-collecting vessel and extracted matter (m_1)				
Mass of the prepared fat-collecting vessel (m_2)				
Mass of the fat-collecting vessel used in the blank test and any extracted matter (m_3)				
Mass of the fat-collecting vessel used in the blank test (m_4)				
Fat content (% m/m)				

Comments:____.

6.3 DETERMINATION OF SNF CONTENT

6.3.1 *INTRODUCTION*

Cream, like milk, has two types of components besides moisture, viz., fat and SNF. SNF comprises of all the components other than fat, such as: protein, lactose, vitamins, and minerals. The method requires prior determination of total solids and fat content of test sample [8].

6.3.2 *PRINCIPLE*

The SNF content is determined by the method of difference. The SNF content of the cream is given by subtracting the fat percentage from the total solids' percentage.

6.3.3 PROCEDURE

6.3.3.1 PREPARATION OF SAMPLE

According to the procedure described in Section 1.4.1 in this chapter.

6.3.3.2 DETERMINATION

- Record the total solids content (TS%) of the specific amount of sample (as described in part I).
- Record the fat content (F%) of that particular amount of that specific sample (as described in part II).

6.3.4 CALCULATION

The SNF content (%) of the cream sample is calculated as follows:

$$\text{SNF}\ (\%) = \text{TS}, \% - \text{F}, \% \qquad (3)$$

where; TS, % is the total solids content of cream sample (%); and F, % is the fat content of the cream sample (%).

6.3.5 DATA COLLECTION SHEET

Title: Determination of SNF content in cream.

Location and Organization:________________________________

Participants: ____

Date (Date/Month/Year): ______________________________

Observations:

Parameter	Repeat 1	Repeat 2	Repeat 3	Average
Total solid content (%)				
Fat content (%)				
SNF content (%)				

Comments:____

6.4 DETERMINATION OF LACTOSE CONTENT IN THE CREAM

6.4.1 INTRODUCTION

Lactose (milk sugar) is a principal component of the cream and other dairy products. It is associated with lactose intolerance, which is a major effect of milk consumption in the world. Lactose is determined by chromatographic or gravimetric methods, which are usually taking a long time. The method described here is based on enzymatic method [9].

6.4.2 PRINCIPLE

Lactose is hydrolyzed in the presence of β-galactosidase and H_2O_2 and broken down to glucose and β-galactose. In the presence of β-galactose dehydrogenase, NAD (nicotinamide adenine dinucleotide) oxidizes β-galactose to galactonic acid. The amount of NADH formed can be measured by absorbance at 340 nm, which is stoichiometric with lactose content.

6.4.3 MATERIALS

6.4.3.1 REAGENTS

- Carrez I solution: Dissolve 3.6 g of $K_4[Fe(CN_6)].3H_2O$ in water and volume is made to 100 ml.
- Carrez II solution: Dissolve 7.2 g of $ZnSO_4.7H_2O$ in water and volume is made to 100 ml.
- NaOH solution (0.1 M).
- Assay reagents: Kits are available containing the following solutions:
- Solution I: 610 mg of lyophilizate consisting of citrate buffer (pH), NAD (35 mg), $MgSO_4$ and stabilizers. Dissolve the lyophilizate in 7 ml distilled water.
- Solution II: 1.7 ml of enzyme suspension of β-galactosidase (100 U).
- Solution III: 34 ml of potassium diphosphate buffer solution (0.51 mol/liter, pH 8.6).
- Solution IV: 1.7 ml enzyme suspension of galactose dehydrogenase (35 U).

6.4.3.2 APPARATUS

- Pipettes.

- Stirrer: glass rod or plastic.
- Spectrophotometer – slit width < 10 nm.
- Cuvettes – glass or plastic (1 cm light path).

6.4.4 PROCEDURE

6.4.4.1 PREPARATION OF SAMPLE FILTRATE

- Take 2 g of cream sample in a volumetric flask having a capacity of 100 ml. Add 60 ml distilled water to it followed by thorough mixing.
- Transfer 5 ml Carrez I solution into the flask, followed by mix properly.
- Then transfer 5 ml Carrez II solution and again mix properly.
- Now add, 10 ml of 0.1 M NaOH.
- Fill the volume with distilled water up to the mark and mix it.
- This is followed by filtration and use the clear filtrate for assay.

6.4.4.2 DETERMINATION

- Get 2 cuvettes, one for blank and one for sample.
- Take 0.2 ml Solution I and 0.05 ml Solution II in both the cuvettes.
- Add 0.1 ml sample solution in the sample cuvette.
- Mix the contents in cuvettes using a stirrer and then keep them in hold for 10 min. at 20–25°C.
- Now add 1 ml of Solution III in each cuvette.
- Then, add 2 ml and 1.9 ml of distilled water in blank and sample cuvettes, respectively.
- Mix and hold for 2 more minutes at 20–25°C.
- Check the absorbance at 340 nm (A_1).
- Now add 0.05 ml of Solution IV, mix and hold at 20–25°C for 10–15 min.
- Again, take the readings for absorbance at 340 nm (A_2).
- Continue taking readings every 2 minutes until the absorbance value (AV) remains constant for about 2 min.

6.4.5 CALCULATION

Convert absorbance (A) to anhydrous lactose (C, g/liter) using the following equation:

$$C = \frac{V \times MW \times \Delta A}{\varepsilon \times d \times v \times 1000} \tag{4}$$

where; *V* is the final volume (ml); *V* is the volume of the sample (ml); MW is the molecular weight of anhydrous lactose (342.3 g/mol); *d* is the light path (in cm); ε is the NADH absorption coefficient at 340 nm [= 6.3 mmol^{-1}cm^{-1})]; ΔA is the absorption differences of blank and sample [(A_{s1} − A_{S2}) − (A_{B1} − A_{B2})].

Introducing values of constants in Eqn. (4), the equation for anhydrous lactose reduces to:

$$C = \frac{3.30 \times 342.3 \times \Delta A}{6.3 \times 1 \times 0.1 \times 1000} = 1.793 \times \Delta A \tag{5}$$

6.4.6 DATA COLLECTION SHEET

Title: Determination of lactose content in cream.

Location and Organization:____________________

Participants: ____________________

Date (Date/Month/Year): ____________________

Observations:

Parameter	Repeat 1	Repeat 2	Repeat 3	Average
Initial absorbance of sample (A_{S1})				
Final absorbance of sample (A_{S2})				
Initial absorbance of blank (A_{B1})				
Final absorbance of blank (A_{B2})				
Lactose content (g/liter)				

Comments:____________________.

6.5 DETERMINATION OF ASH CONTENT

6.5.1 INTRODUCTION

Ash content signifies the incombustible portion of the product, which remains after its complete combustion or incineration. It is the inorganic residue. Ash content is a measure of the mineral content of the milk or milk product.

6.5.2 PRINCIPLE

The principle of determination of the ash content of a product involves charring the test sample by igniting the furnace at 550°C until the point when it gets rid of its carbon content after being preheated in boiling water bath [10–12].

6.5.3 MATERIALS

6.5.3.1 APPARATUS

- Silica crucible;
- Desiccator;
- Boiling water bath;
- Muffle furnace, maintained at 550°C.

6.5.4 METHODOLOGY

6.5.4.1 PREPARATION OF SAMPLE

According to the procedure described in Section 1.4.1 in this chapter.

6.5.4.2 DETERMINATION

- Take 5 g of sample in silica crucible after weighing it and then subject it to drying by evaporation using a water bath;
- Transfer it to a muffle furnace maintained at 550°C to let the contents transform into ash by burning all carbonaceous matter;
- Then keep it inside desiccator and let it cool;
- Finally, take the weight of the silica crucible containing the ash.

6.5.5 CALCULATION

The percentage of ash content in the cream is calculated by the following equation:

$$\text{Ash content (\% by mass)} = \frac{W_1 - W_o}{W} \times 100 \quad (6)$$

where; W is the weight of sample taken (g); W_0 is the weight of the empty silica crucible (g); and W_1 is the weight of the silica crucible containing the ash (g).

6.5.6 DATA COLLECTION SHEET

Title: Determination of ash content in cream.

Location and Organization: ______________________________

Participants: ______________________________

Date (Date/Month/Year): ______________________________

Observations:

Parameter	Repeat 1	Repeat 2	Repeat 3	Average
Weight of the sample taken (W)				
Weight of the empty silica crucible (W_0)				
Weight of the silica crucible containing the ash (W_1)				
Ash content (%)				

Comments: ______________________________

6.6 DETERMINATION OF TOTAL NITROGEN (TN) CONTENT

6.6.1 INTRODUCTION

The total nitrogen (TN) content is estimated by the well-known Kjeldahl method as described in this section [13]. The method lays down the basic step also for quantification of crude protein in food products. This is a reference method for nitrogen and protein content determination in food products.

6.6.2 PRINCIPLE

The determination of the TN content of cream based on the principle of digestion of cream using H_2SO_4 with the help of catalyst ($CuSO_4.5H_2O$) and a boiling point elevator K_2SO_4, helping in nitrogen to be released from protein, thereby its retention as ammonium salt. NH_3 is released by adding concentrated NaOH, following its collection in H_3BO_3 solution and titration.

$$NaCl + AgNO_3 \longrightarrow AgCl + NaNO_3$$
$$K_2CrO_4 + 2AgNO_3 \longrightarrow Ag_2CrO_4 + 2KNO_3$$

6.6.3 MATERIALS

6.6.3.1 REAGENTS

- Sulfuric acid (98% weight): It should be nitrogen-free.
- Copper catalyst solution: 0.05 g/ml in water.
- Potassium sulfate.
- Sodium hydroxide solution: 50% w/w.
- Boiling chips – amphoteric alundum granules (10 mesh size).
- Methyl red/bromocresol green (BCG) indicator – obtained by dissolving 0.2 g methyl red and 1 g BCG in 100 ml and 500 ml 95% ethanol, respectively in separate glassware. Then mix both the solutions in 1: 5 ratio.
- Boric acid solution (4%): Dissolve 40 g H_3BO_3 in one liter of water followed by adding 3 ml methyl red/BCG indicator gives light orange colored solution.
- Hydrochloric acid (HCl) standard solution: 0.1 N.

6.6.3.2 APPARATUS

- Analytical balance;
- Digestion flasks (Kjeldahl);
- Distillation flasks: NH_3 distilled in boric acid solution needs to be completely absorbed, hence trap the condenser outlet;
- Digestion/distillation system;
- Conical flask: capacity 250 ml;
- Burette with stand;
- Measuring cylinder.

$$NaCl + AgNO_3 \longrightarrow AgCl + NaNO_3$$
$$Excess\ AgNO_3 + KSCN \longrightarrow AgSCN + KNO_3$$
$$Fe^{+++} + SCN^{--} \longrightarrow [FeSCN]^{++}$$

6.6.4 METHODOLOGY

6.6.4.1 PREPARATION OF SAMPLE

- Take 15 g of K_2SO_4, 1 ml of $CuSO_4.5H_2O$ and 8–10 boiling chips in digestion flask;
- Warm the cream at 35–40°C in water bath and mix thoroughly and Take 5 g of test sample in the flask;
- Add 25 ml H_2SO_4;
- Prepare a blank by taking 0.5 g of sucrose instead of the test sample.

6.6.4.2 DETERMINATION

6.6.4.2.1 Digestion

- Take 250 ml of water in digestion flask;
- Heat it to 25°C by placing it on a burner;
- Add 3–4 boiling chips to it;
- Heat it from 25°C to rolling boil within 5–6 min;
- Keep the fume ejection system on with the flask placed in inclined position;
- The sample must form foam along the neck of the flask;
- Digest it under low burner setting until white fumes appear at the neck of the flask. It takes approximately 20 min;
- Increase the burner setting to half of maximum and continue heating for 15 min;
- Then increase the burner setting to maximum and heat it for 1–1.5 h, until a clear blue-green colored solution appears;
- Cool the digested mixture to ambient temperature in 25 min (must be in liquid state);
- Now, add 300 ml water to the flask;
- For distillation at a later time, flasks can be stoppered.

6.6.4.2.2 Distillation

- Turn on the water of the condenser.
- Take 50 ml H_3BO_3 solution having indicator added to it.
- Transfer it to 500 ml titration flask.

- The tip should be below H_3BO_3 condenser surface, so place the flask under condenser tip.
- To the diluted digested matter, slowly add 75 ml 50% NaOH along the sides of the flask. A clear layer of NaOH will be seen under diluted digest.
- Connect the flask to the distillation bulb, which is there on the condenser.
- Mix the contents properly by subjecting to vigorous swirling motion.
- Heat until the complete distillation of all the ammonia.
- Put off the heater. Lower down the receiving flask. Let all the liquid flow out from the condenser tip.
- Titrate H_3BO_3 solution with 0.1 N standard HCl solution till the color changes to just pink.
- Note the readings for both sample and blank.

6.6.5 *CALCULATION*

Total nitrogen (%) is calculated with the following equation:

$$\text{Total nitrogen }\% = \frac{1.4007 \times (V_S - V_B) \times N}{w} \tag{7}$$

where; V_S is the volume of HCl which is used for titration of sample (ml); V_B is the volume of HCl which is used for blank titration (ml); N is the normality of HCl solution; and W is the weight of sample (g).

6.6.6 *DATA COLLECTION SHEET*

Title: Determination of total nitrogen content in cream.

Location and Organization:________________________

Participants: ________________________

Date (Date/Month/Year): ________________________

Observations:

Parameter	**Repeat 1**	**Repeat 2**	**Repeat 3**	**Average**
Volume of HCl used for sample (V_S)				
Volume of HCl used for blank (V_B)				
Weight of sample (W)				
Nitrogen content (%)				

Comments:________________________.

6.7 DETERMINATION OF WATER-INSOLUBLE FATTY ACIDS (FAS)

6.7.1 INTRODUCTION

The process described in this section is a titrimetric method, which is used for quantification of FAs (water insoluble) percentage of cream [14–16].

6.7.2 PRINCIPLE

The basic principle consists of extracting the FAs, which are mainly water-soluble and water-insoluble, in cream by solvent, petroleum ether and subjected to washing with water. The water-insoluble portion of FAs is determined by titration with sodium ethylate, using phenolphthalein indicator.

6.7.3 MATERIALS

6.7.3.1 REAGENTS

- Alcohol;
- Congo red paper;
- Ether;
- Glass beads/sand;
- Na_2WO_4;
- NaOH solution (0.1 N);
- Neutral benzene;
- Phenolphthalein indicator solution;
- Sodium ethylates;
- Sulfuric acid (H_2SO_4).

6.7.3.2 APPARATUS

- Centrifuge bottle;
- Siphon;
- Separator;
- Beaker;
- Flask.

6.7.4 METHODOLOGY

6.7.4.1 PREPARATION OF SAMPLE

According to the procedure described in the Section 1.4.1 in this chapter.

6.7.4.2 EXPERIMENTATION

- Take a 250 ml centrifuge bottle. To it, add 50 g sample, 20 ml alcohol and 50 ml ether. Shake it gently to dissolve the fat.
- Keep on adding 0.1 N NaOH to the phenolphthalein added mixture, in 0.2 ml installments until the appearance of pink color. Finally, add 0.5 ml extra NaOH and mix for 2 min.
- Keep on opening the stopper a few times to release the excess pressure.
- Then add 50 ml of petroleum ether which follows centrifugation at 1,200 rpm for 5 min till a clear separation of layers is obtained.
- Carefully siphon off the fat-ether layer.
- To the remnant, again add 25 ml ether and 25 ml petroleum ether followed by mixing after each addition. Again, siphon off the fat-ether layer.
- Add 1 ml of H_2SO_4 and shake thoroughly for a few seconds.
- Next, add 10 ml Na_2WO_4 and shake thoroughly (when you shake Na_2WO_4 treated mixture with ether, emulsions can form, and they may not break even by centrifuging. However, they can be broken by mixing 10–20 ml alcohol and mild centrifuging).
- Then add 75 ml ether, shake for 2 min, centrifuge, and siphon off the ether layer.
- Wash off the residue by shaking with 75 ml of ether, followed by centrifugation and siphoning off.
- Transfer all the siphoned off solvent to a 500 ml separator followed by adding 100 ml alcohol.
- Neutralize with 1N NaOH till pink color is just observed.
- Add 0.5 ml in excess and shake thoroughly for 2 min.
- To it, add 25 ml water, mix, and let it stand until a clear aqueous layer is observed (it may take a few minutes).
- Shift the total aqueous layer to a beaker.
- Add 50 ml alcohol followed by addition of 10 drops of phenolphthalein solution. Use 1N NaOH to neutralize it by shaking it for about 2 min.
- 50 ml and 10 ml of water can be added in the first and second time, respectively, followed by mixing them, letting them stand till clear aqueous layer appearance and draining them into beaker.
- Keep the beaker on steam bath to get rid of ether.
- Transfer it to a beaker having a capacity of 250 ml which is already containing 25 ml of water.

- Now take 5 g of Na_2WO_4 and dissolve the warm solution.
- Bring the temperature down to 20°C by stirring frequently to prevent soaps from forming hard crust on the surface.
- Add H_2SO_4 drop by drop with added indicator Congo red and stir vigorously for thorough liberation of FAs, mashing all pink soap curds.
- Add a 500 mg filter-aid and mix. Perform suction with suitable filter for getting a dry precipitate.
- Wash off the beaker with 15 ml water maintained at 20°C and transfer it to the crucible.
- Use a flask containing few glass beads or grains of sand for receiving flask of filtering apparatus.
- Extract the acid with 15 ml ether by thoroughly mixing and breaking up precipitate.
- Let the ether drip off through filter before applying suction.
- Evaporate the ether extract using water bath and dry acids for 1 h maintained at 100°C.
- Bring the temperature down and record its weight as mg water insoluble acids/100 g cream.
- Dissolve the weighed acids in 10 ml neutral benzene and titrate using 0.1 N sodium ethylate, having phenolphthalein indicator in it.

6.7.5 CALCULATION

$$\text{Molecular weight of fatty acids} = \frac{\text{Observed mg acids}}{NaOH(0.1N)\text{used for titration}} \tag{8}$$

6.7.6 DATA COLLECTION SHEET

Title: Determination of water-insoluble fatty acids content in cream.

Location and Organization: ____________________

Participants: ____________________

Date (Date/Month/Year): ____________________

Observations:

Parameter	Repeat 1	Repeat 2	Repeat 3	Average
Volume of alkali used for titration (ml)				
Fatty acids (water insoluble) content (%)				

Comments:________________________________.

6.8 DETERMINATION OF TITRATABLE ACIDITY OF FAT

6.8.1 INTRODUCTION

The test for determination of titratable acidity of fat in cream is based on the titrimetric method as discussed in this section [17].

6.8.2 PRINCIPLE

Test sample is taken and mixed with a solution containing sodium tetraphosphate and a surface-active agent. Fat is separated by heating the mixture in a boiling water bath. A known amount of extracted fat is dissolved in organic solvent and titrated against alcoholic alkali.

6.8.3 MATERIALS

6.8.3.1 REAGENTS

- Phosphoric acid solution (H_3PO_4): 1 mol/liter.
- BDI (Bureau of Dairy Industries) reagent: Obtained by dissolving 70 g of sodium tetra phosphate in distilled water and make up the volume up to 700 ml and mix properly. Add 30 g octylphenylpoly(ethylene glycol) and again mix it. Phosphoric acid solution can be used to adjust the pH to 6.6. Dilute with water to the volume of 1 liter and then mix it.
- Thymol blue solution ($C_{27}H_{30}O_5S$) (0.1 g/liter in propan-2-ol): Dissolve 0.1 g of sodium salt of thymol blue with 100 ml of propan-2-ol to prepare stock solution. Mix one volume of stock solution with nine volumes of propan-2-ol before use.
- Fat solvent solution: Mix one volume of thymol blue solution with four volumes of petroleum ether (boiling point ranges 60°C to 80°C).
- Potassium hydrogen phthalate solution ($KHC_8H_4O_4$) (0.01 mol/liter); Dissolve 1.0211 g of $KHC_8H_4O_4$ in 500 ml volumetric flask and fill up to the marked volume with water and mix.

- Tetra-n-butylammonium hydroxide solution ($C_{16}H_{37}NO$) (0.01 mol/liter in propan-2-ol and methanol mixture): Dilute one volume of tetra-n-butyl ammonium hydroxide (0.1 mol/liter) in propan-2-ol and methanol mixture with nine volumes of propan-2-ol to get the final concentration of $C_{16}H_{37}NO$ as 0.01 mol/liter.
- Pilot fat: Melt anhydrous milk fat (AMF) (1,000 g) having a fat acidity level between 0.5 mmol and 1 mmol/100 g of fat. Divide the melted AMF sample into subsamples (5 g each). It can be used to check the reproducibility of titration results.
- Reference fat: Milk fat of low-fat acidity (basic fat) spiked with increasing levels of palmitic acid having a fat acidity level between 0.5 mmol and 1.5 mmol/100 g of fat.

6.8.3.2 APPARATUS

- Boiling water bath.
- Calibration syringe (adjustable and capable of delivering a known amount of fat).
- Delivery pipette or syringe (10, 25, and 50 ml).
- Dip-probe connected with colorimeter – suitable enough to measure wavelength in the range of 600–620 nm. It is connected to the titration vessel.
- Fat separation tube or butyrometer.
- Gas wash bottle: Containing petroleum ether (boiling point between 60°C and 80°C).
- Micro-burette (graduated at 0.002 ml).
- Nitrogen supply (CO_2 free).
- Titration vessel: provided with stirring device with capacity ranges from 10 to 100 ml.
- Volumetric flask (volume ranges 100 to 500 ml).
- Water bath (temperature maintained at 45±1°C).

6.8.4 METHODOLOGY

6.8.4.1 PRE-TREATMENT OF SAMPLE

Dilute the cream sample with water or corresponding skim milk to obtain a fat content of between 4% and 6%. Dilution with water results in an underestimation of FAs content compared to skim milk. In such cases, a correction can be added to get correct results.

6.8.4.2 SEPARATION OF FAT

- Mix 3.5 parts of diluted test sample to one part of BDI reagent in the tube for fat separation containing the following steps:
- When Van Gulik butyrometer is used for separation then mix 16±0.5 ml of diluted cream sample and 4.5 ml of BDI reagent.
- When MONED tube is used for separation then mix 31±1 ml of diluted cream sample and 8.9 ml of BDI reagent.
- For other tubes: mix diluted cream sample and BDI reagent in the ratio of 3.5:1.
- Close the fat extraction tube quickly after filling and mix thoroughly.
- Within 5 min, keep it in a boiling water bath, which is maintained at >95°C for 15 min.
- Centrifuge the tubes to achieve better fat separation and the separated fat shall be free from any particles and transparent.
- After the fat separation, put the tubes in a water bath at 45°C and make sure that upper layer of tube content is fully immersed in the water.

6.8.4.3 TITRATION

- Carry out the titration in the titration vessel in a carbon dioxide free atmosphere. Connect it to the nitrogen supply coming from the gas wash bottle. Fill the gas bottle regularly to compensate for the petroleum ether evaporation.
- Transfer fat solvent and pilot fat (0.25 g) to titration vessel and flush with nitrogen to remove the CO_2.
- Wavelength setting of the colorimeter should be controlled and adjust scale at 0% (dark) and 100% (fat solvent with fat sample).
- Adjust the titration end point as 70% on transmission scale. Neutralize the solvent with tetra-n-butylammonium hydroxide solution.
- Add 0.25 g of the pilot fat using calibrated syringe and titrate.
- Transfer a test portion of 0.25 g of prepared fat sample to the titration vessel using calibrated syringe and titrate.
- After 3 titrations are carried out per 2 ml of fat solvent, replace it with a fresh one.

- The end point titration can be estimated by visual observation from yellow to faint greenish (when only a small number of test samples is used). Two titrations can be performed in 5 ml fat solvent.

Note: A titration by manual can also be carried out without using colorimeter with dip-probe and end point of titration is observed by change in color.

6.8.5 *CALCULATIONS*

The fat acidity of the test sample (in millimoles per 100 g of fat) it is calculated by the following equation:

$$\text{Fat acidity (mmol/100 g fat)} = \frac{V_c}{m} \times 100 \qquad (9)$$

where; V_c is the volume of alcoholic alkali used in titration (ml); and m is the mass of test sample (g).

6.8.6 *DATA COLLECTION SHEET*

Title: Determination of titratable acidity of fat in cream.

Location and Organization:______________________________

Participants: ______________________________

Date (Date/Month/Year): ______________________________

Observations:

Parameter	**Repeat 1**	**Repeat 2**	**Repeat 3**	**Average**
Volume of alcoholic alkali used in titration (V_c)				
Mass of test sample (m)				
Acidity of fat in cream (mmol/100 g fat)				

Comments:______________________________.

6.9 DETERMINATION OF PH

6.9.1 *INTRODUCTION*

The pH is defined as the negative logarithm (to the base 10) of the hydrogen ion concentration.

6.9.2 PRINCIPLE

The pH of a sample is determined using a pH meter.

6.9.3 MATERIALS

6.9.3.1 REAGENTS

- Buffer solution (pH 4.0);
- Buffer solution (pH 7.0).

6.9.3.2 APPARATUS

- Glass beaker;
- Glass rod/stirrer;
- pH meter.

6.9.4 METHODOLOGY

6.9.4.1 PREPARATION OF SAMPLE

According to the method described in the Section 1.4.1 in this chapter.

6.9.4.2 CALIBRATION OF PH METER

- Prepare buffer solution of pH 4.0 and 7.0 by dissolving one tablet in 100 ml each in two separate beakers. Use glass rod to properly mix and dissolve it.
- Insert the electrode into buffer solution of pH 4 and check the reading. If the reading shows 4, then it is already adjusted. If it shows some different reading, adjust the reading to 4 by turning the knob.
- Now, insert the electrode into buffer solution of pH 7.0 and check the reading. If the reading shows 7, then it is already adjusted. If it shows some different reading, adjust the reading to 7 by turning the knob.
- Now again check for pH 4.0 and 7.0 until the reading gets perfect.

6.9.4.3 DETERMINATION OF PH

After the pH meter is calibrated properly, insert the electrode (dipped in saturated KCL solution) into the sample. Observe the reading shown in the digital screen. Note down the reading shown in the digital screen of the pH meter. This gives the value of pH of the sample.

6.9.5 DATA COLLECTION SHEET

Title: Determination of pH of cream.

Location and Organization: ______________________________

Participants: ______________________________

Date (Date/Month/Year): ______________________________

Observations:

Parameter	Repeat 1	Repeat 2	Repeat 3	Average
pH of sample				
Temperature of sample (C)				

Comments:______________________________.

KEYWORDS

- **compositional analysis**
- **cream**
- **milk fat**
- **pH**
- **titratable acidity**

REFERENCES

1. AOAC, (1995). *AOAC Official Methods of Analysis (Dairy Products): 33.3.02 Cream –Preparation of Sample* (p. 45). Maryland, United States: AOAC International.

2. ISO. *ISO 707:2008 (IDF 50:208)*: *Milk and Milk Products — Guidance on Sampling* (p. 4). Geneva, Switzerland: International Organization for Standardization (ISO); (2008).
3. ISO. *ISO 5725-1:1994*: *Accuracy (Trueness and Precision) of Measurement Methods and Results, Part 1: General Principles and Definitions* (pp. 1–17). Geneva, Switzerland: International Organization for Standardization (ISO); (1986).
4. Steiger, G., & Mantens, R., (1993). *Cream Powder: Determination of Total Solids Content* (pp. 21–29). Brussels – Belgium: International Dairy Federation (IDF) Bulletin 285.
5. *CODEX. CXS 288-1976: Standard for Cream and Prepared Creams* (p. 3). Rome, Italy: Codex Alimentarius Commission. FAO/WHO. (1986).
6. ISO. *ISO 8262-3:2005 (IDF 124-3:2005)*: *Milk Products and Milk-based Foods — Determination of Fat Content by the Weibull-Berntrop Gravimetric Method (Reference Method) — Part 3: Special Cases* (pp. 1–9). Geneva – Switzerland: International Organization for Standardization; (2005).
7. International Dairy Federation, (1988). *Interlaboratory Collaborative Studies* (Vol. 235, pp. 37–54). Second Series; Bulletin of International Dairy Federation (IDF).
8. AOAC, (1995). *AOAC Official Methods of Analysis (Dairy Products): 33.2.45 Solids-Not-fat in Milk, by Difference Between Total Solids and Fat Contents* (p. 32). Maryland, USA: AOAC International.
9. Biltcliffe, D. O., Kleyn, D. H., Trout, J. R., Azzara, D., Bissonette, J., Bradley, R., Brown, R., et al., (1984). Enzymatic-ultraviolet method for measuring lactose in milk: Collaborative study. *Journal of the Association of Official Analytical Chemists, 67*(3), 637–640.
10. Frary, G. G., & Jordan, B., (1940). Report on total solids and ash in milk and evaporated milk. *Journal of the Association of Official Agricultural Chemists, 23*, 453–457.
11. Frary, G. G., (1945). Report on solids and ash in milk and evaporated milk. *Journal of Association of Official Agricultural Chemists, 28*(2), 211–213.
12. Frary, G. G., (1951). Report on ash and acidity of milk and evaporated milk. *Journal of Association of Official Agricultural Chemists, 34*(2), 239–243.
13. Barbano, D. M., Clark, J. L., Dunham, C. E., & Flemin, R. J., (1990). Kjeldahl method for determination of total nitrogen content of milk: Collaborative study. *Journal of the Association of Official Analytical Chemists, 73*(6), 849–859.
14. Hillig, F., (1947). Determination of water-insoluble fatty acids in cream and butter. *Journal of the Association of Official Agricultural Chemists, 30*, 575–582.
15. Hillig, F., & Ahlmann, S. W., (1948). Water insoluble fatty acids in cream and butter. *Journal of the Association of Official Agricultural Chemists, 31*, 739–749.
16. Hillig, F., Lepper, H. A., & Patterson, W. I., (1949). Water-insoluble fatty acids and butyric acids in cream and butter. *Journal of the Association of Official Agricultural Chemists, 82*, 731–745.
17. ISO, (2012). *ISO/TS 22113: 2012 (IDF/RM 204:2012)*: *Milk and Milk Products — Determination of the Titratable Acidity of Milk Fat* (pp. 1–6). Geneva, Switzerland: International Organization for Standardization (ISO).

CHAPTER 7

COMPOSITIONAL ANALYSIS OF ICE CREAM AND FROZEN DESSERTS

M. P. RAHILA, S. ATHIRA, N. VEENA, CHITRA GUPTA, and RINKU SUNDAR

ABSTRACT

Ice cream and frozen desserts are globally popularized milk products whose industrial demands are rapidly growing day by day. Hence the major share of milk produced across the globe is being utilized for the production and marketing of frozen dairy products. Ice cream is familiar to all age groups of people since decades. The physical properties of the end product are highly influenced by the processing and chemical parameters like pasteurization, homogenization, aeration, and aging of the ice cream mix prior to freezing. Ice cream mixture includes milk components, water, sucrose, dextrose, egg, and egg products, flavoring, and coloring components, stabilizer, emulsifier, and optionally other components. According to the processing parameters and consumer requirements and expectations, composition of ice cream varies from industries to industries which finally get reflected to the quality of the finished product. This chapter deals with the laboratory methods used to estimate the chemical composition of ice cream and frozen dairy foods.

7.1 DETERMINATION OF WEIGHT OF MILK PER UNIT VOLUME

7.1.1 INTRODUCTION

Sometimes, challenges like the overrun of the semi-processed product, which is all due to the air incorporation within the ice cream mixture heading up with

Analytical Methods for Milk and Milk Products, Volume 1: Sampling Methods and Chemical and Compositional Analysis. Megh R. Goyal, N. Veena, and Santosh Kumar Mishra (Eds.)

the freezing process. It enhances both the final product volume and physical properties like melting and the mouthfeel is influenced by the amount of air incorporated. Overrun is nothing but, the percentage (%) increase in ice cream volume obtained in excess to that of the mix volume used to produce the ice cream. Excess air incorporation within the product can be reduced through the uniform supply of air during freezing. Air bubbles can even be stabilized through emulsifiers, stabilizers, fat, and protein in unfrozen mixture [3], which entraps the air through emulsification. The volume of the gas phase greatly differs from a maximum of 50% to a minimum of 10–15%. As per the BIS, the mass of the ice cream and fruit or nuts or chocolate-added ice cream should not be less than 525 g/liter and 540 g/liter, respectively [4]. Determination of the weight/volume of an ice cream indicates the amount of total solids present in ice cream and gives an idea about the extent of overrun.

7.1.2 PRINCIPLE

Overrun is technically expressed as a percentage and is always measured by determining the total mass of a known volume of ice cream. This test is carried out using the volume of ice water, alcohol, or kerosene by replacing the original volume of air present within the ice-cream followed by examining the difference between the sum of these two volumes and the capacity of the flask which is equal to the volume occupied by the sample [1, 4].

7.1.3 METHOD-I: USE OF KEROSENE AS THE IMMERSION FLUID (FOR PACKAGED ICE CREAM)

7.1.3.1 MATERIALS

- Overflow chamber: Use a number 10 or 1 gal can (Figure 7.1).
- Iron bar: Larger than the diameter of the overflow chamber, used to immerse sample in kerosene.
- Analytical balance.
- Measuring cylinder – 500 ml, graduated.
- Kerosene.

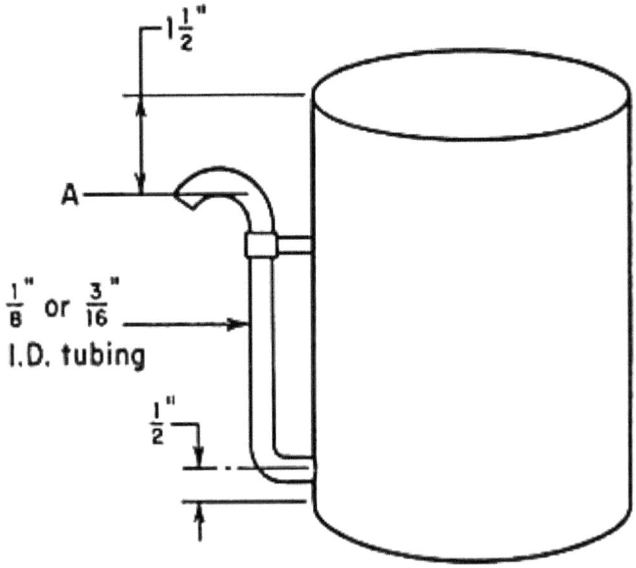

FIGURE 7.1 Overflow can.

7.1.3.2 PROCEDURE

Take the packaged ice cream from the cold room/freezing compartment and immediately carry out the test procedure. Place the overflow can/apparatus on a uniform surface. Use cooled kerosene to fill the can till it starts overflowing via the spout. Once the overflow is ceased, place a 500 ml graduated cylinder (weight measured) under the spout.

Immediately take out the frozen ice cream brick from the carton/package. Weigh the sample appropriately 1–2 g and slowly immerse the sample in a can filled with kerosene, ensuring it is submerged completely by holding it with a spatula until it stops overflowing. The displaced kerosene is collected in a 500 ml graduated cylinder kept beneath the spout. Weigh the cylinder holding the displaced kerosene and calculate the net weight. To obtain the total volume displaced, calculate the net weight by the specific gravity of kerosene [1].

7.1.3.3 CALCULATION

$$\text{Weight/unit volume} = (W \times 8.345)/V \quad (1)$$

where; W is the total sample weight, pounds; and V is the volume of the kerosene displaced, gallon.

7.1.4 METHOD-II: USE OF ALCOHOL AS THE IMMERSION FLUID

7.1.4.1 MATERIALS

7.1.4.1.1 Reagents

- n-Amyl alcohol (Specific gravity – 0.817).

7.1.4.1.2 Apparatus

- Analytical balance – for weighing accurately to 0.001 g;
- Beaker (400 ml);
- Funnel;
- Volumetric flask (250 ml).

7.1.4.2 EXPERIMENTATION

Weigh a unit of given sample and then estimate the weight of sample per liter. For instance, if 200 ml of a full carton sample is given, take the weight of the filled carton first. Later measure the weight of the empty carton by carefully clearing out the ice cream from the carton and then allowing it to dry completely. And weigh the empty dry carton without ice cream. The weight difference between the two cartons, i.e., filled carton and empty carton, gives the accurate mass of the 200 ml frozen product. This mass is five times the weight that is equal to the weight of one liter of sample.

Note down the mass of a clean dry 400 ml beaker. Add 130 g of the accurately weighed sample to the beaker. Followed to this, melt the sample in the beaker by placing in a water bath warmed to 49°C. Note the empty weight of a volumetric flask (250 ml). Finally, transfer 130 g of melted ice-cream using a glass funnel into the 250 ml volumetric flask. Accurately weigh 10 g of n-amyl alcohol and add to the sample containing volumetric flask. To disrupt the surface tension (ST) of the melted sample, give a thorough mix in order to release out the air which is incorporated. About 10 grams of alcohol holds a 12.24 ml volume. Place the mixture containing flask in a cold-water bath to cool up to 15.5°C or can even place it directly under running tap water.

Wash the beaker having ice cream with a small quantity of water and add each washing to the flask. Make up the volume to the mark with water. The lower meniscus must correspond with the 250 ml mark at a temperature of 15.5°C. Note down the mass after drying the flask from outside [5].

7.1.4.2 CALCULATIONS

Determine the weight (g) of the contents and weight (g) of water added to the volumetric flask. Determine the volume in ml of sample under test. Calculate the weight (g)/liter of ice cream mix by multiplying the SG [5].

$$\text{Specific gravity (SG)} = \frac{\text{Weight of mix taken in a flask}}{\text{Volume in ml of samlple occupied}} \quad (2)$$

7.1.5 DATA COLLECTION SHEET

Title: Determination of weight per unit volume of ice cream.

Location and Organization:______________________________

Participants: ______________________________________

Date (Month/Day/Year):________________________________

Observations:

Parameters	**Repeat 1**	**Repeat 2**	**Repeat 3**	**Average**
Weight of the sample (g)				
Volume of the medium (ml)				
Weight of the medium (g)				
Specific gravity of sample				
Weight/volume (g/liter) of ice cream				

Comments:____.

7.2 DETERMINATION OF TOTAL SOLIDS (TS) IN ICE CREAM

7.2.1 INTRODUCTION

Total solids (TS) determination within the dairy products plays a vital role in ensuring the quality parameters. TS are defined as the dry matter that remains after complete removal or extraction of moisture from a food product. Concerned to a food manufacture and a consumer, moisture removal is of great economic importance because water is inexpensive filler as well as a source for spoilage since it is the major constituent in most of the food products especially milk and milk products. For ice creams and frozen dairy products, TS testing is one of a few analytical tests used prior to packaging and distribution.

Traditional method used for the TS or moisture estimation in food products was done using a hot air oven. The obtained results were accurate, but it is a time-consuming method which lasts several hours to dry a sample. Later on, several methods are developed including vacuum, microwave, and infrared drying and the use of several kinds of instruments that exploit ceramic heating elements or halogen lamps. They were often used for on-spot examination of moisture due to their rapid or quick results, but they lack the accuracy and repeatability in results. Because of this reason, traditional gravimetric oven drying method is used commonly to get accurate results.

As per Indian Standard Specification (ISS), ice cream and chocolate ice cream should contain not < 36% of TS; and milk ices or milk lollies should contain not < 20% of TS [5].

7.2.2 PRINCIPLE

In hot air oven drying method, under specified conditions, a known quantity of sample is heated and the weight loss in the sample is used to estimate the TS content. The measured TS content is greatly reliant on the time and temperature of drying, kind of oven and other conditions in the oven.

On drying of ice cream sample, the water is lost leaving the residue behind. To have uniform heating and avoid charring of the solids, the ice cream is dried at 100±2°C after mixing with a known quantity of sand [3, 5].

7.2.3 MATERIALS

7.2.3.1 APPARATUS

- Flat bottomed dishes (7 to 8 cm internal diameter; maximum depth of 2 cm) of aluminum, nickel, or stainless steel;
- Stirring glass rods with widened flat ends;
- Sand which retains on 180 μ and permits through 500 μ IS sieve;
- Hot air oven: Temperature up to 100±2°C and well ventilated;
- Desiccator;
- Weighing balance.

7.2.4 PROCEDURE

7.2.4.1 PREPARATION OF ICE CREAM SAMPLE

For plain product, soften the ice cream at room temperature. Samples should not be heated over flame or water bath since melted fat inclines to separate and comes to the surface. Thoroughly mix by stirring with an eggbeater or spoon or by pouring one beaker to other.

For product containing insoluble particles (like fruits, nuts, and chocolate), use a mixer capable of grinding the ice cream into uniform fine mass. Fill up the mixer cup about 1/3rd of the sample or use 100 to 200 g of sample. Sample must be melted at ambient temperature or in an incubator at 37°C. Transfer the sample to the mixer cup and blend till insoluble particles are uniformly dispersed (around 7 and 3–5 min. for nuts and fruit pieces, respectively). Instead, the sample may be crushed in a pestle and mortar. The sample temperature should not exceed 25°C while softening and mixing steps.

The size of the sample shall not be less than 100 g and it should be taken in a dry, clean container with an airtight closure. The sample containers should be stored at a temperature not more than –15°C.

7.2.4.2 EXPERIMENTATION

Flat bottomed dish having glass rod and 20 g sand is placed in an oven (100±2°C) for around one hour. Then transfer to a desiccator and allow cooling for 30 min. and weigh it. Add 5 g of prepared sample into a dish. A small quantity of water is added to saturate the sand and then uniformly mix ice cream with sand by means of a rod. Spread the mixture over the dish with a glass rod. Keep the dish for 20–30 min on a boiling water bath. Then transfer the dish from the water bath (wipe out the water outside the dish) to well-ventilated oven (100±2°C) for drying for about 4 h.

After 4 h, place the dish to an efficient desiccator for cooling till bearable warm, followed by record weight of the dish. Keep the dish again in oven to re-heat for an additional period of 1 h at 100±2°C, followed by cooling in desiccator and weighing. Repeat the same process till successive weighing agrees within 0.5 mg. Determine the TS content from the observed weight loss [3].

7.2.5 CALCULATIONS

$$\text{Total solids (\%)} = \frac{W_3 - W_1}{W_2 - W_1} \times 100 \qquad (3)$$

where; W_1 is the initial mass of dish, sand, and glass rod (g); W_2 is the mass of dish, sand, and rod + ice cream before drying (g); W_3 is the mass of dish, sand, and rod + ice cream after drying (g).

7.2.6 DATA COLLECTION SHEET

Title: Determination of total solid content of ice cream by oven drying method.

Location and Organization:____________________

Participants: ____________________

Date (Month/Day/Year):____________________

Parameters	Repeat 1	Repeat 2	Repeat 3
Initial mass of dish, sand, and glass rod = W_1 (g)			
Mass of dish, sand, and rod + ice cream before drying = W_2 (g)			
Mass of dish, sand, and rod + ice cream after drying = W_3 (g)			
Total solids (TS), %			
Average total solids (%)			

Comments:____________________.

7.3 DETERMINATION OF FAT CONTENT IN ICE CREAM

7.3.1 INTRODUCTION

Fats and oils are part of a class of biomolecules defined as lipids, which are least soluble in water, but soluble in organic solvents like hexane. Food contains many types of lipids, but the majority of food lipids fall under triacylglycerols or phospholipids.

The internal molecular structure of ice cream is mainly contributed by lipids, and it enhances the buttery and creamy flavor of ice cream. It produces a smooth and creamy texture to the product while consuming by lubricating the palate. During storage, it contributes to the stability of air bubble and helps in prolonging the keeping quality by decreasing water re-crystallization, which therefore helps in making desirable melting properties [9]. The content and type of fat in ice cream are used to classify individual products according to legal standards. 'Dairy ice cream' should have a minimum of 10% milk fat and should not contain any other fat other than milk fat (Table 7.1).

TABLE 7.1 Classification of Ice Cream based on Milk Fat Content [3]

Types	Fat Content (%)
Ice cream	Not < 10.0
Medium fat ice cream	> 2.5 but < 10.0
Low fat ice cream	Not > 2.5

Source: Food Safety Standard Authority of India (FSSAI) [3].

At present, we do have multiple ranges of analytical techniques for estimation of total lipids in foods. Traditional analytical methods hold solvent extraction of lipids and the gravimetric estimation of total fat content. Apart from these, alternative methods like Babcock method, Gerber method, rapid detergent method, automated turbidimetric methods, and infrared spectroscopic methods are scientifically approved for the same. In this section, the crude estimation of total lipids in the ice cream based on the weight of the extracted lipids using the suitable solvents through the Gravimetric approach has been discussed.

7.3.2 PRINCIPLE

Definition of fat clearly points out the solubility in organic solvents and insolubility in water. Hence, water insolubility is a property during analysis which is now the basic property for the separation of fat from other food constituents like water, proteins, and carbohydrates. Commonly used analytical techniques to determine the total fat content of a food is by acid or alkaline hydrolysis followed by Rose-Gottlieb method or by extraction with organic solvents. The accuracy of quantitative estimation of fat from ice cream is ensured only when the protective film surrounding the milk fat globule is broken by using suitable agents. This is done by treating the sample with ammonia and ethyl alcohol. The role of ammonia is to break or cleave the protective layer while the addition of alcohol leads to the easy passage of fat globules from the aqueous to solvent phase. The mixed solvent, petroleum ether and diethyl ether being non-miscible with water results in good extraction of fat from the other solids in the solution to the solvent phase. Then the extract is dried for the removal of residual solvent and the weight of the extracted fat was calculated [8].

7.3.3 MATERIALS

7.3.3.1 REAGENTS

- Ammonium solution (SG of 0.8974 at 16°C);
- Ethyl alcohol (C_2H_5OH; 95–96% v/v);
- Diethyl ether (SG 0.702);
- Petroleum ether (boiling point ranges from 40–60°C);
- Mixed Solvent: Mix 1:1 ratios of diethyl ether and petroleum ether.

7.3.3.2 APPARATUS

- Rose-Gottlieb or Mojonnier fat extraction flask;
- Bark Cork or ground glass stopper;
- Fat collecting vessel: Flat bottom flask (125 ml);
- Boiling water bath;
- Hot air oven (set at temperature of 102±2°C);
- Desiccator;
- Weighing balance;
- Beakers (100 ml);
- Glass beads;
- Measuring cylinders (25 ml).

7.3.4 PROCEDURE

7.3.4.1 PREPARATION OF ICE CREAM SAMPLE

The sample is prepared according to the method mentioned in the sub-chapter on the determination of total solids in ice cream (*refer* to Section 2.4.1).

7.3.4.2 EXPERIMENTATION

4–5 g of prepared ice cream shall be accurately weighed and transferred to a Mojonnier flask (fat extraction tube). Rinse the sides of the flask properly with 2 ml of distilled water and swirl it gently for a uniform mix. Two milliliters of ammonia solution are transferred to the extraction tube and mixed well thoroughly in the lower bulb. Heat the content at 60°C for 20 min in a water bath with occasional stirring and then cooled. Add 10 ml C_2H_5OH to the flask and mix thoroughly between the two bulbs by letting the content to flow forwards and backwards. Followed by dipping the Mojonnier flask in chilled water or allow cooling in cold running water.

Add 25 ml diethyl ether followed by close the tube with a cork (wetted stopper) and mix the contents vigorously for one min. Remove the cork and then 25 ml petroleum ether is added. Close the tube with rewetted cork and shake the tube gently for about half a minute.

Centrifuge at 500–600 rpm for 1–5 min. In case of non-availability of centrifuge, then place the tube in a stand-on position using rack for minimum 30 minutes till the upper solvent layer is clear and noticeably separated from the aqueous layer. If there is a tendency to form emulsion, add few ml of alcohol to aid further separation of the layers.

Take the empty weight of a flat bottom flask (fat collecting vessel) containing a few glass beads. Carefully transfer the supernatant ethereal layer completely (as much as possible) into the vessel. Wash the neck of the tube and cork with a small amount of the mixed solvent when complete transfer is done; collect the rinsing in the vessel.

To extract the remaining fat within the extraction tube, repeat twice the same procedure by using 15 ml each of diethyl ether and petroleum ether in each step. Transfer the ethereal layer carefully from the tube to the same vessel and evaporate off completely.

Heat the vessel containing residual fat for 1 h in an oven (102±2°C), followed by transferring the vessel in a desiccator and record the weight after cooling. Repeat the operations of heating and weighing till the difference between two consecutive weighing should be < 0.5 mg. Note down the minimum weight of the vessel containing residual fat. About 25 milliliter of petroleum ether is added to the vessel so as to check the extracted fat is completely soluble or not. Carry out the same procedure with 10 ml water instead of sample for running blank [3, 5].

7.3.5 CALCULATIONS

Fat content (% w/w) in ice cream sample is calculated as follows:

$$\text{Fat (\%)}\,(w/w) = \left(\frac{W_2 - W_1}{W}\right) \times 100 \tag{4}$$

where; W is the mass of sample taken (g); W_1 is the initial mass of the fat collecting vessel (g); and W_2 is the mass of vessel + extracted fat after drying (g).

7.3.6 DATA COLLECTION SHEET

Title: Determination of fat in ice cream

Location and Organization:____________________________

Participants: ____________________________

Date (Month/Day/Year):____________________________

Observations:

Parameters	**Repeat 1**	**Repeat 2**	**Repeat 3**
Weight of sample taken = W g			
Weight of fat collecting vessel = W_1 g			
Weight of vessel + fat extract after drying = W_2 g			
Fat (%)			
Average fat (%)			

Comments: ____________________________.

7.4 DETERMINATION OF PROTEIN IN ICE CREAM BY KJELDAHL METHOD

7.4.1 INTRODUCTION

Milk protein in ice cream contributes to many important structural functions. They attribute to emulsification of fat to give a stable emulsion during the homogenization of the mix. A milk protein even plays an important role in the aging process by subsequently interacting with the emulsifier and produces a fat emulsion which is capable of partially coalescence in the whipped and frozen product, thus producing a desired phase of fat. During whipping, serum phase proteins aid to the interface formation of an air bubble for preserving small and stable air bubbles. Viscosity of the mix is raised by un-adsorbed proteins mainly in the unfrozen serum portion after hardening. This shows the development of body and texture and reduces the rates of ice re-crystallization. The average protein percentage of ice cream, milk ices and milk lollies fall in 4.1, 3.25, and 0.4%, respectively.

The dairy ice creams are an excellent source of biological value since they are rich with all essential amino acids of milk protein like tryptophan and lysine. Not only these proteins are believed to be complete, but also, they are assimilated 5% to 6% more complete compared to other food sources [13].

The unique structural conformations of proteins could be altered by some agents collectively known as denaturants. They can also alter the functional properties of proteins especially solubility, foaming, and emulsification. The most commonly used denaturants include acids/alkali, urea, heat, guanidine-hydrochloride, organic solvents, and detergents. The determination of total protein content in food products is complicated because of some food constituents (amino sugars, urea, uric acid, some vitamins, etc.) possess similar physicochemical properties. Depending upon the selected method of protein determination, other biomolecules like lipids and carbohydrates may interfere physically.

To determine the total protein content, numerous methods have been developed. All these developed methods focus on the basic principles of either the estimation of nitrogen (N), peptide bonds, amino acids (aromatic), dye-binding ability, light scattering properties and ultraviolet absorptivity of proteins. The most widely accepted method is the Kjeldahl method applicable in quality control and nutritional labeling of dairy products.

7.4.2 PRINCIPLE

The Kjeldahl method was developed by Johann Kjeldahl in 1883. It is truly based on the digestion of organic compounds in the food with a strong acid which results in the release of nitrogen (N). This can be estimated by a titration method and the quantity of total protein present in food is then estimated from the obtained N content. Kjeldahl method is generally regarded as the standard and most accurate method to determine the total protein concentration. However, in this method, it indirectly determines the total protein by measuring the amount of the nitrogen. Since nitrogen is a characteristic of proteins that differentiates proteins from carbohydrates and fats, and the nitrogen content of most proteins is quite similar, it is generally an average value of about 16%. About 6.25 (corresponding to 0.16 g N/g of protein) is the conversion factor (CF) commonly used for several food protein determinations. Each protein has a diverse CF which is dependent on its composition of amino acids. In case of milk proteins, a CF of 6.38 is used [10].

There are three major steps involved in this process.

1. **Digestion of the Sample:** First step involves the digestion of the sample by heating in the presence of concentrated H_2SO_4, anhydrous K_2SO_4 and a catalyst (like copper, selenium, titanium, or mercury). This reaction is enhanced by H_2SO_4, which acts as an oxidizing agent to digest the organic components, whereas anhydrous K_2SO_4 rise the boiling point and thereby accelerates the reaction. The N present in the food gets digested and converted into NH_3 and other organic substances to water and carbon dioxide. Ammonia gas formed as a product is not liberated in an acidic solution because the ammonia thus formed binds to the sulfate ion and forms ammonium sulfate which remains in solution.

$$\text{Protein} + H_2SO_4 \xrightarrow[Heat]{K_2SO_4 + CuSO_4} H_2O + CO_2 + (NH_4)_2SO_4$$

2. **Neutralization and Distillation:** The end product of the first step is diluted with water and further addition of sodium hydroxide (NaOH) converts the acidic solution to alkaline by neutralizing the sulfuric acid. This results in the formation/liberation of ammonia because of the reaction occurred between ammonium sulfate and concentrated NaOH solution.

   Distillation of ammonia is followed by adding an excess of boric acid solution comprised with pH indicators methylene blue

(MB) and methyl red [10]. Distillation lowers the pH and hence the ammonia gas is converted into the ammonium ion, while boric acid is converted into borate ion simultaneously.

$$(NH_4)_2SO_4 + 2NaOH \longrightarrow 2NH_3 + Na_2SO_4 + 2H_2O$$
$$NH_3 + H_3BO_3 \longrightarrow NH_4^+ + H_2BO_3^-$$

3. **Titration:** Borate anion is titrated with standardized hydrochloric acid (HCl) solution. Changing green color to violet color is an indication to stop the titration at the end. The quantity of standard acid used in the titration is equivalent to the nitrogen content in the sample.

$$H_2BO_3^- + H^+ \longrightarrow H_3BO_3$$

7.4.3 MATERIALS

7.4.3.1 REAGENTS

- Concentrated H_2SO_4: Approximately 98% weight and should be nitrogen-free;
- NaOH solution (45%);
- Hydrochloric acid (HCl) solution (0.1 N): Pre-standardized HCl should be used;
- Potassium sulfate (K_2SO_4): It should be nitrogen-free;
- Copper sulfate pentahydrate ($CuSO_4$ $5H_2O$);
- Boric acid solution (4%);
- Mixed indicator. Dissolve 2 g of methyl red and 1 g of MB in 1,000 ml of ethanol.

7.4.3.2 APPARATUS

- Digestion flask (Kjeldahl): Suitable for digestion with capacities 500 or 800 ml;
- Distillation apparatus;
- Conical flask: capacity 250 ml;
- Glass beads (5 mm diameter);
- Burette (50 ml capacity);
- Analytical balance;
- Measuring cylinders.

7.4.4 PROCEDURE

7.4.4.1 PREPARATION OF ICE CREAM SAMPLE

The sample is prepared according to the method mentioned in the sub-chapter on determination of total solids in ice cream (*refer* to Section 2.4.1) [10].

7.4.4.2 EXPERIMENTATION

The prepared ice cream sample weighing about 5–8 g is taken into a Kjeldahl flask of suitable size. Care should be taken while transferring the sample to the flask to avoid no portion gets adheres to the sides of the flask. To this, add 15 g of K_2SO_4, 0.9 to 1.2 g of $CuSO_4.5H_2O$ and 40 ml of concentrated H_2SO_4 and Swirl the contents of Kjeldahl flask thoroughly. Add two to three glass beads prior to the occurrence of the digestion process and then place it on the digestion chamber. During the reaction, moderately heat the flask for the first few minutes to eliminate the possibility of excess foam formation up to the neck of the flask. Then heat moderately (<200°C), mix frequently, until the complete carbonization of the content and the disappearance of foam is seen. Then raise the temperature (400–410°C) to heat more intensively and continue the digestion of sample at the same temperature for at least 1 hour until clear digestion is obtained. Generally, clear digestion indicated with light blue-green color. If clear liquid is not visible as frothing formation on the surface, it may indicate temperature of the chamber is not enough to digest the mixture. Then, digestion time should be increased to 1 to 1.5 h. At the end of the process, the digest must be cooled at room temperature for about 25 minutes. Carefully add 200–300 ml of water in each flask and swirl the content so that any developed crystals will be dissolved.

Turn on the water condenser for distillation which is attached to the tap water circulatory system. To make the above contents strongly alkaline, add enough NaOH solution (about 110 ml) along the side of the flask. The flask is then connected to a distillation setup including flash head and condenser. The reaction of NaOH with a clear digest occurs and ammonia will be liberated. Keep conical flasks containing 100 ml of H_3BO_3 solution and 2–3 drops of mixed indicator in another outlet or delivery tube is inserted. This delivery tube is used for redirecting the liberated ammonia and trap into H_3BO_3 solution.

The flask is subjected for a thorough mixing and heating in such a way that, every 30 min. roughly 150 ml of the distillate is collected in the conical flask receiver. Distillation is continued until all the formed ammonia has passed over to the boric acid, which is indicated by the collected distillate of about 150 ml.

Take out the receiver flask from the distillation unit. Carry out the titration of collected distillate with HCl (0.1 N) solution until the green color changes to violet. Note down the volume of acid used in the titration [3].

Blank test is carried out by taking 0.5 g of sucrose instead of ice cream sample. Rest of procedure is followed as described above [13].

7.4.5 CALCULATIONS

Percent nitrogen and crude protein content is calculated as follows:

Moles of hydrochloric acid = moles of NH_3 = moles of nitrogen in the test portion

$$\%Nitrogen = \frac{N1 \times (A - B) \times 14.007 \times 100}{WX1000} = \frac{1.4007 \times (A - B) \times N1}{W} \quad (5a)$$

where; N1 is the normality of HCl; A is the titer value of HCL (in ml) used for sample; B is the titer value of HCl (in ml) used for blank; W is the weight of ice cream sample; and 14.007 = equivalent weight of nitrogen.

$$\% \text{ Protein} = \% \text{ Nitrogen} \times 6.38 \quad (5b)$$

7.4.6 DATA COLLECTION SHEET

Title: Determination of protein content in ice cream by Kjeldahl method.

Location and Organization:____________________________

Participants: __________________________________

Date (Month/Day/Year):______________________________

Observations:

Parameters	**Repeat 1**	**Repeat 2**	**Repeat 3**
Weight of sample taken = W g			
Titer value of HCL used for sample = A ml			
Titer value of HCl used for blank = B ml			
Total nitrogen (%)			
Total protein (%)			
Average protein (%)			

Comments:____.

7.5 DETERMINATION OF LACTOSE IN ICE CREAM

7.5.1 INTRODUCTION

Lactose is a popularly known as milk sugar which is present in the freeze-concentrated aqueous serum or matrix phase of ice cream along with other components like sugar, protein, polysaccharide, and salts [6]. In ice cream, lactose content is around 6% but it is not contributing much on the sweetness. Whey powder is also used in the ice cream preparation to enhance the body and texture of the final product by increasing the solids-not-fat (SNF) content. In such cases, excessive lactose levels in the mix leads to freezing point (FP) depression which ultimately results in lactose crystallization, development of iciness and softening of ice cream. Certainly, lactose occurs in ice cream as a supersaturated, amorphous state in an extremely viscous mass. Stabilizers retain the lactose molecules in a supersaturated state and prevent crystal formation and growth [14]. Lactose estimation in ice creams and other frozen desserts is determined by the Lane-Eynon method [3].

7.5.2 PRINCIPLE

Lactose can be simply defined as a disaccharide of glucose and galactose linked together by β (1-4) glycosidic linkage. Since the aldehydic group of anomeric carbon of glucose moiety is free, lactose exhibit reducing property. Therefore, it undergoes tautomerization in alkaline solutions and produces an intermediate known as enediols which is highly reactive. The enediol forms of lactose reduce metal ions (Cu, Fe, Ag, etc.).

The Lane-Eynon method is a volumetric analysis of reducing sugar using Fehling solution. Fehling solution contains Fehling-A ($CuSO_4$) and Fehling-B (alkaline $KNaC_4H_4O_6.4H_2O$). When copper sulfate is made alkaline, the cuprous hydroxide [$Cu(OH)_2$] gets precipitated. But in the presence of potassium-sodium tartrate helps to make a soluble complex with copper compound and prevents the precipitation of $Cu(OH)_2$. Before proceeding to reduction reaction, diluted solution of ice cream should be treated with lead acetate and potassium oxalate solution to remove protein and fat. While heating the Fehling solution, cupric oxide (CuO_2) will be formed and subsequently reacts with filtrate. The reducing sugar (lactose) reduces cupric ions (Cu^{2+}) to cuprous ions (Cu^{+}), which forms a brick-red precipitate of cuprous oxide (Cu_2O) and it gets oxidized. The endpoint of titration can be determined by using MB as an indicator which forms a brick-red precipitation of Cu_2O.

$$CuSO_4 + 2NaOH \xrightarrow{Na-K\ tartarate} Na_2SO_4 + Cu(OH)_2$$
$$Cu(OH)_2 \xrightarrow{Heat} CuO + H_2O$$
$$CuO + R-CHO \longrightarrow \underset{Brick\ red\ precipitate}{Cu_2O} + R-COOH$$

7.5.3 MATERIALS

7.5.3.1 REAGENTS

- NaOH (0.1 N).
- Standard lactose solution (5 mg/ml): 5 g of lactose powder is weighed accurately and transferred to a 1-liter volumetric flask containing water and volume made up to the mark.
- Methylene blue (MB) solution: 0.2 g (w/v) in water.
- Neutral lead acetate (10%) solution.
- Potassium Oxalate (10%) solution.
- Fehling solution A: Weigh 34.639 g of $CuSO_4 \cdot 5H_2O$ and dissolve in water. Final volume made up to 500 ml in a volumetric flask. Use asbestos to filter the solution if required.
- Fehling solution B: 173 g of $KNaC_4H_4O_6 \cdot 4H_2O$ (potassium-sodium tartarate, also known as Rochelle salt) and 50 g of NaOH are dissolved separately in water. Mix it thoroughly and final volume is made to 500 ml. Keep this solution undisturbed for 2 days. Use asbestos to filter the solution if required.

7.5.3.2 APPARATUS

- Burette stand.
- Burette: capacity of 50 ml.
- Conical flasks: capacity of 250 ml.
- Funnels.
- Hot plate.
- Measuring cylinder (capacities of 25 and 50 ml).
- Pipettes (capacities of 1, 5, and 10 ml)
- Volumetric flasks (capacities of 100, 250, 500, and 1,000 ml).
- Watch glass.

- Whatman filter paper: Grade 1.
- Wire gage.

7.5.4 PROCEDURE

7.5.4.1 PREPARATION OF ICE CREAM SAMPLE

The sample is prepared according to the method mentioned in the sub-chapter on determination of total solids in ice cream (*refer* to Section 2.4.1) [3].

7.5.4.2 EXPERIMENTATION

1. **Preparation of Ice Cream Sample Filtrate:** A 30 g of ice cream sample should be mixed with 60 ml of neutral lead acetate (10%) solution, and volume made to 250 ml by diluting with water. Filter the solution after allowing it to stand for 15–30 min. Subsequently, add 40 ml of potassium oxalate (10%) solution to 200 ml of the resulting filtrate. Dilute again to 250 ml with water and filter. About 20 ml of the final filtrate should again dilute to 100 ml for the lactose estimation [11].
2. **Standardization of Fehling's Solution:** Take 5 milliliters each of Fehling-A and -B solutions in a 250 ml conical flask and fill up the burette with standard lactose solution. Heat the content of the conical flask over a heater and bring it to boil. Bumping can be prevented by adding 4–5 glass beads to the flask. While boiling, 2–3 drops of MB indicator is added. Titration of the content is performed against the lactose solution till a brick-red color appears. Note the titer value of standard lactose solution used to reduce all the copper. After this preliminary titration, repeat this procedure again by adding almost all the standard solution (1 ml less than the titer value of preliminary titration) before beginning the heating and continue the procedure as mentioned before. Complete the titration within three minutes with continuous boiling by addition of standard lactose solution to prevent oxidation by air [3]. Note down the titer value and denote it as V_1.
3. **Estimation of Lactose in the Filtrate of Ice Cream:** Pipette out 5 milliliters each of Fehling solutions in a conical flask (250 ml) and fill up the burette with filtrate of ice cream. Heat the content of the

conical flask over a burner or heater and boil. Add 4–5 glass beads to avoid bumping. Around 2–3 drops of MB indicator are added while boiling. Volumetric titration is carried out against the filtrated sample from the burette until the color change from blue to a bright brick red color of Cu_2O appears. Repeat the titration to get constant readings and note the titer value as V_2.

7.5.5 CALCULATIONS

Lactose content (%) in ice cream is calculated as follows:

$$\%\text{ Lactose in ice cream} = \frac{V_2 \times 0.005 \times 100}{V_2} \times \frac{250}{W} \times \frac{250}{200} \times \frac{100}{20} \quad (6)$$

where; W is the weight in g of ice cream; C_1 is the concentration of lactose solution (5 mg/ml); V_1 is the ml of lactose solution needed to reduce a Fehling's solution (10 ml); and V_2 is the ml of the sample filtrate needed to reduce a Fehling's solution (10 ml).

7.5.6 DATA COLLECTION SHEET

Title: Determination of lactose in ice cream.

Location and Organization:________________________

Participants: ________________________

Date (Month/Day/Year):________________________

Observations:

Parameter	**Repeat 1**	**Repeat 2**	**Repeat 3**	**Average**
Weight in g of ice cream = W				
ml of standard lactose solution used in titration = V_1				
ml of the filtrated solution of ice cream used in titration = V_2				
% Lactose				

Comments:________________________.

7.6 DETERMINATION OF SUCROSE CONTENT IN ICE CREAM

7.6.1 INTRODUCTION

The most used sweetening agent in ice cream is sucrose. The sweetening agents are added @ 12–17% by weight in the ice cream mix. It is the cheapest source of total solids. The main functionalities of sucrose other than sweetening are to enhance flavor and improve the texture and palatability. It is present in the form of a glassy state with some crystals in ice cream [14].

Sucrose is a non-reducing sugar. Therefore, a direct estimation of sucrose by Lane-Eynon method is not possible. Therefore, a different approach has been adopted to make this sucrose into reducing form by inversion.

7.6.2 PRINCIPLE

Hydrolyzed mixture of sucrose comprising of glucose and fructose, is known as invert sugar. During hydrolysis of sucrose by dilute acid, it liberates one molecule of glucose and fructose. Therefore, the invert sugar has the ability to reduce cupric ions (Cu^{2+}) to cuprous ions (Cu^{+}), which forms a red precipitate of cuprous oxide and itself get oxidized.

7.6.3 MATERIALS

7.6.3.1 REAGENTS

- NaOH solution: 0.1 N and 1 N.
- Hydrochloric acid (HCl) solution: 12 N and 6.34 N.
- Stock invert sugar (IS) solution (2.5 g/100 ml): 23.75 g of sucrose is dissolved in a beaker containing 100 ml of water and mix it thoroughly. The contents are transferred to the 1-liter volumetric flask. To this 10 ml of conc. HCl is added, mix thoroughly, and keep it undisturbed for not less than 3 days at 20°C. Finally, volume is made to one liter with water.
- Standard dilute IS solution (2.5 mg/ml): Take 50 ml of stock solution and neutralize with 1 N NaOH using litmus paper and then dilute to 500 ml immediately before use.
- Neutral lead acetate solution (density 1.25 at 20°C): 41 g of lead acetate is dissolved in cold water and volume made to 100 ml with

water after neutralization (use litmus paper to neutralize the solution using CH_3COOH or NaOH if required).

- Alumina cream: Prepare a saturated aluminum potassium sulfate solution in cold water. To this, enough ammonia solution (specific gravity = 0.88) is added and is stirred continuously, to make the solution alkaline to litmus. Wash with water, by decantation, allowing the gelatinous precipitate to settle thoroughly between each washing, till sulfate remains to be trace in the washings.
- Sodium oxalate ($Na_2C_2O_4$) solution: saturated solution.
- Methylene blue (MB) indicator: 0.2 g (w/v) in water.
- Fehling solution A: Weigh 34.639 g of $CuSO_4 \cdot 5H_2O$ and dissolve in water. Final volume made up to 500 ml. Use asbestos to filter the solution if required.
- Fehling solution B: 173 g of $KNaC_4H_4O_6 \cdot 4H_2O$ (potassium-sodium tartarate, also known as Rochelle salt) and 50 g of NaOH are dissolved separately in distilled water. Mix it and final volume is made to 500 ml. Keep this solution undisturbed for 2 days. Use asbestos to filter the solution if required.

7.6.3.2 APPARATUS

- Burette stand;
- Burette (capacity of 50 ml);
- Conical flasks (capacity of 250 ml);
- Funnels;
- Hot plate;
- Measuring cylinders (capacities of 25 and 50 ml);
- Pipettes (capacities of 1, 5, and 10 ml);
- Volumetric flasks (capacities of 100, 250, 500, and 1,000 ml);
- Watch glass;
- Whatman filter paper: Grade 1;
- Wire gage.

7.6.4 PROCEDURE

7.6.4.1 PREPARATION OF ICE CREAM SAMPLE

The sample is prepared according to the method mentioned in the sub-chapter on determination of total solids in ice cream (*refer* to Section 2.4.1) [14].

7.6.4.2 EXPERIMENTATION

7.6.4.2.1 Standardization of Fehling's Solution

Take 10 ml of each Fehling's solution A and B in a 250 ml conical flask and fill up the burette with standard dilute invert sugar (IS) solution. Heat the content of the conical flask over a burner or heater and boil. To prevent bumping, add 4–5 glass beads to the flask. Contents of the flask are titrated against standard IS solution until it reduces Fehling's solution completely (bright brick red color of Cu_2O appears). Add 1 ml of MB indicator within 2 min from onset of boiling. Continue the titration by adding a small amount of IS solution till the color changes to brick red. Record the volume of IS solution required to reduce all the copper. After this preliminary titration, repeat this procedure again by adding almost all the standard IS solution (1 ml less than the titer value of preliminary titration) before beginning the heating and continues the titration as explained earlier. Complete the titration within three minutes with continuous boiling to prevent oxidation by air. Note down the titer value and denote it as V_1 [5].

7.6.4.2.2 Determination of Original Reducing Sugars

Accurately weighed 10 g of prepared sample is transferred to conical flask (250 ml) and to this add 150 ml of water. Contents of the flask are mixed thoroughly. Drop by drop lead acetate is added to the flask and swirl it by rotating till no precipitate is obtained further. One drop of alumina cream added to it, mix well, and keep it for few min. For complete precipitation of protein, rotate the content of the flask at some intervals. Required quantity of $Na_2C_2O_4$ solution is added to precipitate the surplus of lead. Filter through a filter paper and transferred the filtrate to 250 ml volumetric flak. Wash the precipitate and filter paper with hot water. Collect the washings and transferred to the volumetric flask. Finally, make up the volume to the mark and let this solution be denoted as C.

Pipette 10 milliliters each of Fehling's solution in a 250 ml flask and fill up the burette with filtrate of ice cream (C). Complete the titration with sample filtrate as mentioned under standardization of Fehling's solution. Note down the titer value and denoted it as V_2 [5].

7.6.4.2.3 Determination of Total Reducing Sugar After Inversion

Transfer 50 ml of filtered solution of ice cream (C) into a volumetric flask (200 ml). About 10 ml of 6.34 N HCl and 25 ml of H_2O are added to the flask. Keep it on a water bath set at a temperature of 70°C/5 min. with continuous mixing for the first 3 minutes. Cool the contents of flask immediately to 20°C and use litmus paper to neutralize the contents with 1 N NaOH. Dilute to 200 ml at 20°C and denote it as D.

Pipette 10 milliliters each of Fehling's solution in 250 ml conical flask and fill up a burette with filtrate of sample after inversion (D). Complete the titration with sample filtrate after inversion (D) using the same procedure as described in the standardization of Fehling's solution. Note down the titer value and denoted it as V_3 [5].

7.6.5 CALCULATIONS

1. **Determination of Reducing Sugar:**
   20 ml of Fehling's solution = V_1 ml of IS solution = V_2 ml of sample C
   Therefore, V_2 ml of sample C = V_1 ml of IS solution
   = $V_1 \times C_1$, g of IS
   = $V_1 \times 0.0025$, g of IS

$$\text{Reducing sugar present in ice cream (\%)} = \frac{V_1 \times 0.0025 \times 100}{V_2} \times \frac{250}{W} \quad (7)$$

2. **Determination of Total Reducing Sugar after Inversion:**
   20 ml of Fehling's solution = V_1 ml = V_3 ml
   Therefore, V_3 ml of sample D = V_1 ml of IS solution
   = $V_1 \times 0.0025$ g of IS

Total reducing sugar present in ice cream after inversion (%) =

$$\frac{V_1 \times 0.0025 \times 100}{V_3} \times \frac{250}{W} \times \frac{200}{50} \quad (8)$$

3. **Determination of Sucrose Content in the Ice Cream:**

Sucrose in ice cream (%) = [(percent reducing sugar after inversion – percent reducing sugar) × 0.95] (9) where; W is the weight in g of ice cream (10 g); C_1 is the concentration of standard IS solution (0.0025 g/ml); V_1 is the ml of standard IS solution used in titration; V_2 is the ml of sample C used in titration; and V_3 is the ml of sample D used in titration.

7.6.6 DATA COLLECTION SHEET

Title: Determination of sucrose percentage in ice cream

Location and Organization:______________________________

Participants: ____

Date (Month/Day/Year):________________________________

Observations:

Parameter	**Repeat 1**	**Repeat 2**	**Repeat 3**	**Average**
Weight of ice cream = W g				
Standard IS solution used in titration = V_1 ml				
Sample C used in titration = V_2 ml				
Sample D used in titration = V_3 ml				
Reducing sugar (%)				
Total reducing sugar after inversion (%)				
Sucrose (%)				

Comments:____.

7.7 DETERMINATION OF ASH CONTENT IN ICE CREAM

7.7.1 INTRODUCTION

Ash is the inorganic matter left after removal of organic components either by ignition or by complete oxidation in any food. It is the measure of total minerals that occurs in food. Estimating the ash content is the component of compositional analysis for nutritional labeling [2]. Salts in ice cream are present in the freeze-concentrated aqueous serum or matrix phase along with other components like sugars, proteins, and polysaccharides [6].

7.7.2 PRINCIPLE

When the food sample is incinerated, the organic components get oxidized and burnt away. The leftover residue will be in the form of white ash [7].

7.7.3 MATERIALS

7.7.3.1 APPARATUS

- Silica crucible;
- Muffle furnace;
- Water bath;
- Desiccator;
- Safety tongs.

7.7.4 PROCEDURE

Take the weight of an empty dry silica crucible and denote it as W. Weigh 10 g of prepared sample accurately in a silica crucible and note it as W_1. Keep the sample containing crucible on boiling water bath and evaporate the moisture to dryness. Transfer the crucible to the muffle furnace and incinerate the sample at 550°C for 4 hours to obtain white ash free from carbon. Cool the sample in desiccator. Take the weight of crucible with ash and denote it as W_2. Repeat the heating in a muffle furnace for 30 min followed by cooling and weighing till the two consecutive weighing differences should be < 1 mg [7].

7.7.5 CALCULATIONS

$$\text{Ash content (\%) by mass} = \frac{W_2 - W}{W_1 - W} \times 100 \tag{10}$$

where; W is the mass of empty crucible (g); W_1 is the mass of crucible with ice cream (g); and W_2 is the mass of crucible with ash (g).

7.7.6 DATA COLLECTION SHEET

Title: Determination of ash content in ice cream.

Location and Organization:______________________________

Participants: ______________________________________

Date (Month/Day/Year):________________________________

Observations:

Parameter	Repeat 1	Repeat 2	Repeat 3	Average
Mass of empty crucible = W g				
Mass of crucible containing ice cream = W_1 g				
Mass of crucible containing ash, W_2 g				
% Ash				

Comments: ____.

KEYWORDS

- ash
- fat
- frozen dessert
- ice cream
- lactose
- sucrose
- total solids

REFERENCES

1. AOAC, (2000). AOAC official method, 926.14: Weight per unit volume of packaged ice cream, methods I and II. Chapter 33. In: Hortwiz, W., (ed.), *Official Methods of Analysis of AOAC International* (17th edn., Vol. II, pp. 71, 72). Gaithersburg, MD – USA: AOAC International.
2. Ismail, B. P., (2017). Ash content determination. In: Nielson, S., (ed.), *Food Analysis Laboratory Manual* (pp. 117–119). Cham: Springer.

3. Food Safety Standard Authority of India (FSSAI), (2015). *Lab Manual 1: Manual of Methods of Analysis of Food – Milk and Milk Products* (pp. 58–63). New Delhi: Food Safety Standard Authority of India.
4. Goff, H. D., (2002). Formation and Stabilization of structure in ice cream and related products. *Current Opinion in Colloid and Interface Science, 7,* 432–437.
5. IS. *Specification for Ice Cream (IS: 2802-1964)* (pp. 1–20). New Delhi: Indian Standards Institution; (1964).
6. Goff, H. D., (2003). Ice cream. In: Fox, P. F., & McSweeney, P. L. H., (eds.), *Advanced Dairy Chemistry, – 1 Proteins* (Vol. 1A, pp. 1063–1082). Boston – MA: Springer.
7. FAO, (1986). Food analysis: Quality, adulteration, and tests of identity: Ice cream. In: *Food and Nutrition Paper Series – Manuals of Food Quality Control 14/8* (p. 70). Rome: FAO.
8. AOAC. (2012). Official Method 995.19. Fat in cream: Mojonnier ether extraction method. In: *Official Methods of Analysis of AOAC International* (19th edn., p. 72). Gaithersburg, MD –USA: AOAC International.
9. Granger, C., Leger, A., Barey, P., Langendorff, V., & Cansell, M., (2005). Influence of formulation on the structural networks in ice cream. *International Dairy Journal, 15,* 255–262.
10. AOAC, Official Method 991.2. Kjeldahl estimation of protein. In: *Official Methods of Analysis of AOAC International* (17th edn., p. 72). AOAC International, Gaithersburg, MD– USA; (2002).
11. Peeples, M. L., & Hutcheson, R. M., (1978). Method for determining lactose and sucrose contents in ice cream. *Journal of Food Science, 43*(3), 799–800.
12. *IS 14825: 2000, ISO 5983: 1997, Animal Feeding Stuff – Determination of Nitrogen Content and Calculation of Crude Protein Content -Kjeldahl Method* (pp. 1–10). New Delhi: Bureau of Indian Standards (2000).
13. Shithal, D., Shrikanth, D. K., & Chendraprakash, D. K., (2016). Ice-cream: Composition and Health Effects. *The Encyclopedia of Food and Health, 3,* 385–390.
14. Goff, H. D., (1997). Ice cream. Chapter 12. In: Gunstone, F. D., & Padley, F. B., (eds.), *Lipid Technologies and Applications* (pp. 329–344). New York: Marcel Dekker Inc.

CHAPTER 8

COMPOSITIONAL ANALYSIS OF BUTTER

SATISHKUMAR C. PARMAR, ANURADHA KUMARI, AMIT K. JAIN, AHESANVARISH SHAIKH, and SHRIYESH PATEL

ABSTRACT

Butter is a product produced by the concentration of milk fat by destabilizing cream or milk emulsion. Butter is a fat-rich product and principally a water-in-oil type emulsion. During the churning process of milk or cream, the fat globules coalesce to form a fine granular mass. According to Codex standards, butter should contain 80% fat (minimum); 16% moisture (maximum) and 2% milk SNF (maximum). Ingredients like NaCl, starter culture and water can be permitted as per the regulatory guidelines. To safeguard the interest of the consumers for quality product and assured composition, different regulatory agencies have laid down analytical procedures for compositional analysis of butter. In this chapter, brief principle and procedure for analysis of different parameters of butter like moisture, fat, curd, salt, and acidity are discussed.

8.1 DETERMINATION OF MOISTURE IN BUTTER USING BUTTER MOISTURE BALANCE

8.1.1 INTRODUCTION

Moisture content in butter plays a significant role on the structural behavior of the product and in turn spreadability and consumer acceptance. Butter with less moisture content will make it hard and lack in spreadability. A higher

Analytical Methods for Milk and Milk Products, Volume 1: Sampling Methods and Chemical and Compositional Analysis. Megh R. Goyal, N. Veena, and Santosh Kumar Mishra (Eds.)

percentage of water in butter will lead to problems related to its keeping quality and flavor and texture. Moisture content of butter should be 16% (maximum) [1]. Hence, determination of butter moisture is of importance from standards as well as flavor and texture point of view. According to the Bureau of Indian Standards (BIS), moisture can be determined by moisture tester (method I) or by oven drying method (method II) [3]. Method I is a routine method and method II is a reference method. In this chapter, moisture content is estimated by butter moisture analyzer, which is a very simple, easy, and rapid method.

8.1.2 PRINCIPLE

A known weight of butter is taken in a previously cleaned and dried dish. The dish is heated to expel the moisture from the sample. Then, it is transferred to a desiccator, cooled, and weighed. The moisture lost on heating is read directly as percent moisture.

8.1.3 MATERIALS

8.1.3.1 APPARATUS

- Butter moisture tester;
- Aluminum cups;
- Tongs;
- Spirit lamp/electric hot plate/gas burner;
- Riders and 10 g weight;
- Desiccator.

8.1.4 PROCEDURE

8.1.4.1 PREPARATION OF BUTTER SAMPLE

Gently heat the sample of the container with the stopper in a water bath to a temperature not exceeding 39°C or at the lowest possible temperature in an oven or water bath until by vigorous shakings a homogenous fluid emulsion is obtained [2].

8.1.4.2 EXPERIMENTATION

Clean the aluminum cup, dry it in an oven and allowed to cool in a desiccator. Set the % rider to zero and tare the moisture scale, balance the scale with the empty cup without moving the location of the riders. Take accurately 10 g sample in the cup; hold the cup with tongs and heat over spirit lamp/electric hot plate/gas burner moving constantly in a circular motion to prevent spatting. Heat the sample continuously until the foam gets stopped and the curd at the cup's bottom acquire a specific slight brown color. Cool the cup in a desiccator. Put it on the balance and directly read for moisture percent. The residue is preserved for the analysis of curd [3].

8.1.5 CALCULATIONS

The percent moisture in butter is read directly. Moisture in a given sample of butter by moisture balance is ___________ %.

8.1.6 DATA COLLECTION SHEET

Title: Determination of moisture in butter by moisture balance.

Location and Organization:____________________________________

Participants: ____________________________________

Date (Month/Day/Year):____________________________________

Observations:

Parameters	Repeat 1	Repeat 2	Repeat 3	Average
Moisture (%) from moisture balance method				

Comments:____________________________________.

8.2 DETERMINATION OF MOISTURE IN BUTTER BY GRAVIMETRIC METHOD

8.2.1 INTRODUCTION

Moisture estimation (or total solids) in milk and milk products is performed quite frequently in the laboratory of a composite dairy plant. Results of the analysis are required to serve as a process control parameter, indicator of desired composition of finished product, assessor of storage and transport conditions and index of profit. Here, the moisture content of butter is estimated by standard gravimetric method, which is treated as the reference for other methods. This method is accurate, simple, and easy to perform. However, the time required for analysis is longer as compared to moisture balance method.

8.2.2 PRINCIPLE

When a thin layer of food products is subjected to convecting air at 100°C, the free moisture gets evaporated. The water vapor is taken up by the circulating air and finally thrown out in the atmosphere. The process of evaporation and removal of water vapor lasts till the water activity of the food is higher than that of the circulating air. Removal of the water is taken as loss of weight on drying and is calculated as percent moisture.

8.2.3 MATERIALS

8.2.3.1 APPARATUS

- Aluminum or stainless-steel moisture dish or 250 ml beaker;
- Water bath;
- Hot air oven;
- Glass rods;
- Tong;
- Desiccator.

8.2.4 PROCEDURE

8.2.4.1 PREPARATION OF BUTTER SAMPLE

The sample is prepared according to the method mentioned in the sub-chapter on determination of moisture (*refer* to Section 1.4.1).

8.2.4.2 EXPERIMENTATION

Heat the glass rod and dish for at least 1 h at temperature 100±1°C. Take the weight of the dish after cooling it in the desiccator. Weigh accurately 10 g sample in the dish. Keep the dish on a heater, stirring it frequently until the foaming is stopped and the light brown curd particles are observed at the dish bottom. Transfer the dish to an oven maintained at a temperature of

100±1°C and heat for 1.5 h. Remove the dish from oven using tong and take the weight of the dish after cooling in a desiccator. Heat, cool, and weigh repeatedly until the two consecutive weights agree within 0.1 mg [3].

8.2.5 *CALCULATIONS*

Moisture content (%) in butter is calculated as follows:

$$\text{Moisture (\%)} = 100 \times \frac{W_2 - W_3}{W_2 - W_1} \quad (1)$$

where; W_1 is the weight of the empty dish (g); W_2 is the weight of the dish plus sample (g); and W_3 is the weight of the dish with dried sample (g).

8.2.6 *DATA COLLECTION SHEET*

Title: Determination of moisture in butter by gravimetric method.

Location and Organization:____________________

Participants: ____________________

Date (Month/Day/Year):____________________

Observations:

Parameters	**Repeat 1**	**Repeat 2**	**Repeat 3**	**Average**
Empty dish weight (W_1) g				
Weight of dish with sample (W_2) g				
Weight of dish with dried sample (W_3) g				
Moisture content (% w/w)				

Comments:____________________.

8.3 DETERMINATION OF BUTTER FAT

8.3.1 *INTRODUCTION*

Butter fat can be estimated by direct or indirect methods. In indirect method (fat by difference), the fat content is obtained by subtracting the non-fat portion (moisture, salt, and curd) of butter from 100%, whereas direct method measures the fat directly [4]. Indirect methods are of two types, according to one method; total non-fat portion is physically separated from the sample and volumetrically determined. In another indirect method determines the fat content gravimetrically. The left-over residue, after moisture determination, is used for the estimation of fat content. The fat is extracted from the dried residue using an appropriate solvent and dried. The retained SNF portion represents the salt and curd content. Further, this residue is used for the estimation of salt content. The indirect method has been globally accepted as both routine and reference method. For direct estimation of fat, a number of methods is available. These include; refractometric method, gravimetric separation method, volumetric centrifugation method, solvent extraction method, spectrophotometric method, etc. [4]. BIS has provided three different methods for butter fat estimation; in one method, fat is calculated by extracting the fat using appropriate solvent from the residue left after moisture determination. The second method estimates fat content using a butyrometer (70% scales). These two methods are employed as routine methods. Third, Rose-Gottlieb method is treated as a reference method [3]. In this section, fat content is estimated by Rose-Gottlieb method [3].

8.3.2 PRINCIPLE

Analysis of fat by Rose-Gottlieb method includes extraction of fat from the sample in organic solvents, evaporation of the solvent and residual moisture by heating at controlled temperature and weighing of the fat to quantify it in the given sample.

Fat is present in milk and milk products as a microscopic globule covered with a membrane. Its efficient extraction requires destabilization of the emulsion by destroying the membrane with either concentrated ammonia or HCl. It is then extracted into a mixture of diethyl ether and petroleum ether. Alcohol and sodium chloride (NaCl) are also added to facilitate the destabilization of the emulsion and subsequent separation of aqueous and ethereal layers.

8.3.3 MATERIALS

8.3.3.1 REAGENTS

- Conc. ammonia solution (approx. 35%, Specific gravity 0.88).
- Concentrated HCl (Density 1.122 g/ml at 20°C).

- Ethanol (95–96%, v/v) or spirit.
- Diethyl ether (specific gravity 0.72; it should be peroxide free).
- Petroleum ether (boiling point 40–60°C).
- Solvent mix (mix equal volume of petroleum ether and diethyl ether).
- NaCl solution (0.5% w/v).

8.3.3.2 APPARATUS

- Fat extraction apparatus (fat extraction tube);
- Precision analytical balance (sensitivity ≤ 1 mg);
- Metal dishes or 250 ml glass beakers;
- Electric oven (temperature 100±1°C).

8.3.4 PROCEDURE

8.3.4.1 PREPARATION OF SAMPLE

1. **Butter:** The butter sample is warmed in a water bath (not exceeding 39°C) until a homogenous fluid emulsion is obtained. While sampling, care should be taken to avoid fat separation.
2. **Butter Fat:** A portion of emulsified butter is heated to 50 to 60°C till the fat gets separated. The fatty layer is filtered by a dry filter paper above the solidification point of fat using hot water funnel, if needed. The filtrate is filtered again under similar condition, to make it clear and water free. The fat is completely melted and mixed before taking the sample [2].

8.3.4.2 EXPERIMENTATION

Weigh accurately 1 g of butter in a 50 ml of beaker. Add 9 ml NaCl solution slowly to disperse the content. Add 1 ml conc. ammonia solution and mix it properly. Add 10 ml ethanol to it and transfer the content completely into the fat extraction flask and properly mix. Rinse the beaker used for weighing the sample with 25 milliliters diethyl ether and transfer to the tube. Close the stopper (wet with water before inserting) and vigorously shake for 1 min. Avoid the use of rubber stoppers. Add 25 milliliters of petroleum ether and close the stopper. Shake it again for 30 s. Optimum mixing at each stage ensures complete extraction of fat. Let it stand undisturbed in the rack for 30 min (minimum) until the upper ether layer separated clearly. Gently open the stopper, rinse the stopper and flask inner side with a little mixed solvent (use wash bottle). Hold the small bulb and decant the upper layer into the previously dried and weighed fat collecting vessel with boiling aids. Avoid decanting any of aqueous layers. Wash the outer neck of fat extraction flask with the little solvent mix.

Collect the rinsing in the fat collecting vessel. Add 5 ml ethanol to the fat extraction flask. Repeat the extraction twice by adding 15-milliliter diethyl ether followed by 15-milliliter petroleum ether. Carefully evaporate the solvent and dry the extraction in the oven for 1 h at temperature 98–100°C. Take out the container from the oven and take the weight after cooling in a desiccator. Repeatedly heat the flask in the oven. Weigh it again after cooling in a desiccator. Heat, cool, and weigh repeatedly until the two consecutive weight measurements do not differ by 1 mg maximum. Follow the same procedure with equal quantity of water in place of the sample for running blank.

8.3.5 CALCULATIONS

$$\text{Fat (\% by weight)} = 100 \times \frac{W_1 - W_2}{W} \tag{2}$$

where; W_1 is the weight of the flask with dried fat in g; W_2 is the weight of the empty flask in g; and W is the weight of the sample in g.

8.3.6 DATA COLLECTION SHEET

Title: Determination of fat in butter by Rose-Gottlieb method.

Location and Organization:____________________

Participants: ____________________

Date (Month/Day/Year):____________________

Observations:

Parameters	**Repeat 1**	**Repeat 2**	**Repeat 3**	**Average**
Weight of sample (g)				
Weight of empty dish (g)				
Weight of dish + dried fat (g)				
Fat (%)				

Comments:______________________________________.

8.4 DETERMINATION OF FAT CONTENT IN BUTTER BY GERBER METHOD

8.4.1 INTRODUCTION

The Gerber method is originated from Switzerland in the name of great scientist Dr. Niklaus Gerber. The Gerber acid digests the casein and other organic materials which facilitates fat separation. The wide difference between the specific gravity of fat and other milk constituents, and use of Gerber acid helps in separation of fat during centrifugation. The heat generated by Gerber acid keeps the fat in a liquid state and also reduces the viscosity, which facilitates easy fat separation. Fat estimation by Gerber method is a direct method for fat determination. Different butyrometer scales are used for different types of dairy products as mentioned in Table 8.1. Around 11 types of butyrometer are specified by BIS [6].

TABLE 8.1 Different Butyrometer Scale and Testing Dairy Products

Butyrometer Scale	Dairy Product to be Tested
6% scale	Whole milk and evaporated (unsweetened) milk
8% scale	
10% scale	
0.5% scale	Skim milk
4% scale	Separated milk, partly skim milk and buttermilk
20% scale	Ice cream, condensed milk
20% scale	Dry milk powder
40% scale	Cheese
40% scale	Dry milk powder
70% scale	Cream
90% scale	Butter

8.4.2 PRINCIPLE

The Gerber test employs sulfuric acid and amyl alcohol. The sulfuric acid dissolves the SNF; liberates the fat and generates heat to keep the fat in liquid condition. The amyl alcohol prevents the charring of fat and facilitates a clear fat column. The fat thus liberated from the other milk solids is separated by centrifugal force.

In the case of butter, a known weight of sample is dissolved in a beaker using Gerber H_2SO_4 and then transferred to a clean butyrometer. To have a clear and haze free fat column in a butyrometer, amyl alcohol is added to it. After centrifugation, percent fat in butter is read from butyrometer column [2, 3].

8.4.3 MATERIALS

8.4.3.1 REAGENTS

- Concentration of Gerber H_2SO_4 should be 90 to 91% by mass (density 1.807–1.812 g/ml at 27°C).
- Amyl alcohol.

8.4.3.2 APPARATUS

- Butyrometer (0–70% scale);
- Beaker of 50 ml capacity;
- Pipette or automatic measure (capacity 1 and 10 ml);
- Stirring rods;
- Stoppers for butyrometer;
- Water bath;
- Gerber centrifuge.

8.4.4 PROCEDURE

8.4.4.1 PREPARATION OF THE SAMPLE

The sample is prepared according to the method mentioned in the sub-chapter on determination of fat by Rose-Gottlieb method (*refer* to Section 3.4.1).

8.4.4.2 EXPERIMENTATION

Weigh 2.5 g butter in a 50 ml beaker. Add small proportions of Gerber H_2SO_4 acid (50%) (1:1 ratio) and mix well. Transfer the content into a butyrometer. Using small quantities of H_2SO_4, rinse beaker for about six times to ensure all butter gets transferred to the butyrometer. In a beaker, mix 10 ml of water and 10 ml of Gerber acid and transfer between 15 and 20 ml of this mixture to the butyrometer based on its size. Add 1.0 ml of amyl alcohol, close it with the stopper. Properly mix the butyrometer content until the content is digested. Keep it in the water bath at 65±2°C for not < 5 min. Centrifuge the butyrometer for 4 min. and again transfer it into water bath at 65±2°C between 3 and 10 min. Adjust the fat column in the stem of butyrometer and note down the reading.

8.4.5 CALCULATIONS

Multiply the fat reading with 2 to obtain the fat percentage in the butter.

8.4.6 DATA COLLECTION SHEET

Title: Determination of fat in butter by Gerber method.

Location and Organization:________________________

Participants: _______________________________

Date (Month/Day/Year):__________________________

Observations:

Parameters	Repeat 1	Repeat 2	Repeat 3	Average
Fat reading from butyrometer				
Fat content in Butter				

Comments:____________________________________.

8.5 DETERMINATION OF CURD CONTENT OF BUTTER

8.5.1 INTRODUCTION

The solid not fat (SNF) content, i.e., protein, minerals, lactose, and vitamins contents altogether contribute to curd content in butter. To ensure good storage quality, less quantity of SNF part in the butter is desirable. FSSAI has prescribed maximum 1.5% of curd content in butter. During washing the buttermilk sticking to the butter granules are removed and decreases the curd amount of butter, which improves the butter storage quality. Unwashed butter contains a higher amount of curd content than the washed butter [5]. Amount of curd vary according to the manufacturing method and washing number given. In this section, curd content of butter is analyzed by the official method described by BIS [3].

8.5.2 PRINCIPLE

The residue obtained after moisture determination is made of fat free. This fat free residue represents the curd and salt content in butter. By subtracting the salt content from the weight of the fat-free residue, curd can be calculated.

8.5.3 MATERIALS

8.5.3.1 REAGENTS

- Petroleum ether (boiling range 40–60°C).

8.5.3.2 APPARATUS

- Asbestos;
- Desiccator;
- Flat bottom flask (250 ml);
- Glass funnel;
- Gooch crucible – with filter flask with adopter;
- Pumice stone;
- Whatman filter paper-Grade 1 (12.5 cm).

8.5.4 PROCEDURE

8.5.4.1 PREPARATION OF SAMPLE

The sample is prepared according to the method mentioned in the sub-chapter on determination of moisture (*refer* to Section 1.4.1).

8.5.4.2 EXPERIMENTATION

Prepare an asbestos mat in a crucible, heated in the oven at 100±1°C, cool and weigh. Alternatively, take the weight of funnel with folded filter paper (12.5 cm) after drying. The residue left after moisture determination is used for curd and salt content determination. Keep the funnel with filter paper on of a stand or fit the filter flask with crucible. The melted residue in the dish is added with 25 ml petroleum ether and mixed well. Wet the filter paper or asbestos mat with petroleum ether. Pass the fatty solution through the filter paper or asbestos; leave the sediment in the dish. Dissolve twice the sediment with petroleum ether and pass the fatty solution again through filter paper or asbestos for filtration. The filtrate is collected in a clean, dry, and tarred flat bottom flask (250 ml) containing glass beads. Wash all the sediment and fat of the dish with a wash bottle comprising petroleum ether into filter paper or crucible. Wash the filter paper or crucible and make it fat free; collect the filtrate in the flask. Preserve the filtrate for fat estimation (by gravimetric method). Oven dries the left residue of the filter funnel or crucible at 100±1°C/30 min. Weigh it after cooling it in a desiccator. Dry, cool, and weigh repeatedly until the two consecutive weight measurements do not differ by 0.1 mg maximum.

8.5.5 CALCULATIONS

$$\text{Curd and salt \% by weight} = 100 \times \frac{W_1 - W_2}{W} \tag{3a}$$

$$\text{Curd (\%)} = \text{Curd and salt (\%)} - \text{salt (\%)} \tag{3b}$$

where; W is the weight of the sample in g; W_1 is the weight of filter paper with residue in g; and W_2 is the weight of the filter paper in g.

8.5.6 DATA COLLECTION SHEET

Title: Determination of curd content in butter.

Location and Organization:______________________________

Participants: ____________________________________

Date (Month/Day/Year):________________________________

Observations:

Parameters	Repeat 1	Repeat 2	Repeat 3	Average
Weight of sample (g)				
Weight of filter paper with residue (g)				
Weight of the filter paper (g)				
Curd + Salt (%)				
Curd content (%)				

Comments:____.

8.6 DETERMINATION OF SALT IN BUTTER BY MOHR'S METHOD (DIRECT METHOD)

8.6.1 INTRODUCTION

Salt, usually NaCl, is added in dairy products such as butter, cheese, cheese spread, etc., to enhance taste, flavor, keeping quality and yield. It is important to restrict the amount of salt present in these products for quality control and legal reasons. BIS has described two methods for salt estimation in butter, Mohr's (direct) method and Volhard's (routine) method. In this section, Mohr's method is described for salt estimation [3].

8.6.2 PRINCIPLE

On adding boiling water to the butter sample, the sample is melted. The dissolved chloride is titrated against standard $AgNO_3$ by using K_2CrO_4 (potassium chromate) as an indicator. Since silver ion has more affinity to chloride ion, it does not react with chromate until chloride is totally reacted.

$$NaCl + AgNO_3 \longrightarrow AgCl + NaNO_3$$

$$K_2CrO_4 + 2AgNO_3 \longrightarrow Ag_2CrO_4 + 2KNO_3$$

8.6.3 MATERIALS

8.6.3.1 REAGENTS

- $CaCO_3$ (Analytical grade; free from chloride);

- K_2CrO_4 (5% w/v solution in water);
- Standard $AgNO_3$ solution (0.1 N, standardized against NaCl).

8.6.3.2 APPARATUS

- Conical flask (250 ml);
- Burette (50 ml).

8.6.4 PROCEDURE

8.6.4.1 PREPARATION OF SAMPLE

The sample is prepared according to the method mentioned in the sub-chapter on determination of moisture (*refer* to Section 1.4.1).

8.6.4.2 EXPERIMENTATION

Take 5 g of sample in a conical flask and 100 ml of boiled water is added. Let it stand with random swirl for 5–10 minutes. When the sample solution is cooled to 50–55°C, add 2 ml K_2CrO_4 (indicator) and mix it by shaking. Mix it again after adding 250 mg $CaCO_3$. Titrate it against $AgNO_3$ while hot (50–55°C) with continuous shaking. This titration temperature is maintained to avoid fat coagulation which might affect the end point color. The endpoint is observed by the color change to brown, which persists for 30 s. Follow the same procedure by using the same reagents except the sample for running blank test. The difference between two consecutive results of NaCl concentration should be 0.02% maximum [3].

8.6.5 CALCULATIONS

$$\text{Salt (as NaC1\,\% by weight)} = \frac{5.85 \times N \times (V_1 - V_2)}{W} \tag{4}$$

where; N is the normality of $AgNO_3$; V_1 is the titer value (ml) of 0.1 N $AgNO_3$ used for sample; V_2 is the titer value (ml) of 0.1 N $AgNO_3$ used for blank; and W is the sample weight in g.

8.6.6 DATA COLLECTION SHEET

Title: Determination of salt by Mohr's method (Direct titration method)

Location and Organization:________________________________

Participants: ____

Date (Month/Day/Year): ________________________________

Observations:

Parameters	Repeat 1	Repeat 2	Repeat 3	Average
Weight of sample (g)				
Titre value (ml) of 0.1 N $AgNO_3$ used for sample				
Titre value (ml) of 0.1 N $AgNO_3$ used for blank				
Salt (%)				

Comments:____.

8.7 DETERMINATION OF SALT IN BUTTER BY VOLHARD'S METHOD (INDIRECT METHOD)

8.7.1 INTRODUCTION

Salt, usually NaCl, is added in dairy products such as butter, cheese, cheese spread, etc., to improve/enhance taste, flavor, keeping quality and yield. It is important to maintain the amount of salt present in these products for quality control and legal reasons. Salt is added in butter for the purpose of enhancing the storage stability, taste, and overrun. This chapter depicts the Volhard's indirect method for salt estimation in butter [3].

8.7.2 PRINCIPLE

The salt in butter is obtained by extraction with hot water from dry fat free residue obtained after moisture estimation. On adding excess silver nitrate ($AgNO_3$) solution the chloride gets precipitated and unused $AgNO_3$ is titrated against potassium thiocyanate (KSCN) using ferric ammonium sulfate (FAS) as an indicator. Both, the silver chloride and $AgNO_3$ are sparingly soluble salt. Thus, there are chances for silver chloride to react with the thiocyanate; this is prevented by coating the silver chloride precipitates with nitrobenzene. The KSCN also reacts with the indicator, but the reaction is not favored till $AgNO_3$ is not titrated completely [3].

$$NaCl + AgNO_3 \longrightarrow AgCl + NaNO_3$$
$$\text{Excess } AgNO_3 + KSCN \longrightarrow AgSCN + KNO_3$$
$$Fe^{+++} + SCN^- \longrightarrow [FeSCN]^{++}$$

8.7.3 MATERIALS

8.7.3.1 REAGENTS

- $AgNO_3$ solution (0.05 N, standardized against NaCl);
- Nitric acid (HNO_3; specific gravity, 1.42; approx. 70% w/w);
- Nitric acid (approx. 5 N);
- Nitrobenzene;
- Ferric ammonium sulfate (FAS; $Fe_2(SO_4)_3(NH_4)SO_4{\cdot}24H_2O$) indicator solution: Dissolve 50 g of FAS in 95 ml H_2O containing 5 ml of 5 N HNO_3;
- Standard KSCN solution (0.05 N, standardized against $AgNO_3$).

8.7.3.2 APPARATUS

- Pipette (25 ml);
- Volumetric flask (100 ml);
- Conical flask (250 ml);
- Burette (50 ml, graduated to 0.1 ml).

8.7.4 PROCEDURE

The salt is extracted from the leftover curd and salt residue (*refer* to Section 5.4.2) by repeatedly hot water washing of filter paper or crucible or by keeping the filter paper or crucible in a beaker of hot water. The rinsing is collected in a measuring flask (100 ml) by passing through filter paper and allowed to cool to ambient temperature, make up the volume up to the mark.

Transfer 25 ml of the extract in a 250 ml conical flask. Add an excess of 0.05 N $AgNO_3$ solution (generally 25–30 ml) to the flask. Acidify with

HNO_3; 2 ml indicator and 1 ml nitrobenzene are added. Mix the content and excess $AgNO_3$ is determined by titrating it against KSCN. The end point is observed by color change to slightly orange, which persists for 15 seconds. Follow the same procedure with water instead of sample for running blank.

8.7.5 CALCULATIONS

$$\text{Salt (as NaCl \% by weight)} = \frac{23.38 \times N \times (A - B)}{W} \quad (5)$$

where; N is the normality of KSCN; A is the titer value (ml) of KSCN used for blank; B is the titer value (ml) of KSCN used for sample; and W is the sample weight in g.

8.7.6 DATA COLLECTION SHEET

Title: Determination of salt by Volhard method (Indirect titration method).

Location and Organization:____________________________

Participants: ____________________________

Date (Month/Day/Year):____________________________

Observations:

Parameters	Repeat 1	Repeat 2	Repeat 3	Average
Weight of sample (g)				
Titre value (ml) of KSCN used for blank				
Titre value (ml) of KSCN used for sample				
Salt (as NaCl % by weight)				

Comments:____________________________

8.8 DETERMINATION OF TITRATABLE ACIDITY OF BUTTER

8.8.1 INTRODUCTION

The titratable acidity is, in fact, a measure of the buffering capacity of milk and milk products. It estimates the amount of NaOH required to shift the pH of the sample from the original value to the pH at which phenolphthalein indicator changes its color from colorless to pink. The amount of alkali is customarily converted to equivalent lactic acid and expressed as percent lactic acid by mass. Acidity value of a product serves as a guide to predict its composition, shelf life, sensory property, microbial quality, suitability for further processing, etc. In this section, butter acidity is measured according to the standard method described by BIS [4].

8.8.2 PRINCIPLE

In an acid-base titration, while the aqueous solution of the sample is titrated against standard alkali, the lactic acid and other acidic components are progressively neutralized, leading to an increase in pH. The phenolphthalein indicator, a weak organic acid, turns its color from colorless to light pink when the pH of the sample reached to 8.3 [4].

$$\underset{\text{Lactic acid}}{CH_3CHOHCOOH} + \underset{\text{Sodium Hydroxide}}{NaOH} \longrightarrow \underset{\text{Sodium lactate}}{CH_3CHOHCOONa} + H_2O$$

8.8.3 MATERIALS

8.8.3.1 REAGENTS

- NaOH solution (0.02 N).
- Phenolphthalein indicator solution (0.5%): Dissolve 1.0 g of phenolphthalein in 110 ml of rectified spirit. This solution is neutralized with 0.1 N NaOH solution followed by dilution with water up to 200 ml.

8.8.3.2 APPARATUS

- Burette;
- Porcelain dishes (60 ml) or 250 ml conical flask;
- Stirring rods (glass, flattened one end).

8.8.4 PROCEDURE

8.8.4.1 PREPARATION OF SAMPLE

The sample is prepared according to the method mentioned in the sub-chapter on determination of moisture (*refer* to Section 1.4.1).

8.8.4.2 EXPERIMENTATION

Take 20 g sample in a conical flask (250 ml), add into 90 ml hot water (previously boiled) and mix it. Add 1 ml indicator, mix properly and titrate it against 0.02 N NaOH solution during hot.

8.8.5 CALCULATIONS

$$\text{Acidity(\% lactic acid)} = \frac{9 \times N \times V}{W} \quad (6)$$

where; N is the normality of sodium hydroxide (NaOH); V is the titer value (ml) of 0.02 N sodium hydroxide used for sample; and W is the sample weight in gram.

8.8.6 DATA COLLECTION SHEET

Title: Determination of titratable acidity of butter.

Location and Organization:_____________________________

Participants: __________________________________

Date (Month/Day/Year):______________________________

Observations:

Parameters	Repeat 1	Repeat 2	Repeat 3	Average
Weight of the sample, g				
Volume of 0.02 N NaOH used, ml				
Acidity (% lactic acid)				

Comments:______________________________________.

KEYWORDS

- **acidity**
- **butter**
- **curd**
- **fat**
- **moisture**
- **salt**
- **sampling**

REFERENCES

1. CODEX, (1971). *CXS-279: Standard for Butter* (pp. 1–3). Codex Alimentarius International Food Standards.
2. IS 3507: 1966. *Methods of Sampling and Test for Butter* (pp. 1–31). New Delhi: Bureau of Indian Standards; (1966).
3. IS. SP-18:1981. Butter. In: *ISI Handbook of Food Analysis (Part XI) Dairy Products* (pp. 81–88). New Delhi; Bureau of Indian Standards; (1981).
4. Evers, J. M., (1999). Determination of fat in butter: A review. *Bulletin of the International Dairy Federation, 340*, 3–15.
5. De S., (2004). Butter. Chapter 4. In: *Outlines of Dairy Technology* (12th edn., pp. 143–173). New Delhi: Oxford University Press.
6. IS 1223:2001. *Apparatus for Determination of Milk Fat by Gerber Method – Specification* (pp. 1–19). New Delhi: Bureau of Indian Standards; (2001).

CHAPTER 9

COMPOSITIONAL ANALYSIS OF BUTTER OIL AND CLARIFIED FAT (GHEE)

TANMAY HAZRA, KUNTAL ROY, N. VEENA, MITUL BUMBADIYA, ROHIT G. SINDHAV, and SUDHEENDRA V. K. CHAGANTI

ABSTRACT

Ghee is a fat-rich dairy product, and it constitutes >99% of milk fat. It is also known as butter oil in Western countries, but it lacks the sweet typical aroma of ghee. Butter oil is manufactured by melting butter at <80°C, whereas clarified fat (ghee) is prepared at 105 to 110°C. Ghee has a pleasing flavor while butter oil has a bland flavor. The fatty acid composition of ghee mostly depends on the type of breed, region, season, feed, etc. The establishment of its purity (i.e., free from adulteration) requires elaborate analytical analysis, including physicochemical analysis to check the quality. In this chapter, different analytical techniques for assessment of the quality of ghee have been discussed.

9.1 DETERMINATION OF SPECIFIC GRAVITY OF GHEE

9.1.1 INTRODUCTION

Signature identification of oil and fat can be obtained by analysis of its density or specific gravity. The density of fat or oil is commonly denoted as specific gravity in its liquid form instead of absolute density. Pycnometer (or specific gravity bottle) is used to determine the specific gravity of ghee. Temperature plays an essential role on the physicochemical properties of ghee; therefore, the specific gravity is determined at a specified temperature [3]. The specific

Analytical Methods for Milk and Milk Products, Volume 1: Sampling Methods and Chemical and Compositional Analysis. Megh R. Goyal, N. Veena, and Santosh Kumar Mishra (Eds.)

gravity at 30°C for cow ghee and buffalo ghee ranges from 0.9358 to 0.9443 and 0.9340 to 0.9444, respectively [4]. However, for edible vegetable oils and animal body fats, it varies from 0.910 to 0.927 and 0.894 to 0.906, respectively at 20°C [5]. Several researchers have tried to ascertain the purity of ghee (including cotton tract areas) based on the specific gravity value at 30°C.

9.1.2 PRINCIPLE

Determination of density or specific gravity using a pycnometer is an accurate method. The water is used as a working or standard liquid with a known volume. The spare liquid is discharged after closing fine hole in the top-filled pycnometer, and it permits to detect a given volume of measured and/or standard liquid with high precision [2].

9.1.3 MATERIALS

9.1.3.1 APPARATUS

- Pycnometer (Figure 9.1) fitted with thermometer (0.1 or 0.2°C subdivision);
- Analytical balance;
- Water bath (30±2°C).

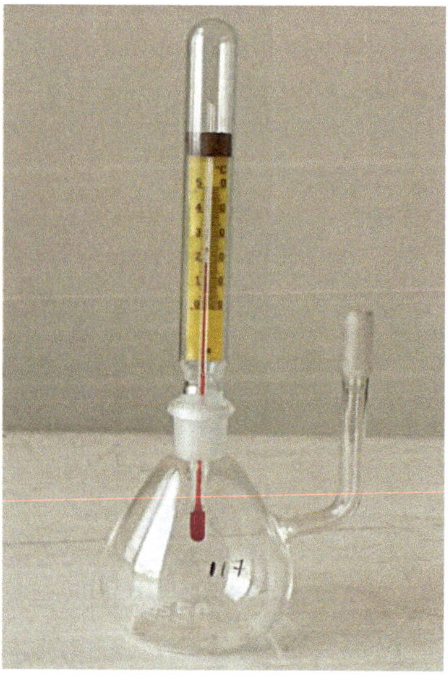

FIGURE 9.1 Pycnometer.

9.1.4 PROCEDURE

9.1.4.1 STANDARDIZATION OF THE PYCNOMETER

Chromic acid is carefully filled in the pycnometer for efficient cleaning followed by rinsing with water. Then, fill the pycnometer with fresh water (boiled) followed by cooling at 20°C and it is kept at 30°C constant temperature in a water bath. Pycnometer is carefully removed from the water bath; it is dried by wiping with a tissue or clean cloth, and it is weighed.

9.1.4.2 PREPARATION OF THE GHEE SAMPLES

Solid ghee sample should be softened (without melting) by mediating on a moderate heat. When the moisture and volatile matter is to be determined, mix the soft sample thoroughly. For the determination of other analysis, melt the sample on water bath that is kept at a temperature not > 50°C. If the sample is clear and transparent, then proceed with the experiment directly. If the sample is turbid or comprises of sediments, then pass the test sample through a Whatman filter paper (#4) using hot water funnel. The fat sample is stored in the dark and cold place to protect it from light/air to avoid rancidity.

9.1.4.3 EXPERIMENTATION

Carefully put the sample into a dry pycnometer so that it is free from air bubbles. After closing with the stopper, the pycnometer should be placed for 30 min at 30±2°C in a water bath. Wash out the oil droplets, which may leak out from the capillary opening. After removing the bottle from the water bath, it should be cleaned and dried. After detaching the cap from the side, the weight of the bottle should be taken by maintaining the temperature at 30°C. Similarly, follow these steps with water instead of a test sample.

9.1.5 CALCULATIONS

$$\text{Specific gravity (sp. gr.) at } 30°\text{C} = \frac{[x-y]}{[z-y]} \qquad (1)$$

where; y is the mass of empty sp. gr. bottle (in g at 30°C); x is the mass of sp. gr. bottle + ghee (in g at 30°C); and z is the mass of sp. gr. bottle + water (in g at 30°C).

9.1.6 DATA COLLECTION SHEET

Title: Determination of specific gravity of ghee.

Location and Organization:________________________________

Participants: ________________________________

Date (Month/Day/Year):________________________________

Observations:

Parameter	Repeat 1	Repeat 2	Repeat 3	Average
Mass of sp. gr. bottle (g) = y				
Mass of sp. gr. bottle + ghee (g) = x				
Mass of sp. gr. bottle + water (g) = z				
Sp. gr. of ghee at 30°C				

Comments:________________________________.

9.2 DETERMINATION OF MOISTURE CONTENT OF GHEE

9.2.1 INTRODUCTION

Moisture content influences the storage quality of the ghee. The amount of moisture is controlled by drying techniques and the weight loss is recognized as water content. The evaporated moisture may contain volatile substances apart from water. Here, the method specified by BIS (IS 3508: 2003) has been discussed [6]. According to Food Safety and Standards Regulations of India (FSSRI: 2021 amendments), moisture content (%, m/m) in ghee, milk fat (butter oil) and anhydrous milk fat (AMF) should not be more than 0.5, 0.4, and 0.1%, respectively.

9.2.2 PRINCIPLE

The moisture content (expressed as percentage by mass) of ghee is loss in the weight when the sample is heated in a drying oven till a constant weight is obtained at 105°C.

9.2.3 MATERIALS

9.2.3.1 APPARATUS

- Aluminum (Al) dish with slip-over cover: Diameter of 7–8 cm and depth of 2–2.5 cm;

- Analytical balance;
- Desiccator;
- Drying oven maintained at 105±1°C.

9.2.4 PROCEDURE

9.2.4.1 SAMPLE PREPARATION

Uniformly mix the test portion in a container, till sample becomes homogeneous. Perform this operation in a cool place, protect it from sun light and finish within the shortest duration. Remix the sample, if any separation occurs in between, during mixing and initiation of moisture estimation.

9.2.4.2 EXPERIMENTATION

Exactly weigh 10 g of test sample in the aluminum dish, which has been previously dried, cooled in the desiccator and then weighed. Keep the dish in a drying oven at 105±1°C for one hour. Place a dish in the desiccator to cool the contents after removing from the oven, followed by recording of the weight of the dish. Repeat this procedure; but place the dish in drying oven for only 30 min each time, till the two consecutive weight differences are not more than 1 mg. Note down the minimum weight of the dish.

Note: Deviations between duplicate analyzes should not be more than 0.1% by mass of moisture.

9.2.5 CALCULATIONS

The moisture content (expressed as % by weight) is calculated by:

$$\text{Moisture content (\%)} = \frac{M_1 - M_2}{M_1 - M_o} \times 100 \tag{2}$$

where; M_0 is the empty dish weight (g); M_1 is the dish plus test sample weight (g); and M_2 is the dish plus dried sample weight (g).

9.2.6 DATA COLLECTION SHEET

Title: Determination of moisture content of ghee.

Location and Organization:________________________________

Participants: ______________________________

Date (Month/Day/Year): ______________________________

Observations:

Parameter	Repeat 1	Repeat 2	Repeat 3	Average
Dish plus test sample weight (g) = M_1				
Dish plus dried sample weight (g) = M_2				
Empty dish weight (g) = M_o				
Moisture content (%)				

Comments: ______________________________.

9.3 DETERMINATION OF REICHERT-MEISSEL (RM) VALUE, POLENSKE VALUE (PV), AND KIRSCHNER VALUE (KV) IN GHEE

9.3.1 *INTRODUCTION*

RM value and PV are expressed in milliliter of dilute alkali (NaOH) solution needed to neutralize water-soluble and water-insoluble steam volatile fatty acids (FAs), respectively, which are distilled from 5 g of sample under defined conditions that are specified in the method. Kirschner value (KV) is expressed in milliliter of dilute alkali solution needed to neutralize the steam volatile water-soluble FAs, which will form silver salts (water-soluble) distilled from 5 g of sample under defined conditions that are specified in the method. The method described here does not determine the total amount of volatile FAs (such as: water soluble and insoluble) that are present in ghee or butter oil. However, the quantity of FAs estimated by the method is reliant on the apparatus dimensions and details of the method in use. The butyric and caproic are major FAs that are present in ghee; and these are estimated by RM value, whereas PV is contributed mainly by caprylic and capric acid. The RM value and PV for ghee usually vary from 17 to 35 and 1.2 to 2.4, respectively. KV indicates only the butyric acid. According to FSSRI (2021 amendments), RM value in ghee, butter oil and AMF should be not less than 24; whereas PV ranges between 0.5 and 2.0.

9.3.2 *PRINCIPLE*

Saponification of ghee with diluted, acidified (with 1 N sulfuric acid) glycerol-potash is followed by immediate steam distillation in the Polenske distillation apparatus (glass apparatus) that is essential. Thereafter, the condensed/cooled distillates are filtrated out. For estimation of RM value, this filtrate is titrated with standard sodium hydroxide (NaOH) solution. To estimate PV (the water insoluble FAs), the condenser and filter paper should be washed with alcohol followed by titration against a standard NaOH solution [7]. The reactions are as follows:

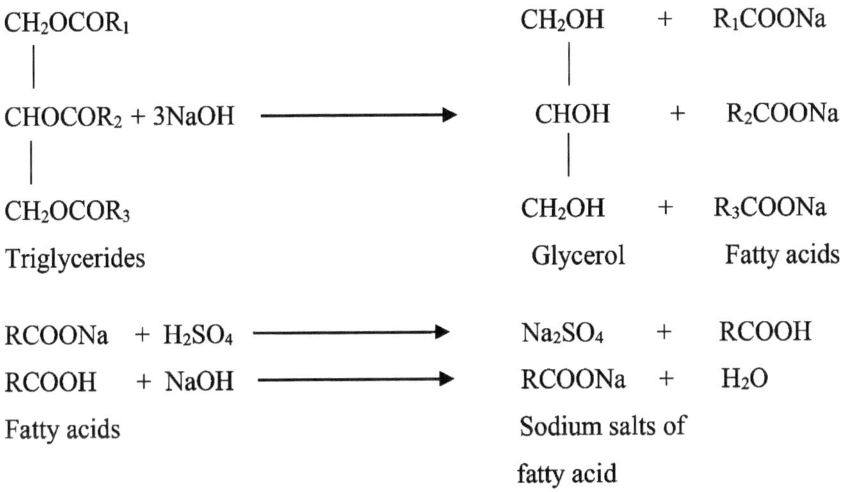

9.3.3 MATERIALS

9.3.3.1 REAGENTS

- Glycerol: purity 98% (m/m).
- Sodium hydroxide (NaOH) solution (50% m/m).
- Dilute H_2SO_4: Dilute 25 ml of H_2SO_4 to one liter in a volumetric flask with water (or dilute in such a way so that 40 milliliters of dilute H_2SO_4 can neutralize 2 ml of 50% (m/m) NaOH solution).
- Neutralized ethanol 95% (v/v).
- Glass Beads (1.5–2.0 mm in diameter) or pumice powder (ground).
- Phenolphthalein indicator: 0.5% solution in ethanol (95% v/v).
- NaOH solution (0.1 N).
- Barium hydroxide [$Ba(OH)_2$] solution (0.1 N).
- Powdered silver sulfate.

9.3.3.2 APPARATUS

- Asbestos board.
- Conical flask (250 ml).
- Graduated cylinders (25 and 100 ml).
- Distillation apparatus (Figure 9.2): Consists of still-head, condenser, flat bottom Polenske flask, receiver (110 ml volumetric flask) and rubber stopper. Strict adherence of dimensions of the constituent apparatus given in the IS: 3508–1966 should be followed [7].
- Glass Funnel: 6 cm diameter (approx.).
- Pipette.
- Whatman filter paper (No 4).
- Gas burner.
- Watch-glass.

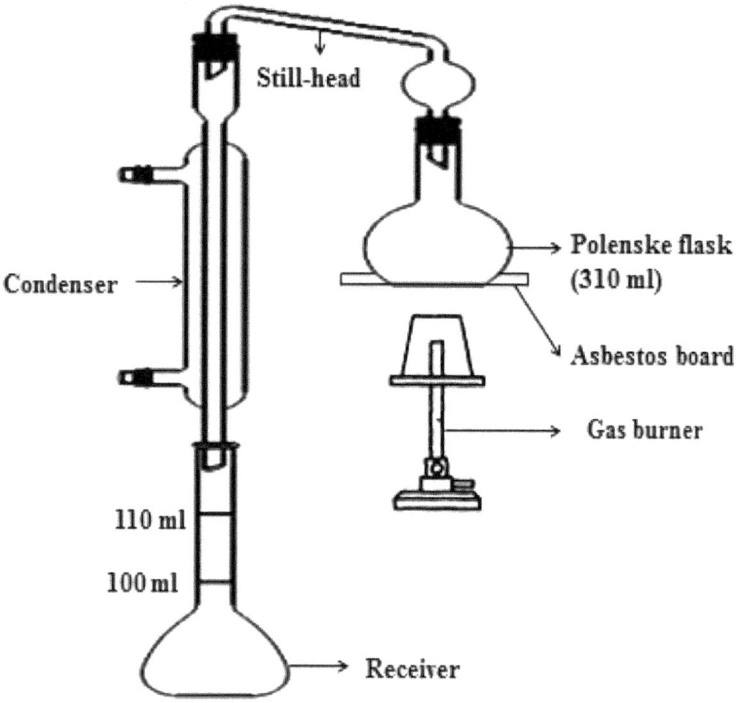

FIGURE 9.2 Distillation apparatus.

9.3.4 PROCEDURE

Exactly 5±0.01 g of prepared sample (as discussed in Section 1.4.2) is taken in a flat bottom flask; sequentially glycerol (20 g) and 50% NaOH (2 ml) solution are added to it; followed by direct heating on Bunsen burner with vigorous swirling of the flask till the ghee sample is completely saponified (forms clear mixture). Special care should be taken to evade overheating. Place a watch-glass above the flask. *Simultaneously*, a blank sample is prepared using the same quantity of reagents but without ghee sample.

The 93 ml of boiled distilled water is measured in a 100 ml cylinder, where water has been vigorously boiled for 15 min. For dissolving the soap, the water should be added before solidifying the mixture and it is sufficiently cooled to avoid its loss.

Note: Incomplete saponification is shown by unclear solution, whereas overheating is represented by dark yellow color. In such a case, repeat the experiment with a fresh sample.

Before attaching the flask to the distillation apparatus, dilute H_2SO_4 (50 ml) solution and two glass beads are added to the flask. Initially, contents in the flask are heated without boiling till it fully melts the insoluble acids in the mixture; thereafter, raise the flame. The 110 milliliters of the distillate should be collected within 19- and 21-minutes. Appropriate speed of flowing water in the condenser is maintained so that the temperature of distillate is in the range of 18–20°C.

Stop heating the flask after collecting 110 ml of distillate in a flask; and replace with a measuring cylinder (25 ml) to collect the draining stuff. Close the flask (110 ml) with stopper and keep it in a water bath without shaking the contents for 10 minutes at 15°C so as to completely submerge it up to the mark (i.e., 110 ml). Take out the flask, dry the outside, and the flask is carefully inverted to avoid drenching of the stopper with insoluble acids. Invert the flask four to five times without vigorous shaking. The mixture should be filtrated through a filter paper (#4) and discard first few drops; and filtrate is collected in a 100 ml volumetric flask. Preserve the filtrate for further titration to determine the RM value and KV.

Remove the still-head; and condenser is washed 3 times with 15 milliliters (ml) of water (cold); each washing is passed separately into a cylinder (15 ml), flask (110 ml) and filter paper; and discard each washing before filtering the next. Similarly carry out 3 washings of condenser, cylinder, flask, and filter paper with 15 ml of ethanol (neutralized) for dissolving the insoluble acids; and drained ethanolic solution (approx. 45 ml) is collected in a flask after each washing. Preserve this solution for further titration to determine the PV.

For determining RM value, transfer 100 ml of above collected filtrate into a conical flask; and add 0.1 ml of indicator (phenolphthalein), mix the contents and titrate these against 0.1 N NaOH [or $Ba(OH)_2$] solution until pink color persists. Record the titer value.

For determining PV, transfer ethanolic extract into a conical flask, add 0.25 ml of indicator, mix the contents and titrate it against NaOH [or $Ba(OH)_2$] solution until pink color persists. Record the titer value.

For determining KV, transfer 110 ml of the above collected filtrate into a conical flask, add 0.1 ml of indicator, mix the contents and titrate it against $Ba(OH)_2$ solution, till there is a pink color. At the end of titration, rinse the 110 ml flask with neutralized solution. Record the volume of $Ba(OH)_2$ solution that was used in the titration. Add 0.5 g of silver sulfate to the neutralized solution. Keep the flask in the dark for 1 h with intermittent mixing. Thereafter, the contents are filtered through filter paper (#4) in the dark.

Collect the filtrate (100 ml) in Polenske flask. Pour 35 ml of cold water to the flask (which has been recently boiled for 15 minutes), 10 ml of diluted H_2SO_4 and 2 glass beads. Attach the flask to the distillation apparatus and follow the same procedure as mentioned above. Collect 110 ml of distillate within 19 to 21 min. Mix the contents (without cooling for 10 min) followed by filtration and titrate 100 ml of filtrate against 0.1 N $Ba(OH)_2$ solution using phenolphthalein. Similarly, run a blank test using all the same reagents without sample. Record the titer value.

9.3.5 *CALCULATIONS*

$$\text{RM value} = [1.10 \times (V_1 - V_2)]$$
$$PV = [(V_3 - V_4)]$$
$$KV = \frac{[121 \times (100 + V_1) \times (V_5 - V_6)]}{10000} \quad (3)$$

where; V_1 is the 0.1 N NaOH [or $Ba(OH)_2$] solution required to titrate water-soluble acids in the test sample (ml); V_2 is the 0.1 N NaOH [or $Ba(OH)_2$] solution required to titrate water-soluble acids in the blank sample (ml); V_3 is the 0.1 N NaOH [or $Ba(OH)_2$] required to titrate water insoluble acids in test sample (ml); V_4 is the 0.1 N NaOH [or $Ba(OH)_2$] required to titrate water insoluble acids in the blank sample (ml); V_5 is the 0.1 N $Ba(OH)_2$ required to titrate water-soluble silver salts in test sample (ml); and V_6 is the 0.1 N $Ba(OH)_2$ required to titrate water-soluble silver salts in blank sample (ml).

9.3.6 *DATA COLLECTION SHEET*

Title: Determination of RM value, PV, and KV in ghee.

Location and Organization: ________________________________

Participants: ________________________________

Date (Month/Day/Year): ________________________________

Observations:

Parameter	Repeat 1	Repeat 2	Repeat 3	Average
0.1 N NaOH [or $Ba(OH)_2$] solution required to titrate water-soluble acids in test sample (ml) = V_1				
0.1 N NaOH [or $Ba(OH)_2$] solution required to titrate water-soluble acids in blank sample (ml) = V_2				
0.1 N NaOH [or $Ba(OH)_2$] required to titrate water insoluble acids in test sample (ml) = V_3				
0.1 N NaOH [or $Ba(OH)_2$] required to titrate water insoluble acids in blank sample (ml) = V_4				
0.1 N $Ba(OH)_2$ required to titrate water-soluble silver salts in test sample (ml) = V_5				
0.1 N $Ba(OH)_2$ required to titrate water-soluble silver salts in blank sample (ml) = V_6				
RM value				
PV				
KV				

Comments: ________________________________.

9.4 DETERMINATION OF SAPONIFICATION VALUE (SV) OF GHEE

9.4.1 *INTRODUCTION*

Saponification value (SV) can identify the characteristics of the FAs present in the specified fat. It is the index of average molecular weight of the FAs that are present in fat. SV represents milligram of KOH required to saponify entirely 1 g of ghee. Saponification equivalent can also be calculated using

SV, which is the quantity of ghee saponified by 1 g equivalent of KOH and is equal to 56,100 divided by SV [8]. According to FSSRI (2021 amendments), SV of ghee should range from 205 to 235.

SV for milk fat, body fat and vegetable oils ranges between 210–233, 192–203, and 170–197, respectively. The SV for olive oil, ground nut oil, sunflower oil, mustard oil, sesame oil is in the range of 187–196, 188–195, 188–194, 169–176, and 188–195, respectively, whereas it is very high for coconut oil (251–263). Along with diverse fatty acid of triglycerides, feeding pattern greatly affects the SV of ghee. For example, feeding of cotton seeds to milch animals affects the SV of milk fat and lowers this value by 7 units [5].

9.4.2 PRINCIPLE

Saponification of ghee sample is proceeded by refluxing it with alcoholic potassium hydroxide (KOH) solution (excess amount) resulting in potassium salts of FAs and glycerol. To determine the amount of alkali required for saponification, titrate the excess KOH with standard HCl solution. The reactions are as follows:

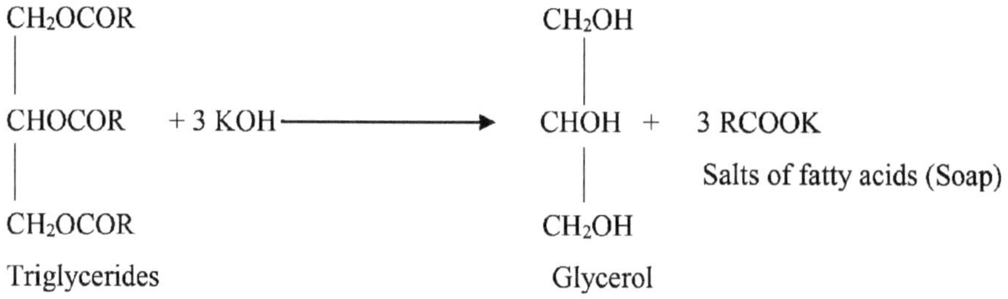

9.4.3 MATERIALS

9.4.3.1 REAGENTS

Alcoholic KOH (0.5 N) solution: Using the reflux condenser, 35 to 40 g of KOH is dissolved in ethanol (95% v/v: aldehyde free) and volume is made up to one liter. It gives pale yellow color to colorless. The resultant solution is stored in the dark.

Note: To prepare aldehyde-free ethanol, 1,200 ml of alcohol is mixed with 10 g of KOH and 6 g of aluminum (Al) foil (or granulated Al) for 1 h. Distill and collect 1,000 ml after rejecting the initial 50 ml.

- Phenolphthalein indicator (1% solution in ethanol (95% v/v)).

- Standard hydrochloric acid (HCl, 0.5 N).

9.4.3.2 *APPARATUS*

- Erlenmeyer flask (flat bottom, 250 ml capacity);
- Reflux condenser (minimum 65 cm length);
- Water bath.

9.4.4 PROCEDURE

The clear liquid ghee sample (2±0.001 g) is taken in a dry Erlenmeyer flask (250 ml). Transfer exactly 25 ml of alcoholic KOH into the flask. Blank test is prepared by measuring the same quantity of alcoholic KOH solution but without ghee sample and is run under the same conditions. Add one or two glass beads in each flask and connect to the reflux condenser. Continuously boil the contents of the flask under a reflux condenser for 30–60 minutes; and swirl the flask at regular interval until clear solution thus obtained indicates complete saponification. Titrate the contents of the flask while still in hot condition against HCl using 0.5 ml of indicator (phenolphthalein) to estimate the leftover KOH [8].

9.4.5 CALCULATIONS

$$SV = \frac{[56.1 \times (V_1 - V_2) \times N]}{W}$$

$$\text{Saponification number} = \frac{56100}{SV} \tag{4}$$

where; V_1 is the amount of HCl (acid) used in blank titration (ml); V_2 is the amount of acid used in sample titration (ml); N is the normality of acid used in titration; and W is the sample weight (g).

9.4.6 DATA COLLECTION SHEET

Title: Determination of saponification value of ghee.

Location and Organization:______________________________

Participants: ____________________________________

Date (Month/Day/Year):________________________________

Observations:

Parameters	Repeat 1	Repeat 2	Repeat 3	Average
Weight of sample (g) = W				
Amount of acid used in blank titration (ml)= V_1				
Amount of acid used in sample titration (ml) = V_2				
SV				
Saponification number				

Comments:____.

9.5 DETERMINATION OF UNSAPONIFIABLE MATTER (USM) IN GHEE

9.5.1 *INTRODUCTION*

The unsaponifiable matter (USM) in ghee comprises of the substance that remains insoluble in water after saponification with the alkali solution. It can be extracted with organic solvents for their estimation. The USM in milk fat is almost 0.45% and it includes fat soluble vitamins, sterols, and pigments, etc. Cholesterol is the highest USM present in ghee. USM analysis is useful to identify the adulteration with vegetable oil in ghee.

9.5.2 *PRINCIPLE*

USM in ghee or butterfat is not soluble in water after saponification. For its determination, it can be extracted with organic solvents followed by washing and drying to a constant weight [9, 10]. The reaction is as below:

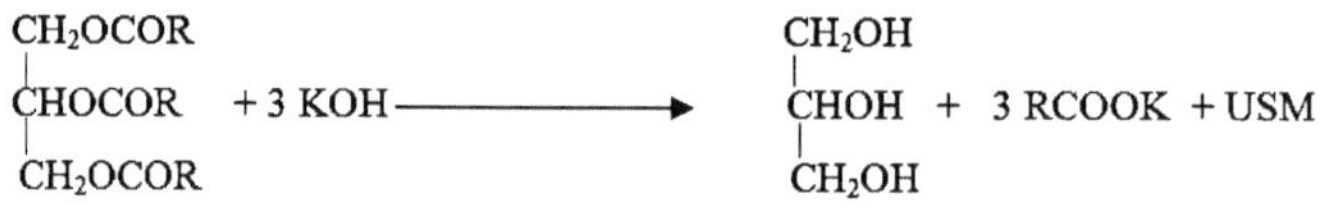

9.5.3 *MATERIALS*

9.5.3.1 *REAGENTS*

- Alcoholic KOH (0.5 N) solution: 35–40 g of KOH is dissolved in 20 ml of H_2O and volume is made up to one liter with ethyl alcohol (95% v/v: aldehyde free). It gives pale yellow color to colorless and approx. strength of not <0.5 N. Let the solution stand for overnight, and the clear upper layer is decanted and stored in the dark place.

- Aqueous KOH (0.5 N) solution.
- Ethyl alcohol (95%).
- Phenolphthalein indicator (1% solution in ethanol (95% v/v)).
- HCl (sp. gr. of 1.18).
- Diethyl ether (sp. gr. of 0.72–0.724 at 15°C).
- NaOH solution (0.1 N) in ethanol.
- Acetone.

9.5.3.2 APPARATUS

- Analytical balance;
- Drying oven;
- Flask (flat bottom, 250 ml capacity);
- Reflux condenser;
- Separating funnel (250 ml);
- Water bath;
- Burette.

9.5.4 PROCEDURE

In a 250 ml round bottom flask, 5±0.01 g of ghee is accurately weighed, followed by addition of 50 ml of alcoholic KOH solution and one to two glass beads. Connect the flask to a reflux condenser, reflux for 1 h on a boiling water bath, and the content is swirled at regular intervals until the clear solution thus obtained indicates complete saponification. Transfer the whole contents to separating funnel (250 ml) followed by rinsing of the flask with water (50 ml). Again, rinse the flask with 50 ml of diethyl ether and transfer carefully to a separating funnel. Close the separating funnel with stopper, mix vigorously and allow it to stand until clear separation of two layers. Remove the lower layer (contains aqueous soap) into the flask that is used for saponification.

Transfer ether layer into a 2nd separating funnel (250 ml) containing 20 ml of water. Repeat the extraction process in the same manner two times, each time using 50 ml of diethyl ether followed by combining of 3 ethereal extracts into a 2nd separating funnel. Wash the ether extract twice by adding 20 ml of water with vigorous shaking followed by sequential washing with 20 ml each of aqueous KOH (0.5 N), water and then with aqueous KOH (0.5 N) and water. Continue washing with water until wash water is free of alkali (no pink color formation on addition of phenolphthalein indicator).

To the previously weighed and dried Erlenmeyer flask, transfer the ether extract and evaporate the ether to a small amount. Then add 2–3 ml of acetone and completely evaporate the solvent by means of an air current; and flask is immersed entirely in boiling water bath, rotated, and held obliquely. Dry the contents of the flask in a drying oven at 80°C till it reaches constant mass.

Pour 10 ml of ethanol (95% v/v, recently boiled and neutralized) in an Erlenmeyer flask to dissolve the contents and then titrate it against alcoholic 0.1 N NaOH solution using phenolphthalein [10].

Note: If the amount of alkali used in the titration is more than 0.1 ml, then reject the test; and repeat the procedure from the start.

9.5.5 CALCULATIONS

Mass of extracted free fatty acids (FFAs) as oleic acid, Y = [0.282 × V × N]

$$\text{USM (\% by weight)} = \frac{[100 \times (X - Y)]}{W} \tag{5}$$

where; V is the amount of 0.1 N NaOH solution used in titration of test sample (ml); N is the normality of NaOH solution; X is the mass of the residue (g); Y is the mass of extracted FFAs (g); and W is the weight of test sample (g).

9.5.6 DATA COLLECTION SHEET

Title: Determination of unsaponifiable matter.

Location and Organization:____________________________________

Participants: ____

Date (Month/Day/Year):______________________________________

Observations:

Parameter	**Repeat 1**	**Repeat 2**	**Repeat 3**	**Average**
Amount of 0.1 N NaOH used in titration of test sample (ml) = V				
Mass of extracted FFAs (g) = Y				
Mass of the residue (g) = X				
Weight of test sample (g) = W				
USM (% by weight)				

Comments:____.

9.6 DETERMINATION OF IODINE VALUE OF GHEE

9.6.1 *INTRODUCTION*

Iodine value gives us the amount of unsaturated FAs or presence of number of double bonds that are present in the milk fat. It is directly proportional to a number of unsaturated FAs in ghee. Ghee or butterfat has generally lower iodine value compared with other vegetable oils except coconut and palm kernel oils. Cotton seed oil as a feed generally increases the iodine value in milk fat. According to FSSRI (2021 amendments), iodine value of ghee should range from 25 to 38.

9.6.2 *PRINCIPLE*

The iodine value is determined based on the absorption of iodine monochloride (ICl) on unsaturated FAs in fat or oil. Initially test sample is dissolved in a solvent followed by addition of excess of ICl. The excess quantity of ICl is estimated by addition of excess KI to the mixture, which will liberate I_2 on reaction. In the presence of starch indicator, sodium thiosulfate ($Na_2S_2O_3$) solution is used to titrate the liberated free iodine. The number of double bonds or amount of unsaturation in the ghee can be calculated by determining the amount of $Na_2S_2O_3$ required to complete the titration [11, 12]. The reactions are as follows:

$$\text{R-CH=CH-R} + \text{ICl (excess)} \longrightarrow \text{R-CHI-CHCl-R} + \text{ICl (remaining)}$$

$$\text{ICl (remaining)} + 2\,\text{KI} \longrightarrow \text{KCl} + \text{KI} + I_2$$

$$\underset{\text{(Blue color)}}{I_2 + \text{starch} + 2\,Na_2S_2O_3} \longrightarrow \underset{\text{(Colorless)}}{2\,\text{NaI} + \text{starch} + Na_2S_4O_6}$$

9.6.3 *MATERIALS*

9.6.3.1 *REAGENTS*

- Potassium dichromate ($K_2Cr_2O_7$).
- Concentrated HCl.
- Potassium iodide (KI, 10%) solution: Prepare fresh and it should be devoid of potassium iodate.
- Starch indicator: 5 g of starch and 0.01 g HgI_2 are dissolved in 30 ml of cold H_2O followed by its transfer to boiling water (1 liter) while

stirring and continuing the boiling for three minutes. Cool the solution to ambient temperature; and collect the clear upper layer of the solution.

- $Na_2S_2O_3$ standard solution (0.1 N): 24.8 g of $Na_2S_2O_3.5H_2O$ is dissolved in the boiled water (free from CO_2) and volume is made to 1 liter. The solution is stored in the dark colored bottle at a cool place before standardization.

The 5 g of powdered $K_2Cr_2O_7$ (dried before at 105±2°C for 1 h) is exactly weighed in a volumetric flask (1,000 ml), is dissolved in water and volume is made up to the mark. Mix the contents thoroughly. For $Na_2S_2O_3$ standardization, transfer exactly 25 ml of $K_2Cr_2O_7$ solution in a stoppered conical flask that has 5 ml of concentrated HCl and 15 ml of KI solution (10%). Place the flask in the dark for 5 min followed by titration against $Na_2S_2O_3$ solution using starch indicator. Note down the endpoint until color changes from blue to green and calculate the normality of $Na_2S_2O_3$ solution as follows:

$$N = \frac{25 \times v}{49.03 \times m} \tag{6}$$

where; N is the normality of $Na_2S_2O_3$ solution; v is the volume of $Na_2S_2O_3$ used in the titration (ml); and m is the mass of $K_2Cr_2O_7$ (g).

- Glacial acetic acid (CH_3COOH, 99%): Melting point 16.6°C and free from reducing impurities.
- Carbon tetrachloride (CCl_4).
- Iodine monochloride (ICl): 98% purity.
- Wij's ICl solution: Mix about 10 ml of ICl in 1,800 ml of glacial CH_3COOH and vigorously shake the mixture. Transfer 5 ml of this solution into a flask; add 10 ml of KI solution followed by titration with $Na_2S_2O_3$ (0.1 N) using starch indicator. Volume of the solution is adjusted until it reaches the strength of 0.2 N. Store this solution in a stoppered bottle in the dark and cool place; and avoid exposure to light.

9.6.3.2 *APPARATUS*

- Conical flask (250 and 500 ml) with ground-glass stopper;
- Burette (50 ml);
- Pipette (5 ml);
- Measuring cylinder.

9.6.4 PROCEDURE

The melted 0.40–0.45 g of ghee is taken in a flask (500 ml) with glass stopper followed by addition of 15 ml of carbon tetrachloride to dissolve the sample, and it is mixed well. Add accurately 25 ml of Wij's solution to the flask by means of burette. Close the stopper, mix the contents well and allow to stand in the dark for 1 h. Transfer 20 ml of KI solution and 150 ml of H_2O into the flask and thoroughly mix the contents. Titrate it against 0.1 N $Na_2S_2O_3$ using starch indicator (2 ml; it shall be added shortly before the end of titration) and swirl the mixture constantly during the titration. Note the end point (as disappearance of blue color) while shaking the contents vigorously. A blank test is also performed in the same manner but without ghee [12].

9.6.5 CALCULATION

$$\text{Iodine value of ghee} = \frac{12.69(X - Y)N}{W} \tag{7}$$

where; X is the amount of $Na_2S_2O_3$ solution used in blank titration (ml); Y is the amount of $Na_2S_2O_3$ solution used in sample titration (ml); W is the mass of sample (g); and N is the normality of $Na_2S_2O_3$ solution.

9.6.6 DATA COLLECTION SHEET

Title: Determination of Iodine value of the ghee.

Location and Organization:______________________________

Participants: ____________________________________

Date (Month/Day/Year): ____________________________

Observations:

Parameter	Repeat 1	Repeat 2	Repeat 3	Average
Weight of test sample (g) = W				
Amount of sodium thiosulfate solution (0.1 N) required for blank titration (ml)= X				
Amount of sodium thiosulfate solution (0.1 N) required for sample titration (ml) = Y				
Iodine value				

Comments:____

9.7 DETERMINATION OF REFRACTIVE INDEX (RI)

9.7.1 *INTRODUCTION*

Refractive index (RI) measures the refraction or bending of light as it passes from one medium to another. It is defined as ratio of sine of angle of incidence to sine of angle of refraction when a light ray of a definite wavelength (589.3 nm; mean of D-lines of Na) passes from air into fat or oil [13]. It is helpful to identify a particular substance, purity confirmation and measure its concentration. Abbe refractometer is used to determine the RI of ghee or butterfat at 40°C. The RI value for ghee usually ranges from 1.4345 to 1.4545.

9.7.2 *PRINCIPLE*

Abbe refractometer is used to measure the RI of ghee sample.

9.7.3 *MATERIALS*

9.7.3.1 *REAGENTS*

- Standard liquid: For checking the accuracy of the refractometer.

9.7.3.2 *APPARATUS*

- Abbe refractometer fitted with a thermometer (40°C to 50°C) (Figure 9.3): Calibrate the instrument using distilled water (RI of 1.3306 at 40°C) or glass prism with a known RI (using bromonaphthalene, which makes an optical contact with prism).
- Hot water rotating device (40±1°C).
- Light source: Sodium lamp (589.3 nm) is commonly used. Day light or diffused white light may also be used, if the instrument is fitted with suitable compensator.

9.7.4 *PROCEDURE*

Before taking the reading of ghee sample, the accuracy of the instrument is verified with liquid of known RI at 40°C. Also, check the refractometer at higher temperature (above 40°C) with a liquid of known temperature coefficient and a correction factor must be applied for the obtained reading from the instrument.

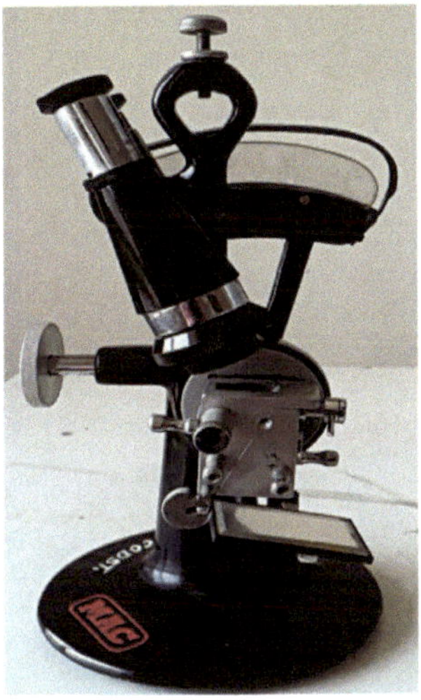

FIGURE 9.3 Abbe refractometer.

Using screw head, open the prism and clean with the tissue paper moistened with ethyl alcohol. Place two to three drops of filtered and melted test sample on the surface of the lower prism. Close the prism with a screw head and the temperature around the prism is maintained at 40°C±1°C by the circulation of water. Focus the light on the prism with the help of reflecting mirror provided in the instrument. Before taking the readings, allow the refractometer to stand still for a few minutes, so that instrument and sample will attain the same temperature.

Measurement of RI is based on the observation of boarder line position of total reflection with respect to the faces of the prism. Adjust the border line (Border line is the line separating dark and light portion and is not actually sharp line but a band of color) into the field of vision of telescope (i.e., separated into light and dark portion) by revolving the double prism via alidade, which is moving forward or backward direction. To remove the colors, rotate the screw head of the compensator till colorless and sharp border line has been attained. Border line shall be adjusted so that it falls on the point of intersection of crosshairs. Note the reading directly from the scale of the sector provided in the instrument. The difference between the duplicate measurements should not be more than 0.0002 unit of RI [13].

Note: The RI of the sample declines with increase in temperature and vice versa. Hence necessary correction shall be made depending on the temperature of measurement.

9.7.5 *CALCULATION*

Read RI of sample directly on scale of sector at 40°C. If the measurement is not done at 40°C, then the temperature correction is necessary using the following formula:

$$R = [R_1 + K_1 (T_1 - T)] \quad (8)$$

where; R is the refractometer value reduced to a specified temperature at T°C (i.e., 40°C); R_1 is the refractometer reading at T_1°C; K is the correction factor (0.000365); T is the specified temperature (40°C); and T_1 is the temperature at which reading R_1 is taken.

Note: Correction factor (K) to the observed value to be deducted for every single degree below or added to every single degree above 40°C.

9.7.6 *DATA COLLECTION SHEET*

Title: Determination of RI.

Location and Organization:______________________________

Participants: ______________________________________

Date (Month/Day/Year):_________________________________

Observations:

Parameter	**1**	**2**	**3**	**Average**
Temperature of test sample				
RI				

Comments:____.

9.8 DETERMINATION OF BUTYRO-REFRACTOMETER (BR) READING

9.8.1 *INTRODUCTION*

BR reading of milk fat or ghee ranges from 40 to 43. This value is comparatively lower than most of the vegetable oils. Hence, adulteration of milk fat with vegetable oil significantly increases the BR reading. According to FSSRI (2021 amendments), BR reading at 40°C of ghee, butter oil or anhydrous butter oil (milk fat) should range from 40 to 44.

9.8.2 PRINCIPLE

BR is used to measure the BR reading of sample on an arbitrary scale at a constant temperature.

9.8.3 MATERIALS

9.8.3.1 REAGENTS

- Standard liquid: For checking the accuracy of the refractometer.

9.8.3.2 APPARATUS

- Butyro-refractometer fitted with a thermometer (Figure 9.4).
- Hot water rotating device (40±1°C).
- Light source: Sodium lamp (589.3 nm) is commonly used. Day light or diffused white light may also be used, if the instrument is fitted with a suitable compensator.

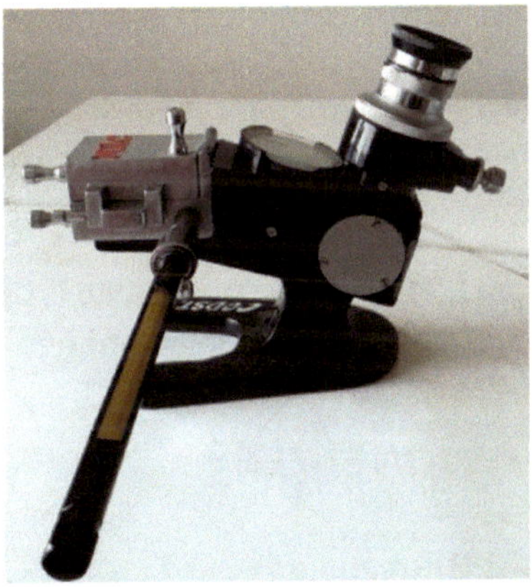

FIGURE 9.4 Butyro-refractometer.

9.8.4 PROCEDURE

Before taking the reading of ghee sample, the accuracy of the instrument is verified with liquid of known BR reading at 40°C. Also check the refractometer at higher temperature (above 40°C) with a liquid of known temperature coefficient and a correction factor must be applied to the obtained reading from the instrument.

Using screw head, open the prism and clean with tissue paper moistened with ethyl alcohol. Place two to three drops of filtered and melted ghee sample on the surface of the lower prism. Close the prism by means of a screw head and the temperature around the prism is maintained at 40°C±1°C by the circulation of water. Focus the light on the prism with the help of reflecting mirror provided in the instrument. Before taking the readings, allow the refractometer to stand still for a few minutes so that instrument and sample will attain the same temperature [13].

Note: The BR value of sample declines with the increase in temperature and vice versa. Hence necessary correction shall be made depending on the temperature of measurement.

9.8.5 CALCULATION

Read BR reading of the sample directly on the scale at 40°C. If the measurement is not done at 40°C, then the temperature correction is necessary using the following formula:

$$R = [R_1 + K_1 (T_1 - T)] \qquad (9)$$

where; R is the refractometer value reduced to a specified temperature at T°C (i.e., 40°C); R_1 is the refractometer reading at T_1°C; K is the correction factor (0.55); T is the specified temperature (40°C); and T_1 is the temperature at which reading R_1 is taken.

Note: Correction factor (K) to the observed value is deducted for every single degree below or added to every single degree above 40°C.

9.8.6 DATA COLLECTION SHEET

Title: Determination of BR reading.

Location and Organization:____________________

Participants: ____________________

Date (Month/Day/Year):____________________

Observations:

Parameter	**1**	**2**	**3**	**Average**
Temperature of ghee sample				
BR reading at 40°C				

Comments:____________________.

9.9 DETERMINATION OF HYDROXYL VALUE IN GHEE

9.9.1 INTRODUCTION

Hydroxyl value is defined as the number of mg of KOH, which is equivalent to the hydroxyl content of one g of ghee. It determines the number of free hydroxyl groups that are present in ghee or specified oils and fats. This value helps to detect rancidity. In the determination of hydroxyl value, the test sample is heated with acetic anhydride and pyridine at 100°C and the excess anhydride is decomposed by heating with water followed by titration with alcoholic KOH solution [15].

9.9.2 PRINCIPLE

To get the hydroxyl content, the amount of KOH is used for acidity titration, reagent blank and the acetylated sample is determined.

9.9.3 MATERIALS

9.9.3.1 REAGENTS

- Acetic anhydride [$(CH_3CO)_2O$].
- Pyridine: Redistilled at 114–115°C.
- Pyridine-acetic anhydride: Prepare the reagent fresh before use. To prepare this reagent, mix one volume of $(CH_3CO)_2O$ with three volumes of pyridine.
- Alcoholic KOH solution: It shall have more than or equal to 0.5 M.
- n-Butyl alcohol: It shall be neutralized with 0.5 M KOH using phenolphthalein indicator.
- Phenolphthalein indicator: 1% solution in ethyl alcohol.

9.9.3.2 APPARATUS

- Pipettes;
- Erlenmeyer flask (250 ml) with glass stopper;
- Taper reflux condenser.

9.9.4 PROCEDURE

According to the expected value of hydroxyl, weigh the sample for acetylation in 250 ml of Erlenmeyer flask. As for the expected hydroxyl value of 0 to 20, 20 to 50, 50 to 100 and 100 to 200, the amount of sample to be used for acetylation is 10, 5, 3, and 2 g, respectively. Transfer 5 ml of pyridine-$(CH_3CO)_2O$ reagent into the flask. Add extra 5 ml of pyridine for the sample having 0 to 20 hydroxyl value. Mix the content by gentle swirling. Keep the flask on a steam bath, connected with a reflux condenser and heat the flask for 1 h. Pass 10 ml water through the condenser and heat it again for 10 min followed by cooling of the contents. Transfer 25 ml of n-Butyl alcohol into the flask (half of the quantity is transferred through the condenser and the rest is used to wash the sides of the flask) and mix thoroughly. Contents in the flask are titrated with 0.5 M KOH in the presence of phenolphthalein indicator (1 ml), until pink color formation indicates the end point. Record the amount of KOH solution used. Prepare a blank test with the same reagents and follow the same procedure but without sample.

For acidity determination, weigh exactly 10 g of the test sample in another flask. Add 10 ml pyridine into the flask followed by neutralization with phenolphthalein. Titrate the contents against 0.5 N KOH solution in the presence of phenolphthalein indicator (1 ml). Note down the titer value [15].

9.9.5 CALCULATION

$$Hydroxyl\ value = \left\{ \frac{\left[B + \left(M \times \frac{T}{A} \right) - S \right]}{M} \times molarity \times 56.1 \right\} \tag{10}$$

where; B is the amount of alkali (KOH) used for blank titration (ml); S is the amount of alkali used for titration of acetylated sample (ml); M is the mass of test sample for acetylation (g); A is the mass of test sample for acidity determination (g); and T is the volume of alkali required for acidity determination (ml).

9.9.6 DATA COLLECTION SHEET

Title: Determination of hydroxyl value of ghee.

Location and Organization:______________________________

Participants: ____

Date (Month/Day/Year): ______________________________________

Observations:

Parameter	Repeat 1	Repeat 2	Repeat 3	Average
Mass of test sample for acetylation (g) = M				
Mass of sample for acidity determination (g) = A				
Amount of alkali (KOH) used for titration of blank (ml)= B				
Amount of alkali used for titration of acetylated sample (ml)=S				
Amount of alkali used for acidity determination (ml) = T				
Hydroxyl value				

Comments: ____.

9.10 DETERMINATION OF ACETYL VALUE OF GHEE

9.10.1 INTRODUCTION

The acetyl value is defined as the quantity of milligrams of KOH that is needed to neutralize the CH_3COOH produced when one g of milk fat or ghee is acetylated with acetic anhydride. The acetyl value indicates a total number of free –OH (hydroxyl groups) in the ghee. An increase in the value indicates more FFAs. Acetyl value help to detect rancidity and also the adulteration of fat. The value is determined by calculating the saponification value (SV) before and after the acetylation [16].

9.10.2 PRINCIPLE

In this method, the ghee is acetylated with an acetic anhydride solution. After acetylation, the SV is determined. The SV of untreated normal ghee is also measured using the method described in Section 9.4 of this chapter. Both values are used to determine the acetyl content of ghee sample.

9.10.3 MATERIAL

9.10.3.1 REAGENTS

- Alcoholic KOH solution: 1.2 liters of ethanol is refluxed for 30 min with 6 g granulated Al (aluminum) and 10 g of KOH. Discard first 50 ml of alcohol and collect one liter. To this one liter of alcohol, 40 g of KOH is added by keeping the temperature at 15°C and it is kept overnight.
- Phenolphthalein indicator solution: 1% solution in rectified spirit.
- Standard HCl (0.5 N).
- Acetic anhydride: It shall be of analytical grade.
- Sodium sulfate: It shall be of analytical grade.

9.10.3.2 APPARATUS

- Pipettes;
- Erlenmeyer flask with stopper (250 ml);
- Reflux condenser (65 cm long) or air condenser;
- Water bath fitted with thermostat;
- Drying oven (100 to 110°C);
- Beakers (500 and 1,000 ml);
- Litmus paper;
- Filter paper.

9.10.4 PROCEDURE

9.10.4.1 PREPARATION OF THE GHEE SAMPLES (SECTION 1.4.2 IN THIS CHAPTER)

9.10.4.2 METHODOLOGY

1. **Acetylation:** Liquid ghee sample (50 ml) is added to freshly distilled acetic anhydride (50 ml) in an Erlenmeyer flask and it is refluxed for 2 h through the condenser. Then, transfer the mixture to a one-liter beaker that is half filled with water (500 ml) and it is boiled for 15

min. To prevent bumping of the contents, air or carbon dioxide is passed through the beaker in the form of bubbles. Siphon off and discard the water portion. Again, add another 500 ml distilled water followed by boiling for another 15 min. and remove the water. Repeat the process one more time. Cool the mixture under ambient conditions. Transfer the acetylated material into a separating funnel. Wash twice with 200 ml warm distilled water. Separate as much water as possible. Now add 5 g of anhydrous sodium sulfate (Na_2SO_4) into it. Keep it undisturbed for 60 min. Filter the contents through a filter paper and keep in an oven at 100 to 110°C. Dry the acetylated material. It should be clear in appearance.

2. **Saponification:** Take 2 g each of acetylated ghee and non-treated ghee into separate Erlenmeyer flasks (250 ml) and then pour 25 ml of alcoholic KOH to each flask. Reflux the contents for 1 h. Titrate the warm solutions with HCl (0.5 N) in the presence of phenolphthalein indicator (1 ml). For the blank test, transfer a same quantity of KOH (alcoholic) in a flask followed by titration with HCl (0.5 N) in the presence of phenolphthalein. Calculate the average volume of acid required for the titration and deduct it from the volume of acid used during titration of acetylated ghee and non-treated ghee. Calculate the saponification number of both samples [16] according to Section 9.4 in this chapter.

9.10.5 *CALCULATION*

$$\text{Acetyl value} = \frac{(P_1 - P_2)}{1-(0.00075 \times P_2)} \quad (11)$$

where; P_1 is the saponification number of acetylated ghee sample, and P_2 is the saponification number of non-treated ghee sample.

9.10.6 *DATA COLLECTION SHEET*

Title: Determination of acetyl value of ghee.

Location and Organization: ____________________________________

Participants: ____________________________________

Date (Month/Day/Year): ____________________________________

Observations:

Parameter	Repeat 1	Repeat 2	Repeat 3	Average
Saponification number of acetylated ghee sample = P_1				
Saponification number of non-treated ghee sample = P_2				
Acetyl value				

Comments:____.

9.11 ESTIMATION OF PEROXIDE VALUE OF GHEE

9.11.1 INTRODUCTION

The peroxide (R-OOH) value helps to determine the content of oxygen (O_2) as peroxide, especially hydroperoxides in a food matrix. This value gives a sign of the initiation of fat oxidation [17]. It is expressed either as milli-molecules or milliequivalents (mEq) of Oxygen/kg of ghee sample. Peroxide value measured by the iodometric method is used to indicate the keeping quality of ghee as shown in Table 9.1. According to FSSRI (2021 amendments), peroxide value of butter oil and AMF should not be more than 0.6 and 0.3 mEq of O_2 per kg of fat, respectively.

TABLE 9.1 Peroxide Value Versus Quality of Ghee

Peroxide Value (Milliliter of 0.002 N $Na_2S_2O_3$ Solution per g)	Quality of Ghee (Butterfat)
Less than 1.5	Very good
1.6 to 2.0	Good
2.1 to 2.5	Fair
2.6 to 3.5	Poor
3.6 to 4.0	Not acceptable

9.11.2 PRINCIPLE

The iodometric method is used for the estimation of the peroxide value of milk fat or ghee. In this method, peroxides interact with KI and thus will liberate iodine (I_2). The released I_2 is titrated with $Na_2S_2O_3$ using starch indicator, until the color changes from blue to colorless. The amount of $Na_2S_2O_3$

indicates the amount of peroxide present in the sample. The reactions are as follows:

$$ROOH + KI \rightarrow ROH + KOH + I_2$$
$$I_2 + starch + 2Na_2S_2O_3 \rightarrow 2NaI + Na_2S_4O_6$$

9.11.3 MATERIALS

9.11.3.1 REAGENTS

- Standard $Na_2S_2O_3$ solution (0.002 N): It shall be prepared from dilution that is freshly prepared 0.1N $Na_2S_2O_3$ solution.
- Potassium Iodate (powdered form).
- KI (5% solution): Prepared fresh.
- Carbon dioxide (CO_2).
- Solvent mixture: Mix 2:1 ratio of glacial CH_3COOH and chloroform.
- Starch indicator: 5 g of starch and 0.01 g of HgI_2 are dissolved in 30 ml of cold water followed by its transfer to boiling water (one liter) while stirring and continuous boiling for 3 min. The solution is cooled to an ambient temperature and clear upper layer of the solution is collected.

9.11.3.2 APPARATUS

- Graduated class-A glass burette (25 ml);
- Water bath;
- Glass rod;
- Conical flask (250 ml);
- Test tube (150 x 25 mm): clean and dry;
- Rubber bung: contains whole at the center through which a glass rod (3–4 mm in diameter) is inserted and it is used to fit the glass tube;
- Weighing balance.

9.11.4 PROCEDURE

Take a required amount of ghee sample in a test tube (the mass of test portion taken for the analysis should be such that the amount of titrate should not exceed 10 ml during titration). The ghee sample should be in a liquid form.

Add 1 g of solid KI and 20 ml mixed solvent into the test tube while the test sample is in the liquid state. Pass the bubbles of CO_2 through the contents in a test tube (this step is not necessary for routine test). Heat the mixture in the test tube to boiling within 30s and boil the contents of the tube vigorously not > 30s till the solvent vapor begins to evaporate. Immediately, close the opening of the rubber bung with glass rod. Then, cool down the contents with a tap water and then these are transferred to a conical flask followed by adding 20 ml of aqueous KOH (5%) solution. Wash the tube twice with about 25–30 ml of water and transfer the washing into the flask. In the presence of starch indicator, titrate the contents of the flask against 0.002 N $Na_2S_2O_3$ solution. Indicator shall be added shortly before the end of titration. Run a blank test using the same reagents but without ghee sample and obtained titer value should not be more than 0.1 ml for blank titration [17].

9.11.5 CALCULATIONS

$$\text{Peroxide value (mEq of } O_2 \text{ per kg)} = 8000 \times \left[\frac{(V_1 - V_2) \times N}{M}\right] \times \frac{1}{8}$$

Or

$$\text{Peroxide value (mmol of } O_2 \text{ per kg)} = 8000 \times \left[\frac{(V_1 - V_2) \times N}{M}\right] \times \frac{1}{16} \quad (12)$$

where; V_1 is the amount of $Na_2S_2O_3$ used during sample titration (ml); V_2 is the amount of $Na_2S_2O_3$ used during blank titration (ml); M is the weight of test sample (g); and N is the normality of $Na_2S_2O_3$.

9.11.6 DATA COLLECTION SHEET

Title: Determination Peroxide value of ghee.

Location and Organization:______________________________

Participants: ____________________________________

Date (Month/Day/Year):________________________________

Observations:

Parameter	Repeat 1	Repeat 2	Repeat 3	Average
Volume of standard sodium thiosulfate solution used during sample titration (ml) = V_1				
Volume of standard sodium thiosulfate solution used during blank titration (ml) = V_2				
Mass of test sample (g) = M				
Peroxide value				

Comments:____.

9.12 DETERMINATION OF ACIDITY OF GHEE

9.12.1 INTRODUCTION

The acidity test for ghee indicates the keeping quality. The acidity of ghee is of two types: natural and developed. The former is responsible for the presence of non-esterified fatty acid in the ghee, and the latter is due to the development or production of FFA due to hydrolysis of fat by lipase action during storage. FFAs are more susceptible to oxidation. Hydrolyzation or oxidation cause formation of off-flavor, which is undesirable to consumers. It is determined in the presence of phenolphthalein indicator: 1 g of sample ghee is titrated with standard KOH or NaOH solution [18]. It is expressed as percent of FFAs in terms of oleic acid percentage. According to FSSRI (2021 amendments), FFA content of butter oil, AMF, and ghee should be not more than 0.4, 0.3, and 2% of oleic acid, respectively.

9.12.2 PRINCIPLE

The procedure of acidity estimation of the ghee sample is done by the titration method. Before titration, the lipid is completely dissolved in the neutral alcohol. The mixture is warmed, followed by titration with NaOH or KOH (0.1 N) solution in the presence of phenolphthalein indicator.

9.12.3 MATERIALS

9.12.3.1 REAGENTS

- Standard alkali solution: KOH/NaOH having normality of 0.1 N;
- Phenolphthalein indicator solution: 1% solution in 95% ethanol;

- Ethyl alcohol (95% v/v: sp. gr. of 0.816): Neutral to phenolphthalein.

9.12.3.2 APPARATUS

- Conical flask (250 ml);
- Burette;
- Weighing balance.

9.12.4 PROCEDURE

Ten g of test sample is transferred to flask (250 ml). In another conical flask, heat ethanol (50 milliliter) to the boiling stage (>70°C), neutralize with NaOH (0.1 N) solution using phenolphthalein (0.5 ml) solution. Add this neutralized ethanol to the sample containing flask and mix the contents thoroughly. Heat the flask and allow these to boil. The sample is titrated while it is hot against standard NaOH or KOH (0.1 N) solution with vigorous shaking. The appearance of pink color (which persists for at least 15 s) indicates the endpoint of the titration [18].

9.12.5 CALCULATIONS

Acid value (milligrams of KOH needed to neutralize FFAs in 1 g of sample) of ghee or butterfat is determined as follows:

$$\text{Acid value} = \frac{5.61 \times V}{M} \tag{13}$$

FFAs content (calculated as oleic acid percentage) of ghee or butterfat is determined as follows:

$$\text{FFAs (\% oleic acid)} = \frac{2.82 \times V}{M} \tag{14}$$

Degree of acidity (total titratable acidity of sample expressed as percentage) of ghee or butterfat is calculated as follows:

$$\text{Degree of acidity (\% oleic acid)} = \frac{100 \times N}{M} \tag{15}$$

where; V is the standard NaOH or KOH used during titration (ml); M is the weight of test sample (g); and N is the amount of NaOH or KOH used (represents ml of one normal solution).

9.12.6 *DATA COLLECTION SHEET*

Title: Determination acid value/FFAs/degree of acidity.

Location and Organization:________________________________

Participants: ____________________________________

Date (Month/Day/Year):__________________________________

Observations:

Parameter	**Repeat 1**	**Repeat 2**	**Repeat 3**	**Average**
Standard NaOH or KOH required during titration (ml) = V				
Weight of test sample (g) = M				
Amount of NaOH or KOH used (represents ml of 1 N solution) = N				
Acid value				
FFAs (% oleic acid)				
Degree of acidity (%)				

Comments:____.

9.13 DETERMINATION OF TOTAL CARBONYL CONTENT IN GHEE

9.13.1 *INTRODUCTION*

Rancidity in ghee mediated by oxygen can also be measured by the total carbonyl content of ghee. Carbonyls give a pleasant flavor to ghee. Volatile carbonyls are produced because of the degradation of hydroperoxides. The cow ghee has a slightly low carbonyl value than the buffalo ghee. The amount of carbonyl is dependent on the temperature of clarification. It increases during storage. The total carbonyl present in ghee is between 4 and 8 μmol per g of ghee. The contents are normally estimated by the flask method [19], where the color produced during the reaction will give the concentration of total carbonyls.

9.13.2 *PRINCIPLE*

When carbonyl reacts with 2,4-dinitrophenylhydrazine (2,4-DNPH), which gives a yellow color compound of dinitrophenyl hydrazones. The intensity of yellow color developed is directly proportional to carbonyl content in a given ghee sample, and absorbance is measured at 340 nm [19].

9.13.3 MATERIALS

9.13.3.1 REAGENTS

- Orthophosphoric acid (OPA): 87%.
- Adsorbent: Cellite – 545.
- 2,4-dinitrophenylhydrazine (2,4-DNPH).
- OPA and 2,4-DNPH mixture: Mix 5 ml of OPA and 0.5 g of 2,4-DNPH and ground the mixture by means of pestle and mortar. Transfer 10 ml of H_2O and reground to get a clear solution.
- Trichloroacetic acid (TCA).
- Benzene: It shall be free from carbonyl content. Transfer one liter of benzene in a flask having round bottom. Transfer 5 g of 2,4-DNPH and 1 g TCA, mix and reflux the contents for more than 3 h. Distill the contents and collect the distillate of carbonyl free benzene.
- Hexane: It shall be free from carbonyl content. Transfer one liter of hexane in a flask having round bottom. Transfer 5 g of 2,4-DNPH and 1 g of TCA, mix and reflux the contents for more than 3 h. Distill the contents and collect the distillate of carbonyl free hexane.

9.13.3.2 APPARATUS

- Glass column;
- Glass rod;
- Round bottom flask;
- Flask (25 ml);
- Mortar and pestle;
- Spectrophotometer;
- Reflux condenser;
- Analytical balance.

9.13.4 PROCEDURE

1. **Preparation of Column Material:** Heat Cellite-545 at 120°C for 12 h to activate the adsorbent. Add the activated adsorbent to the mixture of OPA and 2,4-DNPH mixture and grind it again until homogenous mass has been obtained. Fix the glass column and put some glass wool at the bottom. Add a small amount of hexane and close the opening. Transfer the above prepared mass into the column in three

or four proportions and each time properly packing the material in the glass column using a glass rod. Drain out excess hexane from the column. To the column, add benzene (50 ml) and flush the column using air pressure. Continue the washing until colorless effluent comes out of the column (washing removes excess of 2,4-DNPH). Remove excess hexane and it is ready for the flask method.

2. **Determination:** Transfer 100 mg of liquid ghee in a conical flask (25 ml). Transfer 0.5 ml of hexane (carbonyl free) into the flask and then add 0.5 g of column material. Keep the mixture of flask overnight at ambient temperature. After keeping overnight, add 9.5 ml hexane and mix the contents. Record the absorbance of the solution at 340 nm using a spectrophotometer. Dilute the contents with hexane if the absorbance is more than 0.9. The absorbance reading should be between 0.2 and 0.8. Prepare a blank test in a similar way using the same reagents but without ghee sample.

9.13.5 CALCULATIONS

Calculate the carbonyl content using extinction coefficient (ε) as follows:

$$\varepsilon = \frac{A}{C \times d} \text{ or}$$

$$\text{C, moles/liter} = \frac{A}{\varepsilon \times d}$$

$$\text{Total carbonyl content } (\mu\text{moles/ml}) = \frac{A \times 1 \times 10^{-6}}{22500 \times 1000 \times d} \times \frac{V}{M} \quad (16)$$

where; ε is the extinction coefficient (22,500 for milk fat); C is the concentration of absorbing element; A is the absorbance reading at 340 nm; d is the length of light path in solution; M is the mass of test sample (g); and V is the dilution of test sample (ml).

9.13.6 DATA COLLECTION SHEET

Title: Determination of total carbonyl content in ghee.

Location and Organization:________________________

Participants: ________________________

Date (Month/Day/Year):________________________

Observations:

Parameter	Repeat 1	Repeat 2	Repeat 3	Average
Absorbance = A				
Weight of test sample (g) = M				
Total carbonyl content (μmoles/ml)				

Comments:________________________________.

9.14 ESTIMATION OF CHOLESTEROL IN GHEE

9.14.1 INTRODUCTION

Cholesterol ($C_{27}H_{45}OH$) is the principal sterol that is present in milk. It is mainly present on the membrane of fat globule. Cow ghee contains 302 to 362 mg of cholesterol per 100 g of ghee, whereas in buffalo ghee contains 209 to 312 mg per 100 g of ghee. The 85–95% of cholesterol is present in the free form and the rest is present in the esterified form. Cholesterol in food is usually determined by Liebermann-Burchard (LB) method [20]. Direct ghee sample or pre-extracted USM from ghee can be used for cholesterol estimation by spectrophotometric method at a suitable wavelength. It can also be estimated by the enzymatic technique, gas chromatography or HPLC method.

9.14.2 PRINCIPLE

Cholesterol reacts with H_2SO_4 present in the LB reagent to form a cholesterol dienoid compound, which gives purple color to the solution in the presence of sulfuric acid. The color is estimated by a spectrophotometer at 650 nm. The intensity of developed purple color depends on the presence of amount of cholesterol in the ghee sample. Chloroform is used to stabilize the color for 15 min. or else the color fades due to the formation of polymer compounds [20].

$$\text{Cholesterol} + H_2SO_4 \text{ (present in LB reagent)} \rightarrow \text{Persulfate Cholesterol}$$

$$\text{Persulfate Cholesterol (dehydration by } H_2SO_4) \rightarrow \text{Cholesterol 2,4 – dienoid}$$

$$\text{or}$$

$$\text{Cholesterol 3,4 – dienoid (Purple)}$$

9.14.3 MATERIALS

9.14.3.1 REAGENTS

- Chloroform ($CHCl_3$).
- Concentrated sulfuric acid (H_2SO_4).
- Standard solution of cholesterol (5 mg/ml): 0.5 g standard cholesterol is dissolved in 100 ml of $CHCl_3$.
- Working solution cholesterol: 2 ml of stock solution is transferred to volumetric flask (25 ml) and volume is made to the mark with $CHCl_3$.
- Acetic anhydride.
- LB reagent: Mix one ml of sulfuric acid and 20 ml of acetic anhydride in a conical flask by placing in cold water. Place this reagent at 0°C for a minimum of half an hour.

9.14.3.2 APPARATUS

- Pipette;
- Volumetric flask (25 ml);
- Water bath;
- Spectrophotometer;
- Test tube (10 ml) with glass stopper;
- Conical flask (150 ml);
- Analytical balance.

9.14.4 PROCEDURE

9.14.4.1 PREPARATION OF THE GHEE SAMPLES (SECTION 1.4.2 IN THIS CHAPTER)

9.14.4.2 METHODOLOGY

Transfer 5 g of melted clear anhydrous ghee sample into a volumetric flask (25 ml). Transfer $CHCl_3$ to the flask for dissolving the test sample and then make up the volume to the mark with $CHCl_3$. In a test tube, transfer two milliliters of diluted test solution followed by adding $CHCl_3$ (1 ml) and LB reagent (4 ml). Tubes are incubated at 25°C for 10 min in a water bath. The color developed is analyzed by a spectrophotometer at 650 nm. The absorbance value (AV) of the test sample is recorded.

9.14.4.3 PREPARATION OF STANDARD CURVE

Prepare the standard curve by taking different known quantities (i.e., 0, 0.5, 1.0, 2.0, 2.5, and 3.0 ml) of working standard cholesterol solution in different glass stoppered test tubes and final volume is made to 3 ml with $CHCl_3$. Then transfer 4 ml of LB reagent into each test tube and incubate the tubes in a water bath at 25°C for 10 min. Record the AV in the spectrophotometer at 650 nm. Plot the absorbance (Y-axis) versus concentration (X-axis) to get the standard curve.

Note: Color developed is not stable for long period; therefore, measurement of absorbance should be done within 15 min after addition of LB reagent.

9.14.4.4 CALCULATIONS

Calculate the concentration of cholesterol in the ghee sample from the calibration curve.

$$\text{Cholesterol content } (\mu\text{g per 100 g}) \text{ of ghee sample} = \frac{X \times 25}{2 \times M} \times 100 \quad (17)$$

where; X is the concentration of cholesterol in 2 ml of the prepared test sample obtained from the standard curve (μg); and M is the weight of the test sample (g).

9.14.5 DATA COLLECTION SHEET

Title: Determination of cholesterol content in ghee.

Location and Organization:____________________

Participants: ____________________

Date (Month/Day/Year):____________________

Observations:

Parameter	Repeat 1	Repeat 2	Repeat 3	Average
Weight of test sample (g) = M				
Concentration of cholesterol in 2 ml of prepared test sample obtained from the standard curve (μg) = X				
Cholesterol content (μg/100 g of ghee)				

Comments:____________________.

9.15 DETERMINATION OF FATTY ACIDS (FAS) PROFILE IN GHEE

9.15.1 INTRODUCTION

Milk fat contains > 400 FAs; however, only 15 or 16 FAs are present at concentrations above 1%. Most of the FAs in milk fat are even numbered and contain 4 to 18 carbon atoms, because the FA is elongated by two carbon units (from Acetyl CoA) per cycle through a series of steps and repeated cycles. In general, FAs occurs in cis form. The main saturated, monounsaturated, and polyunsaturated FAs accounts for 62–70%, 25% and 2–6%, respectively of the total FAs in milk fat. The GLC with flame ionization detector (FID) is normally used for the analysis of the FAs in ghee [21, 22]. According to FSSRI (2021: amendment), the product shall meet the requirements of composition of FAs (Table 9.2 for ghee).

9.15.2 PRINCIPLE

Fatty acids methyl esters (FAMEs) are prepared by transesterification of glycerides with methanolic KOH or sodium methoxide ($NaOCH_3$) followed by neutralization with $NaHSO_4$ to prevent saponification of fat. The prepared FAMEs are quantitatively determined by capillary GLC.

9.15.3 MATERIALS

9.15.3.1 REAGENTS

- Methanol (contains H_2O should not be >0.5% by mass).
- Solvent: n-hexane, n-heptane or n-pentane.
- Sodium hydrogen sulfate monohydrate ($NaHSO_4$. H_2O).
- Transesterification reagent: Methanloic KOH or $NaOCH_3$ solution (2 mol/liter). About 11.2 g of KOH or 10.8 g of $NaOCH_3$ is dissolved in 100 ml of methanol. Methanolic $NaOCH_3$ can also be made by adding 4.6 g of metallic Na in absolute methyl alcohol.
- Esterification reagent: Add slowly 1 ml of H_2SO_4 in 15 ml of methanol.
- Standard FAMEs mixture (>90% purity): It is available commercially. It consists of saturated, even-numbered straight-chain FAME from C_4 to C_{22} along with unsaturated ($C_{18:1}$, $C_{18:2}$ and $C_{18:3}$) FAMEs.
- Carrier gas: Hydrogen, He or Nitrogen with 99.99% purity.

- Reference milk fat: Known fatty acid composition used for quantification purpose.

TABLE 9.2 The Composition of FAs (% of Total FAs) of Ghee by GLC

FAs	Composition (%)	FAs	Composition (%)
Saturated FAs		**Monounsaturated FAs**	
Butyric ($C_{4:0}$)	1.0–5.0	Palmitoleic ($C_{16:1}$)	0.9–2.8
Caproic ($C_{6:0}$)	0.5–2.2	Oleic ($C_{18:1}$)	19–32
Caprylic ($C_{8:0}$)	0.4–1.5	**Polyunsaturated FAs**	
Capric ($C_{10:0}$)	0.8–5.0	Linoleic ($C_{18:2}$)	0.5–3.5
Lauric ($C_{12:0}$)	1.5–4.0	α-Linolenic ($C_{18:3}$)	0.3–1.0
Myristic ($C_{14:0}$)	6.0–13.0	–	–
Palmitic ($C_{16:0}$)	23.0–28.0	–	–
Steric ($C_{18:0}$)	8.0–19.0	–	–

9.15.3.2 APPARATUS

- Analytical balance;
- Test tube (16×100 mm) with screw cap;
- Pipettes (1-, 5-, and 10-ml capacity);
- Vortex mixer;
- Centrifuge: Operating at 350±50 g and provided with speed indicator;
- Glass ampoules (5 ml capacity);
- Hot air oven: Operating temperature between 100°C and 110°C;
- Gas-liquid chromatogram (GLC): Comprising of injector, carrier gas supply, Pressure gauge, column (glass or silica capillary column), column oven and FID;
- Micro-syringe (10 μl capacity).

9.15.4 PROCEDURE

9.15.4.1 PREPARATION OF FAMES

Weigh exactly 0.1 g of ghee sample in a test tube. Transfer 5 ml of methanol into a test tube to dissolve the ghee sample. Transfer 0.2 ml of methanolic potassium hydroxide (KOH) or $NaOCH_3$ solution and close the test tube with a screw cap. Vigorously mix the contents of the tube for one minute using a vortex mixer. After five minutes, transfer 0.5 g of $NaHSO_4$ and the contents in the test tube are mixed again. Centrifuge the tube for three minutes at 350 g at ambient temperature. After centrifugation, siphon off an upper portion of clear supernatant for GLC analysis. This method is not suitable for the preparation of methyl esters if the milk fat is partially hydrolyzed or lipolyzed with a fat acidity > 1 mmol of FFA/100 g of fat. In such cases, the following procedure can be adopted [21].

Weigh exactly 2 g of ghee sample into a glass ampoule. Transfer 0.4 ml of methanolic H_2SO_4 and then seal the ampoule with the gas flame. Heat the contents of the ampoule in a drying oven at 100–110°C for 3 h with intermittent mixing. Cool the contents of the ampoule to room temperature and wait for clear phase separation. Open the ampoule and siphon off the upper layer for further GLC analysis. Dilute the prepared ester with suitable solvent, if required [21].

9.15.4.3 GLC ANALYSIS

Following operating conditions to be selected in GLC for separation of FAMEs in test sample:

- Column: J&W DB wax;
- Carrier gas: Hydrogen;
- Column dimensions: diameter 0.32 × length 30 m; thickness 0.25 μm;
- Head pressure of column: 0.4 bar;
- Injection type: split mode;
- Injection volume: 1 μl;
- Injector temperature: 280°C;
- Split ratio of 1:25 measured at 40°C;
- Sample concentration: 2 mg/ml (dissolved in hexane);
- Detector (FID) temperature: 250°C.

Under the same working conditions, analyze the standard FAMEs as used for the ghee sample. Record the retention time (RT) of standard FAMEs. Peaks of test sample are identified by comparing the RT obtained with standard FAMEs and by comparison with chromatogram. Method of normalization is used to determine the percent of the component indicated by a peak in the chromatogram. The sum of the areas under pertinent peaks denotes 100% of components of ghee sample. Estimate the peak areas attributable to FAMEs.

9.15.5 CALCULATION

Calculate the percent peak area (PA) representing the FA(i) as follows:

$$\text{Percentage of peak area } P_{Ai} = \frac{A_i}{\Sigma A_i} \times 100\% \tag{18}$$

where; A_i is the numerical value of PA corresponding to FA(i), and EA_i is the numerical value of the sum of all the PAs corresponding to the component FAMEs.

Calculate the correction factors (f_i, which are used to convert % PA into % mass fraction of the component) as follows:

$$\text{Correction factor (CF) } f_i = \frac{W_i}{A_i} = \frac{W_i \cdot \Sigma A_i}{\Sigma W_i \cdot A_i} \tag{19}$$

where; ΣWi = Sum of mass fraction of various FAs in reference fat (expressed as FFA); and W_i is the Mass fraction of FA(i) in the reference fat (expressed as FFA).

Standardize the values of the individual CF w.r.t. the CF of methyl palmitate to get relative CF of FA(i) by using following formula:

$$f_i' = \frac{f_i}{f_p} \tag{20}$$

where; f_p is the CF value for methyl palmitate ($f_i' = 1$).

Calculate the mass fraction of individual FA(i) in test sample as follows:

$$W_{is} = \frac{f_i' \cdot P_{AiS}}{\Sigma(f_i' \cdot P_{AiS})} \times 100\% \tag{21}$$

where; W_{is} is the percent mass fraction of individual FAs(i) in the test sample (expressed as g of FFAs/100 g of total FAs); and P_{Ais} is the % PA indicating FA(i) in the test sample.

9.15.6 DATA COLLECTION SHEET: COMPOSITION FATTY ACID IN GHEE

Title: Analysis of fatty acid composition of ghee.

Location and Organization: ____________________

Participants: ____________________

Date (Month/Day/Year): ____________________

Observations:

Parameter	Repeat 1	Repeat 2	Repeat 3	Average
Numerical value of PA corresponding to FA(i) = A_i				
Numerical value of sum of all the PAs corresponding to the component FAMEs = ΣA_i				
Relative CF of FA(i) = f_i'				
Percent PA in FA(i) in test sample = P_{Ais}				
Percent mass fraction of individual FAs(i) in test sample = W_{is}				

Comments:______________________________________.

9.16 DETERMINATION OF VITAMIN A IN GHEE

9.16.1 INTRODUCTION

Vitamin A is principally an alcohol derivative of a polymer consisting of four isoprene units. Two molecules of Vitamin A are produced by the cleavage of one molecule of β-carotene (provitamin A). Vitamin A occurs in two forms: A_1 (retinol) and A_2 (3-dehydroretinol). Both vitamin A_1 and A_2 are alcohols. The only difference between these two is that vitamin A_2 contains an additional double bond in the ring structure. Vitamin A estimation is quite essential, because both hyper- and hypo-vitaminosis may cause diseases and even death. As vitamin A is a fat-soluble vitamin, therefore it gets stored in our body, and it can reach to unhealthy levels over time. High consumption of dietary carotenoids is not associated with toxicity. However, the tolerable upper limit of 3 mg or 10,000 IU for adults has been fixed by NIH (National Institute of Health) to prevent its toxicity. According to U.S. Department of Agriculture (USDA), 162 IU of vitamin A is present per 100 g of whole milk, having 3.25% fat. Cow milk ghee contains 19 to 34 IU vitamin A per g of fat [23]. In milk, vitamin A is present as retinol, retinol esters and β-carotene. Among all these, about 97% of retinol is in esterified form, whereas β-carotene accounts for 90% of the provitamin A activity in milk. The current emphasis on nutrient requirements, nutrient labeling and the use of processed foods has generated a demand for the estimation of vitamin A. Vitamin A can be determined by colorimetric, spectrophotometric, fluorometric, GLC, and HPLC methods. In the Colorimetric method, antimony trichloride ($SbCl_3$) is used to measure vitamin A (retinol) in foods.

Spectrophotometric or fluorometric methods are mostly used, when vitamin A content is high.

9.16.2 PRINCIPLE

Vitamins A is a component of USM in ghee. For its estimation, the test sample is saponified with alcoholic KOH and solvent (such as: ether) is used to extract the USM containing sterol, fat soluble vitamins, etc. After solvent evaporation, the dried residue is used for vitamin A determination.

The vitamin A content is measured by spectrophotometric method through Carr-Price reaction. When a solution of $SbCl_3$ in chloroform is added to a dilute chloroform solution of a vitamin A containing sample, a blue color soon reaches a maximum intensity and rapidly fades or changes to reddish brown or other colors, depending on the nature of the sample. Absorbance is measured using a spectrophotometer at 620 nm immediately after the formation of blue color. Using standard curve, the vitamin A concentration in ghee sample can be calculated [24].

$$\text{Vitamin A} + SbCl_3 \rightarrow \text{Blue color}$$

9.16.3 MATERIALS

9.16.3.1 REAGENTS

- Isopropanol or absolute alcohol: It shall show spectral purity with an absorbance not > 0.01 at 350 and 320 nm and not > 0.5 at 300 nm, when measured in 1 cm quartz cell against H_2O.
- Standard Vitamin A: 0.5 g of vitamin A acetate is used.
- Ether: It shall be free from peroxide and shall be redistilled or anesthetic grade.
- KOH solution (concentration of 50% (w/v)).
- Chloroform: It shall be redistilled, and it should be free from water.
- $SbCl_3$ solution: 113.4 g of $SbCl_3$ is dissolved in 300 ml of $CHCl_3$. Transfer around 5 g of $CaCl_2$, mix and the mixture is filtered in hot condition. The volume of the filtrate is made up to 500 ml with $CHCl_3$.
- Na_2SO_4 (granular and anhydrous): Under the specified conditions, it shall not absorb vitamin A; and 10% solution shall not be acidic to methyl red indicator.

9.16.3.2 APPARATUS

- Spectrophotometer (suitable to measure absorbance at 620 nm);
- Analytical balance;
- Volumetric flask (500 ml);
- Filter paper (grade 40);
- Saponification flask;
- Reflux condenser;
- Water bath;
- Separating funnel (500 ml).

9.16.4 PROCEDURE

9.16.4.1 PREPARATION OF UNSAPONIFIABLE FRACTION

Weigh accurately 0.5 g of standard vitamin A and 1 g (vitamin A ranges between 20 and 45 IU) of ghee sample and transfer separately into two saponification flasks. To each flask, transfer 40 ml of ethanol and 7 ml of 50% KOH solution. Reflux the contents in a boiling water bath for 30 to 40 minutes using a reflux condenser. Cool the contents, transfer about 50 ml of water and then transfer to a separating funnel. Extract thrice with 50 ml volumes of diethyl ether and ethereal extracts (these contain vitamin A) are combined in another funnel, which contain 100 ml of distilled water. Remove the aqueous layer after clear separation of two layers in the funnel. Ether extract is washed with water (100 ml in each washing) until the wash water is free from alkali (i.e., no color with phenolphthalein). Concentrate the ether extract in a conical flask to around 50 ml in water bath followed by addition of 5–10 g of Na_2SO_4, mix and allow to settle. Transfer the ether extract to a graduated flask (50 ml). Rinse the Na_2SO_4 with several small portions of ether to completely extract the vitamin A and pour into the flask, volume is made up to the mark with final rinse. Completion of vitamin extraction from Na_2SO_4 can be tested by addition of few drops of $SbCl_3$ solution to the residue.

9.16.4.3 PREPARATION OF A STANDARD CURVE

Evaporate the ether extract to about 5 ml in a boiling water bath. Remaining ether is evaporated under reduced pressure at low temperature. Prepare a stock solution of dried vitamin A acetate residue with a known quantity of chloroform to give an absorbance of 0.8 (or dissolve the residue in 100 ml of $CHCl_3$ to get final concentration of 5 mg/ml for standard). From the stock solution, make a sequence of dilute working solutions with $CHCl_3$, so that it gives an AV of 80, 60, 40, and 20% of the original absorbance. Measure the absorbance (620 nm) of blue color formed when 1 ml each of the diluted solution plus 1 ml of $CHCl_3$ is mixed with 3 ml $SbCl_3$ solution. Prepare a blank test using 2 ml of $CHCl_3$ and 3 ml of $SbCl_3$ solution and adjust to 100% transmittance. Standard curve is drawn by plotting absorbance versus concentration of vitamin A acetate solutions.

9.16.4.4 DETERMINATION

Evaporate the ether completely from the test sample under reduced pressure at low temperature. Dissolve the residue in 2 ml of $CHCl_3$ followed by addition of 3 ml of $SbCl_3$ solution. Set the instrument to 100% transmittance using blank containing 2 ml $CHCl_3$ and 3 ml of $SbCl_3$ solution. Determine the optical density (at 620 nm) of blue color within three seconds.

9.16.5 CALCULATIONS

Estimate the vitamin A content of the test sample from the standard/calibration curve.

$$\text{Vitamin A content (mg/g) of ghee} = \frac{Y}{M} \tag{22}$$

where; Y is the vitamin A content in 2 ml of chloroform solution of sample from graph; M is the weight of test sample (in g).

9.16.6 DATA COLLECTION SHEET

Title: Determination of vitamin A content of ghee.

Location and Organization: ______________________________

Participants: ______________________________

Date (Month/Day/Year): ______________________________

Parameter	Repeat 1	Repeat 2	Repeat 3	Average
Concentration of vitamin A in 2 ml chloroform solution of sample obtained from the standard curve (mg)= Y				
Weight of test sample taken (g) = M				
Vitamin A content (mg/g)				

Comments:____.

9.17 DETERMINATION OF VITAMIN D IN GHEE

9.17.1 INTRODUCTION

Vitamin D (known as calciferol) is a fat-soluble vitamin. It is also known for its pro-hormone properties. It is an integral matter of unsaponifiable matter (USM) present in the fat phase. Vitamin D_2 (ergo-calciferol) and vitamin D_3 (chole-calciferol) are the most important ones among vitamin D. Vitamin D_2 are formed by ultraviolet (UV) irradiation of ergosterol while vitamin D_3 is from irradiation of 7-dehydro-cholesterol. Calciferol has a unique property in the formation of bone and teeth from calcium phosphate by absorbing calcium in the guts. A daily requirement of 400 IU is required for youngsters, adolescents, and pregnant or lactating women. The SACN (Scientific Advisory Committee on Nutrition) recommends that people should have serum concentrations of vitamin D > 25 nmol/liter throughout the year. This concentration is considered sufficient to avoid deficiency diseases in children and adults. The vitamin D in cow ghee is about 8 to 13 IU/g.

Vitamin D is very stable to heat. It is not destroyed by pasteurization, sterilization, drying or condensation treatments. The potency of vitamin D is expressed in international units (IU: one IU is equivalent to 0.025 μg cholecalciferol). Vitamin D is estimated by competitive binding methods, HPLC, and radioimmunoassay. In this section, vitamin D is estimated by spectrophotometric analysis with respect to the standard vitamin D sample [25, 26].

9.17.2 PRINCIPLE

Vitamin D is present in the USM of milk fat. The ghee should be saponified by alcoholic KOH. The USM component is extracted with the solvent ether. The extracted matter is dried, and the residue is dissolved in $CHCl_3$. The chloroform solution of vitamin D is treated with a mixture of antimony trichloride, and acetyl chloride dissolved in methylene chloride (CH_2Cl_2). The developed orange-yellow color is measured at 500 nm using a spectrophotometer [25]. The contents of vitamin D are calculated from the standard curve.

9.17.3 MATERIALS

9.17.3.1 REAGENTS

- Vitamin D standard solution: It shall have concentration of 10 IU per ml of chloroform;
- Chloroform ($CHCl_3$);
- Anhydrous sodium sulfate (Na_2SO_4);
- KOH solution: It shall have a concentration of 50%;
- Ethanol;
- Diethyl ether: It shall be peroxide-free;
- Solution A: To prepare this reagent, mix 20 g of antimony trichloride with 90 ml of CH_2Cl_2;
- Solution B: To prepare this reagent, mix 40 ml of CH_2Cl_2 and 10 ml of acetyl chloride;
- Color reagent: Mix A and B solutions in the ratio of 45:5. It shall be stable for 7 days at ambient temperature.

9.17.3.2 APPARATUS

- Glass funnel;
- Separating funnel (250 and 300 ml);
- Saponification flask;
- Glass beads;
- Air condenser;
- Oven;
- Spectrophotometer: capable to measure absorbance at 500 nm;
- Water bath;
- Analytical balance;
- Whatman filter paper (#40).

9.17.4 PROCEDURE

9.17.4.1 PREPARATION OF UNSAPONIFIABLE FRACTION

Exactly weigh about 5 g of test sample into a flask. Transfer 40 ml of ethanol and 7 ml of 50% solution of KOH. Contents of flask are refluxed in boiling water bath for 30 minutes using a reflux condenser. Cool the contents, transfer about 50 ml of H_2O and then it is poured into a separating funnel. Extract thrice with 50 ml of diethyl ether and ethereal layer (these contain vitamin D) is then combined in another funnel. Ether extract is washed with water (100 ml in each washing) until the wash water is free from alkali (i.e., no color with phenolphthalein). Add a small amount of Na_2SO_4 to ether extract, mix well followed by filtration with filter paper (#40). Add two or three glass beads into the flask containing filtrate, and the solvent is completely evaporated in a boiling water bath. Dried residue is dissolved in 5 ml of $CHCl_3$.

9.17.4.3 DETERMINATION

Mix one milliliter of above prepared chloroform solution of the test sample with four milliliters of color reagent. Set the instrument to 100% transmittance using blank containing 1 and 4 ml of $CHCl_3$ and color reagent, respectively. After 10 minutes, measure the absorbance of orange-yellow color solution at 500 nm.

9.17.4.4 PREPARATION OF CALIBRATION CURVE

From the standard stock solution (SSS), make a sequence of dilute working solutions of vitamin D (range from 2 to 10 IU) using chloroform. Measure the absorbance (500 nm) of orange-yellow color formed when 1 ml each of the diluted $CHCl_3$ solution is mixed with 4 ml color reagent. Prepare a blank test using 1 and 4 ml of $CHCl_3$ and color reagent, respectively, and adjust to 100% transmittance. Prepare the standard or calibration curve; and finally, vitamin D in the test sample can be measured from the calibration curve.

9.17.5 CALCULATIONS

Determine the concentration of vitamin D of the test sample from the calibration graph.

$$\text{Vitamin D content per g of ghee, IU} = \frac{X \times 5}{M} \qquad (23)$$

where; X is the concentration of vitamin D in one ml chloroform solution of test sample obtained from standard curve; and M is the weight of test sample (g).

9.17.6 DATA COLLECTION SHEET

Title: Determination of vitamin D content in ghee.

Location and Organization:______________________________

Participants: ____________________________________

Date (Month/Day/Year):______________________________

Observations:

Parameter	Repeat 1	Repeat 2	Repeat 3	Average
Concentration of vitamin D in 1 ml chloroform solution of sample obtained from graph (IU) = X				
Weight of sample (g) = M				
Vitamin D content per g of ghee				

Comments:____________________________________.

9.18 DETERMINATION OF VITAMIN E IN GHEE

9.18.1 INTRODUCTION

Vitamin E (also known as tocopherol) is an antioxidant and a fat-soluble vitamin. There are eight types of tocopherols in their natural forms. Among all, the most active one is α-tocopherol. The other tocopherols are β-tocopherol, γ-tocopherol and δ-tocopherols. Vitamin E prevents the peroxidation of PUFA (polyunsaturated fatty acids) in tissues and cell membranes. Vitamin E even prevents red blood cells from hemolysis. In cow milk ghee, vitamin E content ranges from 26 to 48 mg/g [23]. Vitamin E is separated and determined by chromatographic methods (such as: paper, column, gas-liquid, thin-layer, and high-pressure liquid chromatography (HPLC)). HPLC is a most common method to estimate different fat-soluble vitamins, such as vitamin E. In this section, vitamin E content in butterfat or ghee is estimated by Emmerie-Engel reaction method [14].

9.18.2 PRINCIPLE

Determination of tocopherol is based on Emmerie-Engel reaction in which tocopherols reduce $FeCl_3$ to $FeCl_2$ that reacts with α, α'-dipyridyl to produce red color whose absorbance is measured at 530 nm [14].

9.18.3 MATERIALS

9.18.3.1 REAGENTS

- Methanol.
- Ethyl alcohol: It shall be aldehyde free.
- Diethyl ether: It shall be peroxide free.
- Light petroleum: Boiling point between 40°C and 60°C.
- Ferric chloride: AR grade.
- KOH solution (aqueous): 2%.
- KOH solution in methanol (2 N): To prepare this solution, mix 112 g of KOH pellets in methanol and make up the volume to one liter.
- HCl (sp. gr. of 1.16).
- Anhydrous Na_2SO_4 (AR grade).
- Benzene (AR grade).
- α, α'-Dipyridyl: 0.5% solution in absolute alcohol.
- Floridin XS column: With purified absorbent, fill a tube having 12 x 30 mm dimensions. To purify the absorbent, digest with HCl for 1 h in a boiling water bath. Repeat the digestion with the fresh volume of HCl at ambient temperature. Thereafter, the absorbent is washed with H_2O till it is free of acid; and it is followed by ethanol and benzene. Later it is dried at ambient temperature.
- Standard tocopherol.

9.18.3.2 APPARATUS

- Pipette (1-, 2-, and 5-ml capacity);
- Water bath;
- Spectrophotometer;
- Conical flask (250 ml);
- Test tube (150 x 25 mm);
- Reflux condenser;
- Glass tube (12 x 30 mm).

9.18.4 PROCEDURE

Transfer exactly 1 g of test sample into a test tube attached with a reflux condenser. Transfer 2 ml of alcoholic KOH solution and reflux it for 10 minutes at 72–74°C in a water bath in an atmosphere of nitrogen. Add 8 ml of methanol and 10 ml of H_2O and the contents of the tube are thoroughly mixed. Extract the USM thrice from the mixture with diethyl ether (50 ml). Wash the combined ethereal extract with water, aqueous KOH solution (2%), with water till the wash water is free of alkali (i.e., no color is observed with phenolphthalein). In a conical flask, concentrate the ether extract to around 50 ml in water bath followed by addition of 5–10 g of Na_2SO_4, mix and allow it to settle. Transfer the ether extract to a graduated flask (50 ml). Rinse the Na_2SO_4 with several small portions of ether to completely extract the USM; and pour into the flask with final rinsing. Evaporate the solvent under vacuum in an atmosphere of CO_2.

To remove the carotene from the obtained USM residue, 5 ml benzene is added to dissolve the residue. Pour this solution to Floridin XS column, which has been previously saturated with benzene. Once the solution has been passed through the column, wash with benzene till the volume of elute has reached to 25 ml. Greenish-blue color is formed on absorbent earth by carotenoids, whereas dark blue is due to vitamin A.

Transfer one ml of carotene-free unsaponifiable extract into the tube. Transfer one ml of alcoholic $FeCl_3$ (0.2%) solution and mix it thoroughly. Add 0.5% of α, α'-dipyridyl solution into the test tube. Mix well and volume is made to 25 ml with ethanol. Make a blank test by following the same procedure. Keep the tubes undisturbed for 10–15 min followed by measuring the optical density at 530 nm using spectrophotometer.

Prepare a working standard solution (WSS) of α-tocopherol in alcohol to obtain a concentration in the range of 20–100 μg. Add the same quantity of reagents as that of the test sample. Prepare the standard curve; and finally, vitamin E in the test sample can be measured from the calibration curve.

9.18.5 CALCULATIONS

Determine the vitamin E content of the test sample from the calibration curve.

$$\text{Vitamin E content per g of test sample, } \mu g = \frac{Y \times 25}{M} \tag{24}$$

where; Y is the concentration of vitamin E in one ml alcoholic solution of test sample (μg); and M is the weight of the sample (g).

9.18.6 DATA COLLECTION SHEET

Title: Determination of vitamin E in ghee.

Location and Organization:______________________________

Participants: ______________________________________

Date (Month/Day/Year):_________________________________

Observations:

Parameter	**Repeat 1**	**Repeat 2**	**Repeat 3**	**Average**
Concentration of vitamin E in one ml alcoholic solution which corresponding to the absorbance of test sample (μg) = Y				
Weight of sample (g) = M				
Vitamin E content (μg/g)				

Comments:____.

KEYWORDS

- **acetyl value**
- **acidity**
- **cholesterol**
- **fatty acid**
- **ghee**
- **hydroxyl value**
- **peroxide value**
- **phospholipids**
- **physicochemical constants**
- **total carbonyl**
- **vitamins A/D/E**

REFERENCES

1. AOAC, (2000). AOAC official method 920.212: Specific gravity (apparent) of oils – pycnometer method. In: Hortwiz, W., (ed.), *Official Methods of Analysis of AOAC International* (17th edn., p. 72). Gaithersburg, MD – USA: AOAC International.
2. FSSAI, (2016). Determination of specific gravity. In: *Manual of Methods of Analysis of Foods: Oils and Fats; Lab Manual – II* (p. 6). New Delhi, India: FSSAI.
3. Karleskind, A., & Wolff, J. P., (1996). *Oils and Fats Manual: A Comprehensive Treatise: Properties, Production, Applications* (Vol. 2, p. 1527). New York: John & Wiley Online.
4. Martin, P. G., (1979). *Manuals of Food Quality Control, 3: Commodities* (Vol. 13, p. 338). Food & Nutrition Report 14/8; Rome, Italy: Food & Agriculture Organization (FAO).
5. Rangappa, K. S., & Achaya, K. T., (1974). *Indian Dairy Products* (p. 386). Bombay – India: Asia Publishing House.
6. IS 3508-1966. Determination of moisture content. In: *Methods of Sampling and Test for Ghee (Butterfat)* (pp. 10, 11). New Delhi: Bureau of Indian Standards (BIS); (2003).
7. IS 3508–1966. Determination of soluble and insoluble volatile acids (Reichert-Meissl, Polenske and Kirschner Values). In: *Methods of Sampling and Test for Ghee (Butterfat)* (pp. 22–29). New Delhi: Bureau of Indian Standards (BIS); (1997).
8. IS 3508-1966 (Reaffirmed 1997). Determination of saponification value. In: *Methods of Sampling and Test for Ghee (Butterfat)* (pp. 29, 30). New Delhi: Bureau of Indian Standards; (1997).
9. FSSAI, (2012). *Manual of Methods of Analysis of Foods: Oils and Fats* (pp. 20–22). New Delhi – India: Food Safety and Standard Authority of India (FSSAI).
10. IS 3508-1966. Determination of unsaponifiable matter. In: *Methods of Sampling and Test for Ghee (Butterfat)* (pp. 34, 35). New Delhi – India: Bureau of Indian Standards (BIS); (1997).
11. FSSAI, (2016). Determination of iodine value. In: *Manual of Methods of Analysis of Foods: Oils and Fats* (pp. 26–30). New Delhi, India: FSSAI.
12. IS 3508-1966. Determination of iodine value. In: *Methods of Sampling and Test for Ghee (Butterfat)* (pp. 30–33). New Delhi – India: Bureau of Indian Standards (BIS); (1997).
13. IS 3508-1966. Determination of refractive index. In: *Methods of Sampling and Test for Ghee (Butterfat)* (pp. 13–15). New Delhi – India: Bureau of Indian Standards (BIS); (1997).
14. IS 3508-1966: Determination of tocopherol. In: *Methods of Sampling and Test for Ghee (Butterfat)* (pp. 39–41). New Delhi: Bureau of Indian Standards; Reaffirmed; (2003).
15. AOAC, (2019). AOAC official method, 965.32: Hydroxyl value of oils and fats. Chapter 41. In: Latmer Jr. G. W., (ed.), *Official Methods of Analysis of AOAC International* (21st edn., Vol. III, p. 6). Rockville, Maryland – USA: AOAC International.
16. AOAC, (2000). AOAC official method, 921.09 acetyl value of oils and fats. In: *Official Methods of Analysis of AOAC International* (17th edn., p. 26). Gaithersburg, Maryland–USA: AOAC International.
17. IS 3508-1966: Determination of the peroxide value. In: *Methods of Sampling and Test for Ghee (Butterfat)* (pp. 56–59). New Delhi–India: Bureau of Indian Standards (BIS); (2003).

18. IS 3508-1966: Determination of acidity. In: *Methods of Sampling and Test for Ghee (Butterfat)* (pp. 21, 22). New Delhi: Bureau of Indian Standards (BIS); (2003).
19. Kumazawa, H., & Oyama, T., (1965). Estimation of total carbonyl content in oxidized oil by 2, 4-dinitrophenylhydrazine. *Journal of Japan Oil Chemists' Society, 14*(4), 167–171.
20. Kenny, A. P., (1952). The determination of cholesterol by the Liebermann-Burchard reaction. *Biochemical Journal, 52*(4), 611–619.
21. ISO 15884 (IDF 182): 2002. *Milk Fat – Preparation of Fatty Acid Methyl Esters* (p. 9). Switzerland: International Organization for Standardization (ISO); (2002).
22. ISO 15885 (IDF 184): 2002. *Milk Fat – Determination of the Fatty Acid Composition by Gas-Liquid Chromatography* (p. 8). Switzerland: International Organization for Standardization (ISO); (2002).
23. Aneja, R. P., Mathur, B. N., Chandan, R. C., & Banerjee, A. K., (2002). *Technology of Indian Milk Products: Handbook on Process Technology Modernization for Professionals, Entrepreneurs and Scientists* (p. 187). Anand – India: National Dairy Development Board (NDDB).
24. IS 3508-1966: Determination of Vitamin A, (2003). *Methods of Sampling and Test for Ghee (Butterfat)* (pp. 36–38). New Delhi: Bureau of Indian Standards (BIS).
25. Oser, B. L., & Hawk, P. B., (1965). *Hawk's Physiological Chemistry* (p. 723). New York: McGraw-Hill Book Co.
26. Parrish, D. B., & Richter, E. F., (1979). Determination of vitamin D in foods: A review. *Critical Reviews in Food Science & Nutrition, 12*(1), 29–57.

INDEX

D

E

T

W

Z